# Extreme Financial Risks

Yannick Malevergne   Didier Sornette

# Extreme Financial Risks

From Dependence to Risk Management

 Springer

Yannick Malevergne

Institut de Science Financière et d'Assurances
Université Claude Bernard Lyon 1
50 Avenue Tony Garnier
69366 Lyon Cedex 07
France

and

EM Lyon Business School
23 Avenue Guy de Collongue
69134 Ecully Cedex
France
E-mail: yannick.malevergne@univ-lyon1.fr

Didier Sornette

Institute of Geophysics and Planetary Physics
and Department of Earth and Space Science
University of California, Los Angeles
California 90095
USA

and

Laboratoire de Physique de la Matière
Condensée, CNRS UMR6622
and Université des Sciences
Parc Valrose
06108 Nice Cedex 2
France
E-mail: sornette@moho.ess.ucla.edu

Library of Congress Control Number: 2005930885

ISBN-10  3-540-27264-X Springer Berlin Heidelberg New York
ISBN-13  978-3-540-27264-9 Springer Berlin Heidelberg New York

Springer is a part of Springer Science+Business Media
springeronline.com
© Springer-Verlag Berlin Heidelberg 2006
Printed in The Netherlands

Typesetting: by the authors and TechBooks using a Springer LATEX macro package
Cover design: *design & production* GmbH, Heidelberg

Printed on acid-free paper     SPIN: 10939901     54/TechBooks     5 4 3 2 1 0

An error does not become truth by reason of multiplied propagation, nor does truth become error because nobody sees it.

M.K. Gandhi

# Preface: Idiosyncratic
and Collective Extreme Risks

Modern western societies have a paradoxical relationship with risks. On the one hand, there is the utopian quest for a zero-risk society [120]. On the other hand, human activities may increase risks of all kinds, from collaterals of new technologies to global impacts on the planet. The characteristic multiplication of major risks in modern society and its reflexive impact on its development is at the core of the concept of the "Risk Society" [47]. Correlatively, our perception of risk has evolved so that catastrophic events (earthquakes, floods, droughts, storms, hurricanes, volcanic eruptions, and so on) are no more systematically perceived as unfair outcomes of an implacable destiny. Catastrophes may also result from our own technological developments whose complexity may engender major industrial disasters such as Bhopal, Chernobyl, AZT, as well as irreversible global changes such as global warming leading to climatic disruptions or epidemics from new bacterial and viral mutations. The proliferation of new sources of risks imposes new responsibilities concerning their determination, understanding, and management. Government organizations as well as private institutions such as industrial companies, insurance companies, and banks which have to face such risks, in their role of regulators or of risk bearers, must ensure that the consequences of extreme risks are supportable without endangering the institutions in charge of bearing these risks.

In the financial sector, crashes probably represent the most striking events among all possible extreme phenomena, with an impact and frequency that has been increasing in the last two decades [450]. Consider the worldwide crash in October 1987 which evaporated more than one thousand billion dollars in a few days or the more recent collapse of the internet bubble in which more than one-third of the world capitalization of 1999 disappeared after March 2000. Finance and stock markets are based on the fluid convertibility of stocks into money and vice versa. Thus, to work well, money is requested to be a reliable standard of value, that is, an effective store of value, hence the concerns with the negative impacts of inflation. Similarly, investors look at the various financial assets as carriers of value, like money, but with additional return potentials (accompanied with downturn risks). But for this view to hold so as to promote economic development, fluctuations in values need to be tamed to minimize the risk of losing a lifetime of savings, or to avoid the risks of losing the investment potential of companies, or even to prevent economic and social recessions in whole countries (consider the situation of California after 2002 with a budget gap representing more than one-fourth of the entire State budget resulting essentially from the losses of financial and tax incomes following the collapse of the internet bubble). It is thus highly desirable to have the tools for monitoring, understanding, and limiting the extreme risks of financial markets. Fully aware of these problems, the worldwide banking organizations have promoted a series of advices and norms, known as the recommendations of the Basle committee [41, 42]. The Basle committee has proposed models for the internal management of risks and the imposition of minimum margin requirements commensurate with the risk exposures. However, some criticisms [117, 467] have found these recommendations to be ill-adapted or even destabilizing. This controversy underlines the importance of a better understanding of extreme risks, of their consequences and ways to prevent or at least minimize them.

In our opinion, tackling this challenging problem requires to decompose it into two main parts. First, it is essential to be able to accurately quantify extreme risks. This calls for the development of novel statistical tools going significantly beyond the Gaussian paradigm which underpins the standard framework of classical financial theory inherited from Bachelier [26], Markowitz [347], and Black and Scholes [60] among others. Second, the existence of extreme risks must be considered in the context of the practice of risk management itself, which leads to ask whether extreme risks can be diversified away similarly to standard risks according to the mean-variance approach. If the answer to this question is negative as can be surmised for numerous concrete empirical evidences, it is necessary to develop new concepts and tools for the construction of portfolios with minimum (but unavoidable) exposition of extreme risks. One can think of mixing equities and derivatives, as long as derivatives themselves do not add an extreme risk component and can really provide an insurance against extreme moves, which has been far from true in recent dramatic instances such as the crash of October 1987. Another approach could involve mutualism as in insurance.

Risk management, and to the same extent portfolio management, thus requires a precise and rigorous analysis of the distribution of the returns of the portfolio of risks. Taking into account the moderate sizes of standard portfolios (from tens to thousands of assets typically) and the non-Gaussian nature of the distributions of the returns of assets constituting the portfolios, the distributions of the returns of typical portfolios are far from Gaussian, in contradiction with the expectation from a naive use of the central limit theorem (see for instance Chap. 2 of [451] and other chapters for a discussion of the deviations from the central limit theorem). This breakdown of universality then requires a careful estimation of the specific case-dependent distribution of the returns of a given portfolio. This can be done directly using the time series of the returns of the portfolio for a given capital allocation. A more constructive approach consists in estimating the joint distribution of the returns of all assets constituting the portfolio. The first approach is much simpler and rapid to implement since it requires solely the estimation of a monovariate distribution. However, it lacks generality and power by neglecting the observable information available from the basket of all returns of the assets. Only the multivariate distribution of the returns of the assets embodies the general information of all risk components and their dependence across assets. However, the two approaches become equivalent in the following sense: the knowledge of the distribution of the returns for all possible portfolios for all possible allocations of capital between assets is equivalent to the knowledge of the multivariate distributions of the asset returns. All things considered, the second approach appears preferable on a general basis and is the method mobilizing the largest efforts both in academia and in the private sector.

However, the frontal attack aiming at the determination of the multivariate distribution of the asset returns is a challenging task and, in our opinion, much less instructive and useful than the separate studies of the marginal distributions of the asset returns on the one hand and the dependence structure of these assets on the other hand. In this book, we emphasize this second approach, with the objective of characterizing as faithfully as possible the diverse origins of risks: the risks stemming from each individual asset and the risks having a collective origin. This requires to determine (i) the distributions of returns at different time scales, or more generally, the stochastic process underlying the asset price dynamics, and (ii) the nature and properties of dependences between the different assets.

The present book offers an original and systematic treatment of these two domains, focusing mainly on the concepts and tools that remain valid for large and extreme price moves. Its originality lies in detailed and thorough presentations of the state of the art on (i) the different distributions of financial returns for various applications (VaR, stress testing), and (ii) the most important and useful measures of dependences, both unconditional and conditional and a study of the impact of conditioning on the size of large moves on the measure of extreme dependences. A large emphasis is thus put on the theory of copulas, their empirical testing and calibration, as they offer intrinsic and complete measures of dependences. Many of the results presented here are novel and have not been published or have been recently obtained by the authors or their colleagues. We would like to acknowledge, in particular, the fruitful and inspiring discussions and collaborations with J.V. Andersen, U. Frisch, J.-P. Laurent, J.-F. Muzy, and V.F. Pisarenko.

Chapter 1 describes a general framework to develop "coherent measures" of risks. It also addresses the origins of risks and of dependence between assets in financial markets, from the CAPM (capital asset pricing model) generalized to the non-Gaussian case with heterogeneous agents, the APT (arbitrage pricing theory), the factor models to the complex system view suggesting an emergent nature for the risk-return trade-off.

Chapter 2 addresses the problem of the precise estimation of the probability of extreme events, based on a description of the distribution of asset returns endowed with heavy tails. The challenge is thus to specify accurately these heavy tails, which are characterized by poor sampling (large events are rare). A major difficulty is to neither underestimate (Gaussian error) or overestimate (heavy tail hubris) the extreme events. The quest for a precise quantification opens the door to model errors, which can be partially circumvented by using several families of distributions whose detailed comparisons allow one to discern the sources of uncertainty and errors. Chapter 2 thus discusses several classes of heavy tailed distributions: regularly varying distributions (*i.e.*, with asymptotic power law tails), stretched-exponential distributions (also known as Weibull or subexponentials) as well as log-Weibull distributions which extrapolate smoothly between these different families.

The second element of the construction of multivariate distributions of asset returns, addressed in Chaps. 3–6, is to quantify the dependence structure of the asset returns. Indeed, large risks are not due solely to the heavy tails of the distribution of returns of individual assets but may result from a collective behavior. This collective behavior can be completely described by mathematical objects called *copulas*, introduced in Chap. 3, which fully embody the dependence between asset returns.

Chapter 4 describes synthetic measures of dependences, contrasting and linking them with the concept of copulas. It also presents an original estimation method of the coefficient of tail dependence, defined, roughly speaking, as the probability for an asset to lose a large amount knowing that another asset or the market has also dropped significantly. This tail dependence is of great interest because it addresses in a straightforward way the fundamental question whether extreme risks can be diversified away or not by aggregation in portfolios. Either the tail dependence coefficient is zero and the extreme losses occur asymptotically independently, which opens the possibility of diversifying them away. Alternatively, the tail dependence coefficient is non-zero and extreme losses are fundamentally dependent and it is impossible to completely remove extreme risks. The only remaining strategy is to develop portfolios that minimize the collective extreme risks, thus generalizing the mean-variance to a mean-extreme theory [332, 336, 333].

Chapter 5 presents the main methods for estimating copulas of financial assets. It shows that the empirical determination of a copula is quite delicate with significant risks of model errors, especially for extreme events. Specific studies of the extreme dependence are thus required.

Chapter 6 presents a general and thorough discussion of different measures of conditional dependences (where the condition can be on the size(s) of one or both returns for two assets). Chapter 6 thus sheds new light on the variations of the strength of dependence between assets as a function of the sizes of the analyzed events. As a startling concrete application of conditional dependences, the phenomenon of contagion during financial crises is discussed in detail.

Chapter 7 presents a synthesis of the six previous chapters and then offers suggestions for future work on dependence and risk analysis, including time-varying measures of extreme events, endogeneity versus exogeneity, regime switching, time-varying lagged dependence and so on.

This book has been written with the ambition to be useful to (a) the student looking for a general and in-depth introduction to the field, (b) financial engineers, economists, econometricians, actuarial professionals and researchers, and mathematicians looking for a synoptic view comparing the pros and cons of different modeling strategies, and (c) quantitative practitioners for the insights offered on the subtleties and many dimensional components of both risk and dependence. The content of this book will also be useful to the broader scientific community in the natural sciences, interested in quantifying the complexity of many physical, geophysical, biophysical etc. processes, with a mounting emphasis on the role and importance of extreme phenomena and their non-standard dependences.

Lyon, Nice and Los Angeles                          *Yannick Malevergne*
August 2005                                              *Didier Sornette*

# Contents

**1  On the Origin of Risks and Extremes** .................... 1
  1.1  The Multidimensional Nature of Risk
      and Dependence ........................................ 1
  1.2  How to Rank Risks Coherently? .......................... 4
    1.2.1  Coherent Measures of Risks ....................... 4
    1.2.2  Consistent Measures of Risks and Deviation Measures .. 7
    1.2.3  Examples of Consistent Measures of Risk ............. 10
  1.3  Origin of Risk and Dependence .......................... 13
    1.3.1  The CAPM View ................................. 13
    1.3.2  The Arbitrage Pricing Theory (APT)
         and the Fama–French Factor Model ................. 18
    1.3.3  The Efficient Market Hypothesis ................... 20
    1.3.4  Emergence of Dependence Structures
         in the Stock Markets ............................. 24
    1.3.5  Large Risks in Complex Systems ................... 29
  Appendix ................................................. 30
    1.A  Why Do Higher Moments Allow
        us to Assess Larger Risks? ....................... 30

**2  Marginal Distributions of Returns** ....................... 33
  2.1  Motivations ........................................... 33
  2.2  A Brief History of Return Distributions ................. 37
    2.2.1  The Gaussian Paradigm ........................... 37
    2.2.2  Mechanisms for Power Laws in Finance ............. 39
    2.2.3  Empirical Search for Power Law Tails
        and Possible Alternatives ......................... 42
  2.3  Constraints from Extreme Value Theory ................. 43
    2.3.1  Main Theoretical Results on Extreme Value Theory ... 45
    2.3.2  Estimation of the Form Parameter and Slow
        Convergence to Limit Generalized Extreme Value
        (GEV) and Generalized Pareto (GPD) Distributions ... 47

          2.3.3   Can Long Memory Processes Lead to Misleading
                  Measures of Extreme Properties? . . . . . . . . . . . . . . . . . . . .   51
          2.3.4   GEV and GPD Estimators of the Distributions
                  of Returns of the Dow Jones and Nasdaq Indices . . . . . .   52
     2.4   Fitting Distributions of Returns with Parametric Densities. . . .   54
          2.4.1   Definition of Two Parametric Families . . . . . . . . . . . . . . .   54
          2.4.2   Parameter Estimation Using Maximum Likelihood
                  and Anderson-Darling Distance . . . . . . . . . . . . . . . . . . . . .   60
          2.4.3   Empirical Results on the Goodness-of-Fits . . . . . . . . . . .   62
          2.4.4   Comparison of the Descriptive Power
                  of the Different Families. . . . . . . . . . . . . . . . . . . . . . . . . . .   69
     2.5   Discussion and Conclusions . . . . . . . . . . . . . . . . . . . . . . . . . . . . . .   76
          2.5.1   Summary. . . . . . . . . . . . . . . . . . . . . . . . . . . . . . . . . . . . . . .   76
          2.5.2   Is There a Best Model of Tails? . . . . . . . . . . . . . . . . . . . .   76
          2.5.3   Implications for Risk Assessment . . . . . . . . . . . . . . . . . . .   78
     Appendix . . . . . . . . . . . . . . . . . . . . . . . . . . . . . . . . . . . . . . . . . . . . . . . .   80
          2.A    Definition and Main Properties of Multifractal Processes   80
          2.B    A Survey of the Properties
                  of Maximum Likelihood Estimators . . . . . . . . . . . . . . . . . .   87
          2.C    Asymptotic Variance–Covariance of Maximum
                  Likelihood Estimators of the SE Parameters . . . . . . . . . .   91
          2.D    Testing the Pareto Model versus
                  the Stretched-Exponential Model . . . . . . . . . . . . . . . . . . . .   93

3   Notions of Copulas . . . . . . . . . . . . . . . . . . . . . . . . . . . . . . . . . . . . . .   99
     3.1   What is *Dependence*? . . . . . . . . . . . . . . . . . . . . . . . . . . . . . . . . . .  101
     3.2   Definition and Main Properties of Copulas . . . . . . . . . . . . . . . . .  103
     3.3   A Few Copula Families. . . . . . . . . . . . . . . . . . . . . . . . . . . . . . . . . .  107
          3.3.1   Elliptical Copulas . . . . . . . . . . . . . . . . . . . . . . . . . . . . . . .  107
          3.3.2   Archimedean Copulas . . . . . . . . . . . . . . . . . . . . . . . . . . . .  111
          3.3.3   Extreme Value Copulas . . . . . . . . . . . . . . . . . . . . . . . . . . .  116
     3.4   Universal Bounds for Functionals
           of Dependent Random Variables. . . . . . . . . . . . . . . . . . . . . . . . . . .  118
     3.5   Simulation of Dependent Data with a Prescribed Copula . . . . .  120
          3.5.1   Simulation of Random Variables Characterized
                  by Elliptical Copulas. . . . . . . . . . . . . . . . . . . . . . . . . . . . . .  120
          3.5.2   Simulation of Random Variables Characterized
                  by Smooth Copulas . . . . . . . . . . . . . . . . . . . . . . . . . . . . . . .  122
     3.6   Application of Copulas . . . . . . . . . . . . . . . . . . . . . . . . . . . . . . . . . .  124
          3.6.1   Assessing Tail Risk . . . . . . . . . . . . . . . . . . . . . . . . . . . . . .  124
          3.6.2   Asymptotic Expression of the Value-at-Risk . . . . . . . . . .  128
          3.6.3   Options on a Basket of Assets. . . . . . . . . . . . . . . . . . . . . .  131
          3.6.4   Basic Modeling of Dependent Default Risks . . . . . . . . . . .  137
     Appendix . . . . . . . . . . . . . . . . . . . . . . . . . . . . . . . . . . . . . . . . . . . . . . . .  138

3.A    Simple Proof of a Theorem on Universal Bounds
       for Functionals of Dependent Random Variables . . . . . . . 138
3.B    Sketch of a Proof of a Large Deviation Theorem
       for Portfolios Made of Weibull Random Variables . . . . . . 140
3.C    Relation Between the Objective
       and the Risk-Neutral Copula . . . . . . . . . . . . . . . . . . . . . . . 143

4    **Measures of Dependences** . . . . . . . . . . . . . . . . . . . . . . . . . . . . . . . . . 147
   4.1    Linear Correlations . . . . . . . . . . . . . . . . . . . . . . . . . . . . . . . . . 147
      4.1.1    Correlation Between Two Random Variables . . . . . . . . . 147
      4.1.2    Local Correlation . . . . . . . . . . . . . . . . . . . . . . . . . . . . . . . . 151
      4.1.3    Generalized Correlations Between $N > 2$ Random
            Variables . . . . . . . . . . . . . . . . . . . . . . . . . . . . . . . . . . . . . . 152
   4.2    Concordance Measures . . . . . . . . . . . . . . . . . . . . . . . . . . . . . . . 154
      4.2.1    Kendall's Tau . . . . . . . . . . . . . . . . . . . . . . . . . . . . . . . . . . . 154
      4.2.2    Measures of Similarity Between Two Copulas . . . . . . . . . 158
      4.2.3    Common Properties of Kendall's Tau,
            Spearman's Rho and Gini's Gamma . . . . . . . . . . . . . . . . 161
   4.3    Dependence Metric . . . . . . . . . . . . . . . . . . . . . . . . . . . . . . . . . . 162
   4.4    Quadrant and Orthant Dependence . . . . . . . . . . . . . . . . . . . . 164
   4.5    Tail Dependence . . . . . . . . . . . . . . . . . . . . . . . . . . . . . . . . . . . 168
      4.5.1    Definition . . . . . . . . . . . . . . . . . . . . . . . . . . . . . . . . . . . . . . 168
      4.5.2    Meaning and Refinement of Asymptotic Independence . 168
      4.5.3    Tail Dependence for Several Usual Models . . . . . . . . . . . . 170
      4.5.4    Practical Implications . . . . . . . . . . . . . . . . . . . . . . . . . . . . 177
   Appendix . . . . . . . . . . . . . . . . . . . . . . . . . . . . . . . . . . . . . . . . . . . . . 182
      4.A    Tail Dependence Generated by Student's Factor Model . 182

5    **Description of Financial Dependences with Copulas** . . . . . . . 189
   5.1    Estimation of Copulas . . . . . . . . . . . . . . . . . . . . . . . . . . . . . . . 190
      5.1.1    Nonparametric Estimation . . . . . . . . . . . . . . . . . . . . . . . . 190
      5.1.2    Semiparametric Estimation . . . . . . . . . . . . . . . . . . . . . . . 195
      5.1.3    Parametric Estimation . . . . . . . . . . . . . . . . . . . . . . . . . . . 200
      5.1.4    Goodness-of-Fit Tests . . . . . . . . . . . . . . . . . . . . . . . . . . . . 203
   5.2    Description of Financial Data in Terms of Gaussian Copulas . . 204
      5.2.1    Test Statistics and Testing Procedure . . . . . . . . . . . . . . . 204
      5.2.2    Empirical Results . . . . . . . . . . . . . . . . . . . . . . . . . . . . . . . 207
   5.3    Limits of the Description in Terms of the Gaussian Copula . . . 212
      5.3.1    Limits of the Tests . . . . . . . . . . . . . . . . . . . . . . . . . . . . . . 212
      5.3.2    Sensitivity of the Method . . . . . . . . . . . . . . . . . . . . . . . . . 213
      5.3.3    The Student Copula: An Alternative? . . . . . . . . . . . . . . . 215
      5.3.4    Accounting for Heteroscedasticity . . . . . . . . . . . . . . . . . . 217
   5.4    Summary . . . . . . . . . . . . . . . . . . . . . . . . . . . . . . . . . . . . . . . . . . 219
   Appendix . . . . . . . . . . . . . . . . . . . . . . . . . . . . . . . . . . . . . . . . . . . . . 221

5.A    Proof of the Existence of a $\chi^2$-Statistic
for Testing Gaussian Copulas ....................... 221
5.B    Hypothesis Testing with Pseudo Likelihood .......... 222

**6    Measuring Extreme Dependences** ........................ 227
6.1    Motivations ............................................ 230
6.1.1    Suggestive Historical Examples ..................... 230
6.1.2    Review of Different Perspectives ................... 231
6.2    Conditional Correlation Coefficient ....................... 233
6.2.1    Definition ......................................... 234
6.2.2    Influence of the Conditioning Set ................... 234
6.2.3    Influence of the Underlying Distribution
for a Given Conditioning Set ....................... 237
6.2.4    Conditional Correlation Coefficient on Both Variables . . 239
6.2.5    An Example of Empirical Implementation ............ 240
6.2.6    Summary ........................................... 246
6.3    Conditional Concordance Measures ...................... 247
6.3.1    Definition ......................................... 248
6.3.2    Example .......................................... 249
6.3.3    Empirical Evidence ................................ 251
6.4    Extreme Co-movements ................................ 254
6.5    Synthesis and Consequences ........................... 256
Appendix ..................................................... 261
6.A    Correlation Coefficient for Gaussian Variables
Conditioned on Both $X$ and $Y$ Larger Than $u$ ........ 261
6.B    Conditional Correlation Coefficient for Student's
Variables ......................................... 266
6.C    Conditional Spearman's Rho ....................... 270

**7    Summary and Outlook** ................................. 271
7.1    Synthesis ............................................. 271
7.2    Outlook and Future Directions .......................... 274
7.2.1    Robust and Adaptive Estimation of Dependences ...... 274
7.2.2    Outliers, Kings, Black Swans and Their Dependence . . . 276
7.2.3    Endogeneity Versus Exogeneity ..................... 276
7.2.4    Nonstationarity and Regime Switching in Dependence . 279
7.2.5    Time-Varying Lagged Dependence ................... 280
7.2.6    Toward a Dynamical Microfoundation of Dependences . 281

**References** ..................................................... 283

**Index** ......................................................... 309

# 1

## On the Origin of Risks and Extremes

### 1.1 The Multidimensional Nature of Risk and Dependence

In finance, the fundamental variable is the return that an investor accrues from his investment in a basket of assets over a certain time period. In general, an investor is interested in maximizing his gains while minimizing uncertainties ("risks") on the expected value of the returns on his investment, at possibly multiple time scales – depending upon the frequency with which the manager monitors the portfolio – and time periods – depending upon the investment horizon. From a general standpoint, the return-risk pair is the unavoidable duality underlying all human activities. The relationship between return and risk constitutes one of the most important unresolved questions in finance. This question permeates practically all financial engineering applications, and in particular the selection of investment portfolios. There is a general consensus among academic researchers that risk and return should be related, but the exact quantitative specification is still beyond our comprehension [414].

Uncertainties come in several forms, which we cite in the order of increasing aversion for most human beings:

(i) stochastic occurrences of events quantified by known probabilities;
(ii) stochastic occurrences of events with poorly quantified or unknown probabilities;
(iii) random events that are "surprises," *i.e.*, that were previously thought to be impossible or unthinkable until they happened and revealed their existence.

Here we address the first form, using the mathematical tools of probability theory.

Within this class of uncertainties, one must still distinguish several branches. In the simplest traditional theory exemplified by Markowitz [347], the uncertainties underlying a given set of positions (portfolio) result from the interplay of two components: risk and dependence.

(a) Risk is embedded in the amplitude of the fluctuations of the returns. its simplest traditional measure is the standard deviation (square-root of the variance).

(b) The dependence between the different assets of a portfolio of positions is traditionally quantified by the correlations between the returns of all pairs of assets.

Thus, in their most basic incarnations, both risk and dependence are thought of, respectively, as one-dimensional quantities: the standard deviation of the distribution of returns of a given asset and the correlation coefficient of these returns with those of another asset of reference (the "market" for instance). The standard deviation (or volatility) of portfolio returns provides the simplest way to quantify its fluctuations and is at the basis of Markowitz's portfolio selection theory [347]. However, the standard deviation of a portfolio offers only a limited quantification of incurred risks (seen as the statistical fluctuations of the realized return around its expected – or anticipated – value). This is because the empirical distributions of returns have "fat tails" (see Chap. 2 and references therein), a phenomenon associated with the occurrence of non-typical realizations of the returns. In addition, the dependences between assets are only imperfectly accounted for by the covariance matrix [309].

The last few decades have seen two important extensions.

- First, it has become clear, as synthesized in Chap. 2, that the standard deviation offers only a reductive view of the genuine full set of risks embedded in the distribution of returns of a given asset. As distributions of returns are in general far from Gaussian laws, one needs more than one centered moment (the variance) to characterize them. In principle, an infinite set of centered moments is required to faithfully characterize the potential for small all the way to extreme risks because, in general, large risks cannot be predicted from the knowledge of small risks quantified by the standard deviation. Alternatively, the full space of risks needs to be characterized by the full distribution function. It may also be that the distributions are so heavy-tailed that moments do not exist beyond a finite order, which is the realm of asymptotic power law tails, of which the stable Lévy laws constitute an extreme class. The Value-at-Risk (VaR) [257] and many other measures of risks [19, 20, 73, 447, 453] have been developed to account for the larger moves allowed by non-Gaussian distributions and non-linear correlations.

- Second and more recently, the correlation coefficient (and its associated covariance) has been shown to only be a partial measure of the full dependence structure between assets. Similarly to risks, a full understanding of the dependence between two or more assets requires, in principle, an infinite number of quantifiers or a complete dependence function such as the copulas, defined in Chap. 3.

These two fundamental extensions from one-dimensional measures of risk and dependence to infinitely dimensional measures of risk and dependence constitute the core of this book. Chapter 2 reviews our present knowledge and the open challenges in the characterization of distribution of returns. Chapter 3 introduces the notion of copulas which are applied later in Chap. 5 to financial dependences. Chapter 4 describes the main properties of the most important and varied measures of dependence, and underlines their connections with copulas. Finally, Chap. 6 expands on the best methods to capture the dependence between extreme returns.

Understanding the risks of a portfolio of $N$ assets involves the characterization of both the marginal distributions of asset returns and their dependence. In principle, this requires the knowledge of the full (time-dependent) multivariate distribution of returns, which is the joint probability of any given realization of the $N$ asset returns at a given time. This remark entails the two major problems of portfolio theory: (1) to determine the multivariate distribution function of asset returns; (2) to derive from it useful measures of portfolio risks and use them to analyze and optimize the performance of the portfolios. There is a large literature on multivariate distributions and multivariate statistical analysis [363, 468, 282]. This literature includes:

- the use of the multivariate normal distribution on density estimation [428];
- the corresponding random vectors treated with matrix algebra, and thus on matrix methods and multivariate statistical analysis [173, 371];
- the robust determination of multivariate means and covariances [297, 298];
- the use of multivariate linear regression and factor models [160, 161];
- principal component analysis, with excursions in clustering and classification techniques [276, 254];
- methods for data analysis in cases with missing observations [133, 310];
- detecting outliers [249, 250];
- bootstrap methods and handling of multicollinearity [461];
- methods of estimation using the plug-in principles and maximum likelihood [144];
- hypothesis testing using likelihood ratio tests and permutation tests [398];
- discrete multivariate distributions [253];
- computer-aided geometric design, geometric modeling, geodesic applications, and image analysis [464, 105, 426];
- radial basis functions [86], scattered data on spheres, and shift-invariant spaces [139, 433];
- non-uniform spline wavelets [139];
- scalable algorithms in computer graphics [76];
- reverse engineering [139], and so on.

The growing literature on (1) non-stationary processes [85, 210, 222, 361] and (2) regime-switching [172, 180, 215, 269] is not covered here. Nor do we address the more complex issues of embedding financial modeling within economics and social sciences. We do not cover either the consequences for risk assessment coming from the important emerging field of behavioral finance, with its exploration of the impact on decision-making of imperfect bounded subjective probability perceptions [36, 206, 437, 439, 474]. Our book thus uses objective probabilities which can be estimated (with quantifiable errors) from suitable analysis of available data.

## 1.2 How to Rank Risks Coherently?

The question on how to rank risks, so as to make optimal decisions, is recurrent in finance (and in many other fields) but has not yet received a general solution.

Since the middle of the twentieth century, several paths have been explored. The pioneering work by Von Neuman and Morgenstern [482] has given birth to the mathematical definition of the expected utility function, which provides interesting insights on the behavior of a rational economic agent and has formalized the concept of risk aversion. Based upon the properties of the utility function, Rothschild and Stiglitz [419, 420] have attempted to define the notion of increasing risks. But, as revealed by Allais [4, 5], empirical investigations have proven that the postulates chosen by Von Neuman and Morgenstern are actually often violated by humans. Many generalizations have been proposed for curing the so-called Allais' Paradox, but until now, no generally accepted procedure has been found.

Recently, a theory due to Artzner *et al.* [19, 20] and its generalization by Föllmer and Schied [174, 175] have appeared. Based on a series of postulates that are quite natural, this theory allows one to build *coherent* (resp., convex) measures of risks that provide tools to compare and rank risks [383]. In fact, if this theory seems well-adapted to the assessment of the needed economic capital, that is, of the fraction of capital a company must keep as risk-free assets in order to face its commitments and thus avoid ruin, it seems less natural for the purpose of quantifying the *fluctuations* of the asset returns or equivalently the deviation from a predetermined objective. In fact, as will be exposed in this section, it turns out that the two approaches consisting in assessing the risk in terms of economic capital on the one hand, and in terms of deviations from an objective on the other hand, are actually the two sides of the same coin as recently shown in [407, 408].

### 1.2.1 Coherent Measures of Risks

According to Artzner *et al.* [19, 20], the risk involved in the variations of the values of a market position is measured by the amount of capital invested in a risk-free asset, such that the market position can be prolonged in the future. In other words, the potential losses should not endanger the future actions of the fund manager of the company, or more generally, of the person or structure which underwrites the position. In this sense, a risk measure constitutes for Artzner *et al.* a measure of economic capital. The risk measure $\rho$ can be either positive, if the risk-free capital must be increased to guarantee the risky position, or negative, if the risk-free capital can be reduced without invalidating it.

A risk measure is said to be *coherent* in the sense of Artzner *et al.* [19, 20] if it obeys the four properties or axioms that we now list. Let us call $\mathcal{G}$ the space of risks. If the space $\Omega$ of all possible states of nature is finite, $\mathcal{G}$ is isomorphic to $\mathbb{R}^N$ and a risky position $X$ is nothing but a vector in $\mathbb{R}^N$. A risk measure $\rho$ is then a map from $\mathbb{R}^N$ onto $\mathbb{R}$. A generalization to other spaces $\mathcal{G}$ of risk has been proposed by Delbaen [123].

Let us consider a risky position with terminal value $X$ and a capital $\alpha$ invested in the risk-free asset at the beginning of the time period. At the end of the time period, $\alpha$ becomes $\alpha \cdot (1 + \mu_0)$, where $\mu_0$ is the risk-free interest rate. Then,

**Axiom 1 (Translational Invariance)**

$$\forall X \in \mathcal{G} \quad \text{and} \quad \forall \alpha \in \mathbb{R}, \quad \rho(X + \alpha \cdot (1 + \mu_0)) = \rho(X) - \alpha . \tag{1.1}$$

This simply means that an investment of amount $\alpha$ in the risk-free asset decreases the risk by the same amount $\alpha$. In particular, for any risky position $X$, $\rho(X + \rho(X) \cdot (1 + r)) = 0$, which expresses that investing an amount $\rho(X)$ in the risk-free asset enables one to exactly make up for the risk of the position $X$.

Let us now consider two risky investments $X_1$ and $X_2$, corresponding to the positions of two traders of an investment house. It is important for the supervisor that the aggregated risk of all traders be less than the sum of risks incurred by all traders. In particular, the risk associated with the position $(X_1 + X_2)$ should be smaller than or equal to the sum of the separated risks associated with the two positions $X_1$ and $X_2$.

**Axiom 2 (Sub-additivity)**

$$\forall (X_1, X_2) \in \mathcal{G} \times \mathcal{G}, \qquad \rho(X_1 + X_2) \le \rho(X_1) + \rho(X_2) . \tag{1.2}$$

The condition of sub-additivity encourages a portfolio managers to aggregate her different positions by diversification to minimize her overall risk. This axiom is probably the most debated among the four axioms underlying the theory of coherent measures of risk (see [131] and references therein). As an example, the VaR is well known to lack sub-additivity. At the same time, VaR is comonotonically additive, which means that the VaR of two comonotonic assets equals the sum of the VaR of each individual asset. But, since the comonotonicity represents the strongest kind of dependence (see Chap. 3), it is particularly disturbing to imagine that one can find situations where a portfolio made of two comonotonic assets is less risky than a portfolio with assets whose marginal risks are the same as in the previous situation but with a weaker dependence. Here is the rub with sub-additivity.

**Axiom 3 (Positive Homogeneity)**

$$\forall X \in \mathcal{G} \text{ and } \forall \lambda \geq 0, \qquad \rho(\lambda \cdot X) = \lambda \cdot \rho(X) . \qquad (1.3)$$

This third axiom stresses the importance of homogeneity. Indeed, it means that the risk associated with a given position increases with its size, here proportionally with it. Again, this axiom is controversial. Obviously, one can assert that the risk associated with the position $2 \cdot X$ is naturally twice as large as the risk of $X$. This is true as long as we can consider that a large position can be cleared as easily as a smaller one. However, it is not realistic because of the limited liquidity of real markets; a large position in a given asset is more risky than the sum of the risks associated with the many smaller positions which add up to the large position.

Eventually, if it is true that, for all possible states of nature, the risk of $X$ leads to a loss larger than that of $Y$ (*i.e.*, all components of the vector $X$ in $\mathbb{R}^N$ are always less than or equal to those of the vector $Y$), the risk measure $\rho(X)$ must be larger than or equal to $\rho(Y)$ :

**Axiom 4 (Monotony)**

$$\forall X, Y \in \mathcal{G} \text{ such that } \quad X \leq Y, \qquad \rho(X) \geq \rho(Y) . \qquad (1.4)$$

These four axioms define the coherent measures of risks, which admit the following general representation:

$$\rho(X) = \sup_{\mathbb{P} \in \mathcal{P}} E_{\mathbb{P}} \left[ \frac{-X}{1 + \mu_0} \right] , \qquad (1.5)$$

where $\mathcal{P}$ denotes a family of probability measures. Thus, any coherent measure of risk appears as the expectation of the maximum loss over a given set of scenarios (the different probability measures $\mathbb{P} \in \mathcal{P}$). It is then obvious that the larger the set of scenarios, the larger the value of $\rho(X)$ and thus, the more conservative the risk measure.

It is particularly interesting that expression (1.5) is very similar to the result obtained in the theory of utility with non-additive probabilities [202, 203]. Indeed, in such a case, the utility of position $X$ is given by

$$U(X) = \inf_{\mathbb{P} \in \mathcal{P}} E_{\mathbb{P}} \left[ u(X) \right] , \qquad (1.6)$$

where $u(\cdot)$ is a usual utility function.

When the coherent risk measure is invariant in law and comonotonically additive, an alternative representation holds in terms of the spectral measure of risk [285, 471]

$$\rho(X) = p \int_0^1 \text{VaR}_\alpha(X) \, dF(\alpha) + (1 - p)\text{VaR}_1(X) , \qquad (1.7)$$

where $F$ is a continuous convex distribution function on $[0, 1]$, $p$ is any real in $[0, 1]$ and $\text{VaR}_\alpha$ is the Value-at-Risk defined in (3.85) page 125. Therefore, most coherent measures of risk appear as a convex sum of $\text{VaR}_\alpha$ (a non-coherent risk measure) at different probability levels. The weighting function $F$ can be interpreted as a distortion of the objective probabilities, as underlined in the non-expected utility context [431, 495].

Coherent measures of risk can be generalized to define the so-called *convex* measures of risk by replacing the controversial axioms 2–3, by a single axiom of convexity of the risk measure [174, 175]. In the case where the risk measure is still positively homogeneous, this requirement is equivalent to the sub-additivity, but it becomes less restrictive when Axiom 3 is discarded. Then, one obtains the following representation of the convex risk measures:

$$\rho(X) = \sup_{\mathbb{P} \in \mathcal{M}} E_\mathbb{P} \left[ \frac{-X}{1 + \mu_0} - \alpha(\mathbb{P}) \right], \tag{1.8}$$

where $\mathcal{M}$ is the set of all probability measures on $(\Omega, \mathcal{F})$, $\mathcal{F}$ denotes a $\sigma$-algebra on the state space $\Omega$. More generally, $\mathcal{M}$ is the set of all finitely additive and non-negative set functions $\mathbb{P}$ on $\mathcal{F}$ satisfying $\mathbb{P}(\Omega) = 1$ and the functional

$$\alpha(\mathbb{P}) = \sup_{X \in \mathcal{G} | \rho(X) \leq 0} E_\mathbb{P} \left[ \frac{-X}{1 + \mu_0} \right] \tag{1.9}$$

is a penalty function that fully characterizes the convex measure of risk. In the case of a coherent risk measure, the set $\mathcal{P}$ (in (1.5)) is in fact the class of set functions $\mathbb{P}$ in $\mathcal{M}$ such that the penalty function vanishes: $\alpha(\mathbb{P}) = 0$.

Another alternative leads one to replace Axiom 4 by the following:

## Axiom 5 (Expectation-Boundedness)

$$\forall X \in \mathcal{G} \quad \rho(X) \geq \frac{E[-X]}{1 + \mu_0}, \tag{1.10}$$

where the equality holds if and only if $X$ is certain.[1] Then, together with axioms 1–3, it allows one to define the *expectation-bounded* risk measures [407]. They are particularly interesting insofar as they enable one to capture the inherent relationship existing between the assessment of risk in terms of economic capital and the measure of risk in terms of deviations from a target objective, as we shall see hereafter.

## 1.2.2 Consistent Measures of Risks and Deviation Measures

We now present a slightly different approach, which we think offers a suitable complement to coherent (and/or convex) risk measures for financial investments, and in particular for portfolio risk assessments. These measures are

---

[1] We say that $X$ is *certain* if $X(\omega) = a$, for some $a \in \mathbb{R}$, for all $\omega \in \Omega$, such that $\mathbb{P}(\omega) \neq 0$, where $\mathbb{P}$ denotes a probability measure on $(\Omega, \mathcal{F})$ and $\mathcal{F}$ is a $\sigma$-algebra so that $(\Omega, \mathcal{F}, \mathbb{P})$ is a usual probability space.

called "*consistent* measures of risks" in [333] and "*general deviation measures*" in [407]. As before, we consider the future value of a risky position denoted by $X$, and we call $\mathcal{G}$ the space of risks.

Let us first require that the risk measure $\tilde{\rho}(\cdot)$, which is a functional on $\mathcal{G}$, always remains positive:

**Axiom 6 (Positivity)**

$$\forall X \in \mathcal{G}, \qquad \tilde{\rho}(X) \geq 0, \tag{1.11}$$

where the equality holds if and only if $X$ is certain. Let us now add to this position a given amount $\alpha$ invested in the risk-free asset whose return is $\mu_0$ (with therefore no randomness in its price trajectory) and define the future wealth of the new position $Y = X + \alpha(1 + \mu_0)$. Since $\mu_0$ is non-random, the fluctuations of $X$ and $Y$ are the same. Thus, it is desirable that $\tilde{\rho}$ enjoys a property of *translational invariance*, whatever $X$ and the non-random coefficient $\alpha$ may be:

$$\forall X \in \mathcal{G}, \ \forall \alpha \in \mathbb{R}, \qquad \tilde{\rho}(X + (1 + \mu_0) \cdot \alpha) = \tilde{\rho}(X). \tag{1.12}$$

This relation is obviously true for all $\mu_0$ and $\alpha$; therefore, we set

**Axiom 7 (Translational Invariance)**

$$\forall X \in \mathcal{G}, \ \forall \kappa \in \mathbb{R}, \qquad \tilde{\rho}(X + \kappa) = \tilde{\rho}(X). \tag{1.13}$$

We also require that the risk measure increases with the quantity of assets held in the portfolio. This assumption reads

$$\forall X \in \mathcal{G}, \ \forall \lambda \in \mathbb{R}_+, \qquad \tilde{\rho}(\lambda \cdot X) = f(\lambda) \cdot \tilde{\rho}(X), \tag{1.14}$$

where the function $f : \mathbb{R}_+ \longrightarrow \mathbb{R}_+$ is increasing and convex to account for liquidity risk, as previously discussed. In fact, it is straightforward to show[2] that the only functions satisfying this requirement are the functions $f_\zeta(\lambda) = \lambda^\zeta$ with $\zeta \geq 1$, so that Axiom 3 can be reformulated in terms of positive homogeneity of degree $\zeta$:

**Axiom 8 (Positive Homogeneity)**

$$\forall X \in \mathcal{G}, \ \forall \lambda \in \mathbb{R}_+, \qquad \tilde{\rho}(\lambda \cdot X) = \lambda^\zeta \cdot \tilde{\rho}(X). \tag{1.15}$$

Note that the case of liquid markets is recovered by $\zeta = 1$ for which the risk is directly proportional to the size of the position, as in the case of the *coherent* risk measures.

These axioms, which define the so-called consistent measures of risk [333] can easily be extended to the risk measures associated with the return on the

---

[2] Using the trick $\tilde{\rho}(\lambda_1 \lambda_2 \cdot X) = f(\lambda_1) \cdot \tilde{\rho}(\lambda_2 \cdot X) = f(\lambda_1) \cdot f(\lambda_2) \cdot \tilde{\rho}(X) = f(\lambda_1 \cdot \lambda_2) \cdot \tilde{\rho}(X)$ leading to $f(\lambda_1 \cdot \lambda_2) = f(\lambda_1) \cdot f(\lambda_2)$. The unique increasing convex solution of this functional equation is $f_\zeta(\lambda) = \lambda^\zeta$ with $\zeta \geq 1$.

risky position. Indeed, a one-period return is nothing but the variation of the value of the position divided by its initial value $X_0$. One can thus easily check that the risk defined on the risky position is $[X_0]^\zeta$ times the risk defined on the return distribution. In the following, we will only consider the risk defined on the return distribution and, to simplify the notations, the symbol $X$ will be used to denote both the asset price and its return in their respective context without ambiguity.

Now, restricting to the case of a perfectly liquid market ($\zeta = 1$) and adding a sub-additivity assumption

**Axiom 9 (Sub-additivity)**

$$\forall (X, Y) \in \mathcal{G} \times \mathcal{G}, \qquad \tilde{\rho}(X + Y) \le \tilde{\rho}(X) + \tilde{\rho}(X), \tag{1.16}$$

one obtains the so-called general deviation measures [407]. Again, this axiom is open to controversy and its main *raison d'être* is to ensure the well-posedness of optimization problems (such as minimizing portfolio risks). It could be weakened along the lines used previously to derive the convex measures of risk from the coherent measures of risk.

One can easily check that the deviation measures defined in (1.16) correspond one-to-one to the expectation-bounded measures of risk defined in (1.10) through the relation

$$\rho(X) = \tilde{\rho}(X) + \frac{\mathrm{E}\left[-X\right]}{1 + \mu_0} \iff \tilde{\rho}(X) = \rho\left(X + \mathrm{E}\left[-X\right]\right). \tag{1.17}$$

It follows straightforwardly that minimizing the risk of a portfolio (measured either by $\rho$ or by $\tilde{\rho}$) under constraints on the expected return is equivalent, as long as the constraints on the expected return are active. Indeed, in such a case, searching for the minimum of $\tilde{\rho}$ or of $\tilde{\rho}(X) + \frac{\mathrm{E}[-X]}{1+\mu_0}$ is the same problem since the value of the expected return is fixed by the constraints.

Additionally, it can be shown that the expectation-bounded measure of risk $\rho$ defined by (1.17) is coherent if (and only if) the deviation measure $\tilde{\rho}$ satisfies [407]

$$\forall X \in \mathcal{G}, \qquad \tilde{\rho}(X) \le \mathrm{E}\left[X\right] - \inf X. \tag{1.18}$$

The general representation of the deviation measures satisfying this restriction can be easily derived from the representation of coherent risk measures. When such a requirement is not fulfilled, one can still have the following representation:[3]

---

[3] Strictly speaking, this representation only holds for lower semicontinous deviation measures, *i.e.*, deviation measures such that the sets $\{X | \tilde{\rho}(X) \le \epsilon\}$ are closed in $\mathcal{L}^2(\Omega)$, for all $\epsilon > 0$. This condition is fulfilled by most of the deviation measures of common use: the standard deviation, the semi-standard deviation, the absolute deviation, and so on.

$$\tilde{\rho}(X) = \sup_{Y \in \mathcal{Y}} \text{E}\left[Y \cdot (\text{E}\left[X\right] - X)\right] = \sup_{Y \in \mathcal{Y}} \text{Cov}(-X, Y), \tag{1.19}$$

where $\mathcal{Y}$ is a closed and convex subset of $\mathcal{L}^2(\Omega)$ such that

1. $1 \in \mathcal{Y}$,
2. $\forall Y \in \mathcal{Y}, \quad \text{E}\left[Y\right] = 1$,
3. $\forall X \in \mathcal{L}^2(\Omega)$, $\exists Y \in \mathcal{Y}$, such that $\text{E}\left[Y \cdot X\right] < \text{E}\left[X\right]$.

When the random variables in $\mathcal{Y}$ are all positive, they can be interpreted as density functions relative to some reference probability measure $\mathbb{P}_0$ on $(\Omega, \mathcal{F})$ (the objective probability measure). Thus, the term $\text{E}\left[Y \cdot X\right]$ is nothing but the expectation of $X$ under the probability measure $\mathbb{P}$, such that its Radon density $\frac{d\mathbb{P}}{d\mathbb{P}_0} = Y$. Therefore, one obtains a deviation measure associated with a coherent measure of risk.

These derivations show that deviation measures of risk on the one hand and coherent (or convex/expectation-bounded) measures of risk on the other hand are inextricably entangled. In fact, they are the two sides of the same coin, as mentioned in the introduction to this section. The various representation theorems show that, in most cases, these risk measures can be interpreted as worst-case scenarios, which rationalizes the use of *stress-testing* procedures as a sound practice for risk management.

In the more general case when the exponent $\zeta$ defined in Axiom 8 is no more equal to 1, and more precisely, when we only require that Axioms 6–8 hold, there is no general representation for the consistent risk measures to the best of our knowledge. The risk measures $\tilde{\rho}$ obeying Axioms 7 and 8 are known as the *semi-invariants* of the distribution of returns of $X$ (see [465, pp. 86–87]). Among the large family of semi-invariants, we can cite the well-known centered moments and cumulants of $X$ (including the usual variance). They are interesting cases that we discuss further below.

### 1.2.3 Examples of Consistent Measures of Risk

The set of risk measures obeying Axioms 7–8 is huge since it includes all the homogeneous functionals of $(X - \text{E}[X])$, for instance. The centered moments (or moments about the mean) and the cumulants are two well-known classes of semi-invariants. Then, a given value of $\zeta$ can be seen as nothing but a specific choice of the order $n$ of the centered moments or of the cumulants.[4] In this case, the risk measure defined via these semi-invariants fulfills the two following conditions:

$$\tilde{\rho}(X + \mu) = \tilde{\rho}(X), \tag{1.20}$$
$$\tilde{\rho}(\lambda \cdot X) = \lambda^n \cdot \tilde{\rho}(X). \tag{1.21}$$

---

[4] The relevance of the moments of high order for the assessment of large risks is discussed in Appendix 1.A.

In order to satisfy the positivity condition (Axiom 6), one needs to restrict the set of values taken by $n$. By construction, the centered moments of even order are always positive while the odd order centered moments can be negative. In addition, a vanishing value of an odd order moment does not mean that the random variable, or risk, $X \in \mathcal{G}$ is certain in the sense of footnote 1, since for instance any symmetric random variable has vanishing odd order moments. Thus, only the even-order centered moments seem acceptable risk measures. However, this restrictive constraint can be relaxed by first recalling that, given any homogeneous function $f(\cdot)$ of order $p$, the function $f(\cdot)^q$ is also homogeneous of order $p \cdot q$. This allows one to decouple the order of the moments to consider, which quantifies the impact of the large fluctuations, from the influence of the size of the positions held, measured by the degree of homogeneity of the measure $\tilde{\rho}$. Thus, considering any even-order centered moments, we can build a risk measure $\tilde{\rho}(X) = \mathrm{E}\left[(X - \mathrm{E}[X])^{2n}\right]^{\zeta/2n}$, which accounts for the fluctuations measured by the centered moment of order $2n$ but with a degree of homogeneity equal to $\zeta$.

A further generalization is possible for odd-order moments. Indeed, the *absolute* centered moments satisfy the three Axioms 6–8 for any odd or even order. So, we can even go one step further and use non-integer order absolute centered moments, and define the more general risk measure

$$\tilde{\rho}(X) = \mathrm{E}\left[|X - \mathrm{E}[X]|^{\gamma}\right]^{\zeta/\gamma}, \tag{1.22}$$

where $\gamma$ denotes any positive real number.

Due to the Minkowski inequality, these risk measures are convex for any $\zeta$ and $\gamma$ larger than 1 (and for $0 \le u \le 1$) :

$$\tilde{\rho}(u \cdot X + (1 - u) \cdot Y) \le u \cdot \tilde{\rho}(X) + (1 - u) \cdot \tilde{\rho}(Y), \tag{1.23}$$

which ensures that aggregating two risky assets diversifies their risk. In fact, in the special case $\gamma = 1$, these measures enjoy the stronger sub-additivity property, and therefore belong to the class of general deviation measures.

More generally, any discrete or continuous (positive) sum of these risk measures with the same degree of homogeneity is again a risk measure. This allows us to define "spectral measures of fluctuations" in the spirit of Acerbi [2]:

$$\tilde{\rho}(X) = \int d\gamma \, \phi(\gamma) \, \mathrm{E}\left[|X - \mathrm{E}[X]|^{\gamma}\right]^{\zeta/\gamma}, \tag{1.24}$$

where $\phi$ is a positive real-valued function defined on any subinterval of $[1, \infty)$, such that the integral in (1.24) remains finite. It is sufficient to restrict the construction of $\tilde{\rho}(X)$ to normalized functions $\phi$, such that $\int d\gamma \, \phi(\gamma) = 1$, since the risk measures are defined up to a global normalization factor. Then, $\phi(\gamma)$ represents the relative weight of the fluctuations measured by a given moment order and can be considered as a measure of the risk aversion of the risk manager with respect to large fluctuations.

The situation is not so clear for the cumulants, since the even-order cumulants, as well as the odd-order ones, can be negative (even if, for a large class of distributions, even-order cumulants remain positive, especially for fat-tailed distributions – even though there are simple but somewhat artificial counterexamples). In addition, cumulants suffer from another problem with respect to the positivity axiom. As for the odd-order centered moments, they can vanish even when the random variable is not certain. Just think of the cumulants of the Gaussian law. All but the first two (which represent the mean and the variance) are equal to zero. Thus, the strict formulation of the positivity axiom cannot be fulfilled by the cumulants. Should we thus reject them as useful measures of risks? It is important to emphasize that the cumulants enjoy a property which can be considered as a natural requirement for a risk measure. It can be desirable that the risk associated with a portfolio made of independent assets is exactly the sum of the risk associated with each individual asset. Thus, given $N$ independent assets $\{X_1, \ldots, X_N\}$, and the portfolio $S_N = X_1 + \cdots + X_N$, we would like to have

$$\tilde{\rho}(S_N) = \tilde{\rho}(X_1) + \cdots + \tilde{\rho}(X_N) \ . \tag{1.25}$$

This property is verified for all cumulants, while it does not hold for centered moments excepted the variance. In addition, as seen from their definition in terms of the characteristic function

$$\mathrm{E}\left[e^{ik \cdot X}\right] = \exp\left(\sum_{n=1}^{+\infty} \frac{(ik)^n}{n!} C_n\right) \ , \tag{1.26}$$

cumulants $C_n$ of order larger than 2 quantify deviations from the Gaussian law and therefore measure large risks beyond the variance (equal to the second-order cumulant).

What are the implications of using the cumulants as *almost* consistent measures of risks? In particular, what are the implications on the preferences of the agents employing such measures? To address this question, it is informative to express the cumulants as a function of the centered moments. For instance, let us consider the fourth-order cumulant:

$$C_4 = \mu_4 - 3 \cdot \mu_2^2 = \mu_4 - 3 \cdot C_2^2 \ , \tag{1.27}$$

where $\mu_n$ is the centered moment of order $n$. An agent assessing the fluctuations of an asset with respect to $C_4$ exhibits an aversion for the fluctuations quantified by the fourth central moment $\mu_4$ – since $C_4$ increases with $\mu_4$ – but is attracted by the fluctuations measured by the variance – since $C_4$ decreases with $\mu_2$. This behavior is not irrational because it remains globally risk-averse. Indeed, it depicts an agent which tries to avoid the larger risks but is ready to accept the smallest ones. This kind of behavior is characteristic of any agent using the cumulants as risk measures. In such a case, having $C_4 = 0$ does not mean that the agent considers that the position is not risky (in the sense that the position is certain) but that the agent is indifferent between the large risks of this position measured by $\mu_4$ and the small risks quantified by $\mu_2$.

To summarize, centered moments of even orders possess all the minimal properties required for a suitable portfolio risk measure. Cumulants only partially fulfill these requirements, but have an additional advantage compared with the centered moments, that is, they fulfill the condition (1.25). For these reasons, we think it is interesting to consider both the centered moments and the cumulants in risk analysis and decision making. Finally let us stress that the variance, originally used in Markowitz's portfolio theory [347], is nothing but the second centered moment, also equal to the second-order cumulant (the three first cumulants and centered moments are equal). Therefore, a portfolio theory based on the centered moments or on the cumulants automatically contains Markowitz's theory as a special case, and thus offers a natural generalization encompassing large risks of this masterpiece of financial science. It also embodies several other generalizations where homogeneous measures of risks are considered, as for instance in [241].

We should also mention the measure of attractiveness for risky investments, the gain–loss ratio, introduced by Bernardo and Ledoit [50]. The gain (loss) of a portfolio is the expectation, under a benchmark risk-adjusted probability measure, of the positive (negative) part of the portfolio's excess payoff. The gain–loss ratio constitutes an improvement over the widely used Sharpe ratio (average return over volatility). The advantage of the gain–loss ratio is that it penalizes only downside risk (losses) and rewards all upside potential (gains). The gain–loss ratio has been show to yield useful bounds for asset pricing in incomplete markets that gives the modeler the flexibility to control the trade-off between the precision of equilibrium models and the credibility of no-arbitrage methods. The gain–loss approach is valuable in applications where the security returns are not normally distributed. Bernardo and Ledoit [50] cite the following domains of application: (i) valuing real options on non-traded assets; (ii) valuing executive stock options when the executive cannot trade the options or the underlying due to insider trading restrictions; (iii) evaluating the performance of portfolio managers who invest in derivatives; (iv) pricing options on a security whose price follows a jump-diffusion or a fat-tailed Pareto–Lévy diffusion process; and (v) pricing fixed income derivatives in the presence of default risk.

## 1.3 Origin of Risk and Dependence

### 1.3.1 The CAPM View

Our purpose is not to review the huge literature on the origin of risks and their underlying mechanisms, but to suggest guidelines for further understanding. For enticing introductions and synopses, we refer to the very readable books of Bernstein [51, 52]. In [51], Bernstein reviews the history, since ancient times, of those thinkers who showed how to quantify risk:

The capacity to manage risk, and with it the appetite to take risk and make forward-looking choices, are key elements [...] that drive the economic system forward.

The concept of risks in economics and finance is elaborated in [52], starting with the origins of the Cowles foundation as the consequence of Cowles's personal interest in the question: Are stock prices predictable? In the words of J.L. McCauley (see his customer review on www.amazon.com),

this book is all about heroes and heroic ideas, and Bernstein's heroes are Adam Smith, Bachelier, Cowles, Markowitz (and Roy), Sharpe, Arrow and Debreu, Samuelson, Fama, Tobin, Samuelson, Markowitz, Miller and Modigliani, Treynor, Samuelson, Osborne, Wells-Fargo Bank (McQuown, Vertin, Fouse and the origin of index funds), Ross, Black, Scholes, and Merton. The final heroes (see Chap. 14, The Ultimate Invention) are the inventors of (synthetic) portfolio insurance (replication/synthetic options).

One of these achievements is the capital asset pricing model (CAPM), which is probably still the most widely used approach to relative asset valuation, although its empirical roots have been found to be weaker in recent years [59, 160, 223, 287, 306, 401]. Its major idea was that priced risk cannot be diversified and cannot be eliminated through portfolio aggregation. This asset valuation model describing the relationship between expected risk and expected return for marketable assets is strongly entangled with the Mean-Variance Portfolio Model of Markowitz. Indeed, both of them fundamentally rely on the description of the probability distribution function (pdf) of asset returns in terms of Gaussian functions. The mean-variance description is thus at the basis of the Markowitz portfolio theory and of the CAPM and its inter-temporal generalization (see for instance [359]).

The CAPM is based on the concept of economic equilibrium between rational expectation agents. Economic equilibrium is itself the self-organized result of complex nonlinear feedback processes between competitive interacting agents. Thus, while not describing the specific dynamics of how self-organization makes the economy converge to a stable regime [10, 18, 280], the concept of economic equilibrium describes the resulting state of this dynamic self-organization and embodies all the hidden and complex interactions between agents with infinite loops of recurrence. This provides a reference base for understanding risks.

We put some emphasis on the CAPM and its generalized versions because the CAPM is a remarkable starting point for answering the question on the origin of risks and returns: in economic equilibrium theory, the two are conceived as intrinsically entangled. In the following, we expand on this class of explanation before exploring briefly other directions.

Let us now show how an equilibrium model generalizing the original CAPM [308, 364, 429] can be formulated on the basis of the coherence measures adapted to large risks. This provides an "explanation" for risks from the point of view of the non-separable interplay between agents' preferences and their collective organization. We should stress that many generalizations have already been proposed to account for the fat-tailness of the assets return distributions, which led to the multimoments CAPM. For instance, Rubinstein [421], Krauss and Litzenberger [278], Lim [306] and Harvey and Siddique [223] have underlined and tested the role of the asymmetry in the risk premium by accounting for the skewness of the distribution of returns. More recently, Fang and Lai [162] and Hwang and Satchell [241] have introduced a four-moments CAPM to take into account the leptokurtic behavior of the assets return distributions. Many other extensions have been presented such as the VaR-CAPM [3] or the Distributional-CAPM [389]. All these generalizations become more complicated but unfortunately do not necessarily provide more accurate predictions of the expected returns.

Let us assume that the relevant risk measure is given by any measure of fluctuations previously presented that obey the Axioms 6–8 of Sect. 1.2.2. We will also relax the usual assumption of a homogeneous market to give to the economic agents the choice of their own risk measure: some of them may choose a risk measure which puts the emphasis on the small fluctuations, while others may prefer those which account for the larger ones. In such an heterogeneous market, we will recall how an equilibrium can still be reached and why the excess returns of individual stocks remain proportional to the market excess return, which is the fundamental tenet of CAPM.

For this, we need the following assumptions about the market:

- H1: We consider a one-period market, such that all the positions held at the beginning of a period are cleared at the end of the same period.
- H2: The market is perfect, *i.e.*, there are no transaction costs or taxes, the market is efficient and the investors can lend and borrow at the same risk-free rate $\mu_0$.

Of course, these standard assumptions are to be taken with a grain of salt and are made only with the goal of obtaining a normative reference theory. We will now add another assumption that specifies the behavior of the agents acting on the market, which will lead us to make the distinction between homogeneous and heterogeneous markets.

### Equilibrium in a Homogeneous Market

The market is said to be homogeneous if all the agents acting on this market aim at fulfilling the same objective. This means that:

- H3-1: All the agents want to maximize the expected return of their portfolio at the end of the period under a given constraint of measured risk, using the same measure of risks $\rho_\zeta$ for all of them (the subscript $\zeta$ refers to the degree of homogeneity of the risk measure, see Sect. 1.2).

In the special case where $\rho_\zeta$ denotes the variance, all the agents follow a Markowitz's optimization procedure, which leads to the CAPM equilibrium, as proved by Sharpe [429]. When $\rho_\zeta$ represents the centered moments, this leads to the market equilibrium described in [421]. Thus, this approach allows for a generalization of the most popular asset pricing in equilibrium market models.

When all the agents have the same risk function $\rho_\zeta$, whatever $\zeta$ may be, we can assert that they have all a fraction of their capital invested in the same portfolio $\Pi$ (see, for instance [333] for the derivation of the composition of the portfolio), and the remaining in the risk-free asset. The amount of capital invested in the risky fund only depends on their risk aversion and/or on the legal margin requirement they have to fulfill.

Let us now assume that the market is at equilibrium, i.e., supply equals demand. In such a case, since the optimal portfolios can be any linear combinations of the risk-free asset and of the risky portfolio $\Pi$, it is straightforward to show that the market portfolio, made of all traded assets in proportion of their market capitalization, is nothing but the risky portfolio $\Pi$. Thus, as shown in [333], we can state that, whatever the risk measure $\rho_\zeta$ chosen by the agents to perform their optimization, the excess return of any asset $i$ over the risk-free interest rate $(\mu(i) - \mu_0)$ is proportional to the excess return of the market portfolio $\Pi$ over the risk-free interest rate:

$$\mu(i) - \mu_0 = \beta_\zeta^i \cdot (\mu_\Pi - \mu_0), \tag{1.28}$$

where

$$\beta_\zeta^i = \left. \frac{\partial \ln \left( \rho_\zeta^{\frac{1}{\zeta}} \right)}{\partial w_i} \right|_{w_1^*, \cdots, w_N^*}, \tag{1.29}$$

where $w_1^*, \ldots, w_N^*$ are the optimal allocations of the assets in the following sense:

$$\begin{cases} \inf_{w_i \in [0,1]} \rho_\zeta(\{w_i\}) \\ \sum_{i \geq 0} w_i = 1 \\ \sum_{i \geq 0} w_i \mu(i) = \mu, \end{cases} \tag{1.30}$$

In other words, the set of normalized weights $w_i^*$ define the portfolio with minimum risk as measured by any convex[5] measure $\rho_\zeta$ of risk obeying Axioms 6–8 of Sect. 1.2.2 for a given amount of expected return $\mu$.

When $\rho_\zeta$ denotes the variance, we recover the usual $\beta^i$ given by the mean-variance approach:

$$\beta^i = \frac{\text{Cov}(X_i, \Pi)}{\text{Var}(\Pi)}. \tag{1.31}$$

---

[5] Convexity is necessary to ensure the existence and the unicity of a minimum.

Thus, the relations (1.28) and (1.29) generalize the usual CAPM formula, showing that the specific choice of the risk measure is not very important, as long as it follows the Axioms 6–8 characterizing the fluctuations of the distribution of asset returns.

## Equilibrium in a Heterogeneous Market

Does this result hold in the more realistic situation of an heterogeneous market? A market will be said to be heterogeneous if the agents seek to fulfill different objectives. We thus consider the following assumption:

- H3-2: There exist $N$ agents. Each agent $n$ is characterized by her choice of a risk measure $\rho_\zeta(n)$ so that she invests only in the mean-$\rho_\zeta(n)$ efficient portfolios.

According to this hypothesis, an agent $n$ invests a fraction of her wealth in the risk-free asset and the remaining in $\Pi_n$, the mean-$\rho_\zeta(n)$ efficient portfolio, only made of risky assets. Again, the fraction of wealth invested in the risky fund depends on the risk aversion of each agent, which may vary from one agent to another.

The composition of the market portfolio $\Pi$ for such a heterogeneous market is found to be nothing but the weighted sum of the mean-$\rho_\zeta(n)$ optimal portfolio $\Pi_n$ [333]:

$$\Pi = \sum_{n=1}^{N} \gamma_n \Pi_n \,, \tag{1.32}$$

where $\gamma_n$ is the fraction of the total wealth invested in the fund $\Pi_n$ by the $n^{\text{th}}$ agent.

Moreover, for every asset $i$ and for any mean-$\rho_\zeta(n)$ efficient portfolio $\Pi_n$, for all $n$, the following equation holds

$$\mu(i) - \mu_0 = \beta_n^i \cdot (\mu_{\Pi_n} - \mu_0) \,, \tag{1.33}$$

where $\beta_n^i$ is defined in (1.29). Multiplying these equations by $\gamma_n/\beta_n^i$, we get

$$\frac{\gamma_n}{\beta_n^i} \cdot (\mu(i) - \mu_0) = \gamma_n \cdot (\mu_{\Pi_n} - \mu_0) \,, \tag{1.34}$$

for all $n$, and summing over the different agents, we obtain

$$\left( \sum_n \frac{\gamma_n}{\beta_n^i} \right) \cdot (\mu(i) - \mu_0) = \left( \sum_n \gamma_n \cdot \mu_{\Pi_n} \right) - \mu_0 \,, \tag{1.35}$$

so that

$$\mu(i) - \mu_0 = \beta^i \cdot (\mu_\Pi - \mu_0) \,, \tag{1.36}$$

with

$$\beta^i = \left( \sum_n \frac{\gamma_n}{\beta_n^i} \right)^{-1}. \tag{1.37}$$

This allows us to conclude that, even in a heterogeneous market, the expected excess return of each individual stock is directly proportional to the expected excess return of the market portfolio, showing that the homogeneity of the market is not required for observing a linear relationship between individual excess asset returns and the market excess return.

The above calculations miss the possibility stressed by Rockafellar et al. [408] that two kinds of efficient portfolios $\Pi_n$ may exist in a heterogeneous market: long optimal portfolios which correspond to a net long position, and short optimal portfolios which correspond to a net short position. If the existence of the second kind of portfolio is not compatible with an equilibrium in a homogeneous market,[6] their existence is not precluded in a heterogeneous market. Indeed, the net short positions of a certain class of agents can be compensated by the net long position of another class of agents. Thus, as long as a market portfolio $\Pi$ corresponding to an overall long position exists, an equilibrium can be reached, and the results derived in this section still hold.

### 1.3.2 The Arbitrage Pricing Theory (APT) and the Fama–French Factor Model

The CAPM proposed a solution for what Roll [414] called

> perhaps the most important unresolved problem in finance, because it influences so many other problems, (which) is the relation between risk and return. Almost everyone agrees that there should be some relation, but its precise quantification has proven to be a conundrum that has haunted us for years, embarrassed us in print, and caused business practitioners to look askance at our scientific squabbling and question our relevance.

Indeed, past and recent tests cast strong doubts on the validity of the CAPM. The recent Fama–French analysis [160] shows basically no support for the CAPM's central result of a positive relation between expected return and global market risk (quantified by the so-called beta parameter). In contrast, other variables, such as market capitalization and the book-to-market ratio,[7] present some weak explanatory power.

---

[6] An equilibrium cannot be reached if all investors want to sell stocks.

[7] Ratio of the book value of a firm to its market value. Typically, the book-to-market is used to identify undervalued companies. If the book-to-market is less than one the stock is overvalued, while it is undervalued otherwise.

## The Arbitrage Pricing Theory (APT)

The empirical inadequacy of the CAPM has led to the development of more general models of risk and return, such as Ross's Arbitrage Pricing Theory (APT) [418]. Quoting Sargent [427],

> Ross posited a particular statistical process for asset returns, then derived the restrictions on the process that are implied by the hypothesis that there exist no arbitrage possibilities.

Like the CAPM, the APT assumes that only non-diversifiable risk is priced. But it differs from the CAPM by accounting for multiple causes of such risks and by assuming a sufficiently large number of such factors so that almost riskless portfolios can be constructed. Reisman recently presented a generalization of the APT showing that, under the assumption that there exists no asymptotic arbitrage (*i.e.*, in the limit of a large number of factors, the market risk can be decreased to almost zero), there exists an approximate multi-beta pricing relationship relative to any admissible proxy of dimension equal to the number of factors [402]. Unlike the CAPM which specifies returns as a linear function of only systematic risk, the APT is based on the well-known observations that multiple factors affect the observed time series of returns, such as industry factors, interest rates, exchange rates, real output, the money supply, aggregate consumption, investor confidence, oil prices, and many other variables [414]. However, while observed asset prices respond to a wide variety of factors, there is much weaker evidence that equities with larger sensitivity to some factors give higher returns, as the APT requires.

## The Fama–French Three Factor Model

This empirical weakness in the APT has led to further generalizations of factor models, such as the Fama–French three-factor model [160], which does not use an arbitrage condition anymore. Fama and French started with the observation that two classes of stocks show better returns than the average market: (1) stocks with small market capitalization ("small caps") and (2) stocks with a high book-value-to-price ratio (often "value" stocks as opposed to "growth" stocks). They added the overall market return to obtain the three factors: (i) the overall market return (Rm), (ii) the performance of small stocks relative to big stocks (SMB, small minus big), and (iii) the performance of value stocks relative to growth stocks (HML, high minus low). See the website of Professor K.R. French[8] which updates every quarter the benchmark factors and also presents the performance of several benchmark portfolios using different combinations of weights on the three factors. An important observation must be made concerning Fama and French's approach to risk in

---

[8]  http://mba.tuck.dartmouth.edu/pages/faculty/ken.french/data_library.html

their factor decomposition: they still see, as in the CAPM and APT, a large return as a reward for taking a high risk. For instance suppose that returns are found to increase with book/price. Then those stocks with a high book/price ratio must be more risky than average. This is in a sense the opposite to the traditional interpretation of a financial professional analyst, who would say that high book/price indicates a buying opportunity because the stock looks cheap. In contrast, according to the efficient market theory, a stock, which is cheap, can only be so because investors think it is risky.

Actually, the relationship between return and risk is not automatically positive. Diether *et al.* [124] have recently documented that firms with more uncertain earnings (as measured by the dispersion of analysts' forecasts) have smaller stock returns. As stressed by Johnson [255], this finding is important because it directly links asset returns with information, but the relation is apparently in contradiction with standard economic wisdom: the larger the risks, the smaller the return! Actually, Johnson proposes a simple explanation reconciling this new anomaly with the standard asset pricing theory, which is based on the following ingredients: (i) the equity value of the leveraged firm (*i.e.*, with non-zero debt) is equivalent to a call option on the firm's value, following Merton's model of credit risk [358]; (ii) the dispersion of analysts' forecasts is a measure of idiosyncratic risk, which is not priced. Then, by the Black–Merton–Scholes price for the equity-call option, the firm expected excess return (*i.e.*, relative variation of the equity price) has its risk premium amplified by a factor reflecting the effective exposure of the equity price to the real firm value. This factor turns out to decrease with increasing volatility, because more unpriced risk raises the option value, which has the consequence of lowering its exposure to priced risks. It is important to stress that this effect increases with the firm leverage and vanishes if the firm has no debt, as verified empirically with impressive strength in [255]. This new anomaly is thus fundamentally due to the impact of the volatility in the option pricing of the firm equity value in the presence of debt, together with the existence of a non-priced component of volatility.

### 1.3.3 The Efficient Market Hypothesis

The efficient market hypothesis (EMH) has a long history in finance and offers a mechanism for the trade-off between risk and return [158, 159]. Similarly to the concept of economic equilibrium, it must be understood as the result of repetitive feedback interactions between investors, and thus provides a top–down answer to the question on the origin of risk and return.

#### Origin of Possible Efficiency of Stock Markets

Roll uses an illuminating biological analogy to explain the principle leading to the EMH, in terms of the model of the hawks and the doves [414], which has been introduced to illustrate the concept of an evolutionary stable equilibrium (see also [118] for a seminal presentation of the concept of an evolutionary stable equilibrium in the genetic and biological context):

> Biologists note that competition for food results in a stable evolution-ary equilibrium characterized by multiple strategies. When competi-tors meet at a food site, they can either fight over the prize and risk injury – the "hawk" strategy – or withdraw and lose the food – the "dove" strategy. If every individual fights, a mutant who withdraws would eventually have a greater probability of procreating than the average fighter because of the risk of injury and the fact that only one fighter can win. (The dove occasionally finds uncontested food.) On the other hand, if every individual followed the dove strategy, a single fighter would gain a lot of food. The evolutionary equilibrium can be shown to involve either (a) part of the population always follows the hawk strategy and the complementary part follows the dove strategy or (b) every individual follows a randomized strategy, sometimes be-having as a hawk and sometimes as a dove. We can definitely rule out a world in which everyone follows the same fixed strategy.
>
> The analogy to market efficiency is immediate: investors compete for the most "undervalued" asset. The hawk strategy is conducting se-curity analysis. The dove strategy is passive investing: expending no effort on information analysis. Clearly, if everyone analyses securities, the benefits will be less than the costs. If everyone is passive, the benefits of analysis will be tremendous. The equilibrium is that some analyze, some don't. Does it sound familiar? Note that the final equi-librium is characterized by a situation in which it is not worthwhile for the marginal passive investor to begin analyzing nor for the marginal active investor to cease conducting security analysis.

The EMH is an idealization of a self-consistent dynamical state of the market resulting from the incessant actions of the traders (arbitragers). It is not the out-of-fashion equilibrium approximation sometimes described but rather embodies a very subtle cooperative organization of the market. A grow-ing number of academic studies and many practitioners have questioned the EMH on the basis of the non-rationality of individuals. Studies in psychology and behavioral sciences show indeed that people cannot be represented faith-fully by the Von Neumann/Morgenstern axioms of expected utility, especially in their limited intelligence, partial memory of the past and finite processing abilities, in their overconfidence and their biased assessments of probabilities [469]. However, interestingly, there are many works that demonstrate that "zero-intelligence" agents (to use the term of Farmer *et al.* [166]), who are very inefficient individually, often collectively provide efficient solutions. In-deed, the relevant question for understanding stock markets is not so much to focus on these irrationalities but rather to study how they aggregate in the complex, long-lasting, repetitive, and subtle environment of the market. This extension requires to abandon the emphasis on the description of the individual in favor of the search for emerging collective behaviors. Three fields of research highlight this idea and suggest a reconciliation, while enlarging significantly the perspective of the EMH.

## Collective Phenomena in Statistical Physics

In statistical physics, the fight between order (through the interaction between elementary constituents of matter) and disorder (modeled by thermal fluctuations) gives rise to the spontaneous occurrence of "spontaneous symmetry breaking" also called phase transitions in this context [451]. The understanding of the large-scale organization as well as the sudden macroscopic changes of organization due to small variations of a control parameter has led to powerful concepts such as "emergence" [9]: the macroscopic organization has many properties not shared by its constituents. For the market, this suggests that its overall properties can only be understood through the study of the transformation from the microscopic level of individual agents to the macroscopic level of the global market. In statistical physics, this can often be performed by the very powerful tool called the "renormalization group" [490, 489].

## Collective Phenomena in Biological Systems

Biology has clearly demonstrated that an organism has greater abilities than its constituent parts. This is true for multiorgan animals as well as for insect societies for instance (see E. O. Wilson's book [488]). More recently, this has led to the concept of "swarm intelligence" [67, 68, 70, 135]: the collective behaviors of (unsophisticated) agents interacting locally with their environment may cause coherent functional global patterns to emerge. Swarm intelligence is being used to obtain collective (or distributed) problem solving without centralized control or the provision of a global model in many practical industrial applications [69]. The importance of evolution, competition, and ecologies to understand stock markets has been stressed by Farmer [164].

## Collective Phenomena in Agent-Based Models

Agent-based models (also called multi-agent games) are composed of collections of synthetic, autonomous, interacting entities. They are used to explore how structure and interactions control the emergence of macroscopic behaviors [24]. Ultimately, the goal is to produce faithful synthetic models of reality by capturing the salient structure and strategies of real agents. The so-called minority game is perhaps the simplest in the class of multi-agent games of interacting inductive agents with limited abilities competing for scarce resources. Many published works on minority game have motivated their study by their relevance to financial markets, because investors exhibit a large heterogeneity of investment strategies, investment horizons, risk aversions and wealth, and have limited resources and time to dedicate to novel strategies, and the minority mechanism is found in markets. For an introduction to the Minority Game see [92, 251] and the Web page on the Minority Game by D. Challet at www.unifr.ch/econophysics/minority/minority.html. An important outcome of this work is the discovery of different market regimes, depending on the value of a control parameter, roughly defined as the ratio of the number of effective strategies available to agents divided by the total number of agents. In the minority game, agents choose their strategies according to the condition of the market so as on average to minimize their chance of being in the majority. When the "control" parameter is large, the recent history of the game contains some useful information that strategies can exploit and the market is not efficient. Below a critical value of the control parameter (*i.e.*, for sufficiently many agents), reasonable measures of predictability suggest that the market is efficient and cannot be predicted. These two phases are characterized by different risks, which can be quantified as a function of the control parameter. However, even in the "efficient market" phase, large and extreme price moves occur, which may be preceded by distinct patterns that allow agents in some cases to forecast them [289, 7].

**Self-Organization During Bubbles and Crashes**

A particular type of organization which requires special mention in this book is found in the occurrence of crashes. Market crashes exemplify in a dramatic way the spontaneous emergence of extreme events in self-organizing systems. Stock market crashes are indeed remarkable vehicles of important ideas needed to deal and cope with our risky world, as explained in [450]. By studying the frequency distribution of drawdowns, or runs of successive losses, Johansen and Sornette have shown that large financial crashes are "outliers" [249]: they form a class of their own which is characterized by its specific statistical signatures. An important consequence derives from this property: if large financial crashes are "outliers," they are special and thus requires a special explanation, a specific model, a theory of their own. In addition, their special properties may perhaps be used for their prediction. The main mechanism at work in bubbles and then in their destabilization during crashes is the existence of positive feedbacks, *i.e.*, self-reinforcement. Positive feedbacks have many sources both technical and behavioral, a dominant one being imitative behavior and herding between investors [450], which has been associated with behavioral "irrational exuberance" [438]. Positive feedbacks provide the fuel for the development of speculative bubbles, preparing the instability for a major crash. The understanding of financial bubbles and crashes requires a synthesis between the theory of collective behavior combined with the economic theory of anticipating agents who can change the future by their forecasts and the actions based on them. During a time of market instabilities, the tools of economic and financial theory break down; for instance, the idea of portfolio insurance breaks down as no portfolio can be perfectly insured against extreme deviations, especially those that occurred in October 1987 and wiped out confidence in the methods of so-called portfolio insurance of Leland-O' Brien-Rubinstein Associates. Similarly, the assumptions of near-normal distributions and stable covariance broke down during the failure of LTMC (Long-Term Capital Management) in October 1998 [394].

### 1.3.4 Emergence of Dependence Structures in the Stock Markets

#### Factors and Large Eigenvalues of Correlation Matrices

As mentioned above, factor models are nowadays the approaches most often used for extracting regularities in and for explaining the vagaries of stock market prices. Factor models conceptually derive from and generalize the CAPM and APT models. Factors, which are often invoked to explain prices, are the overall market factor and the factors related to firm size, firm industry and book-to-market equity, thought to embody most of the relevant dependence structure between the studied time series [160, 161]. Indeed, there is no doubt that observed equity prices respond to a wide variety of unanticipated factors, but there is much weaker evidence that expected returns are higher for equities that are more sensitive to these factors, as required by Markowitz's mean-variance theory, by the CAPM and the APT [414]. This severe failure of the most fundamental finance theories could conceivably be attributed to an inappropriate proxy for the market portfolio, but nobody has been able to show that this is really the correct explanation. This remark constitutes the crux of the problem: the factors invoked to model the cross-sectional dependence between assets are not known in general and are either postulated based on the economic intuition in financial studies, or obtained as black-box results in the recent analyses using the random matrix theory to large financial covariance matrices [392, 288]. In other words, explanatory factors emerge endogenously.

Here, we follow [337] to show that the existence of factors have a natural bottom-up explanation: they can be seen to result from a collective effect of the assets, similar to the emergence of a macroscopic self-organization of interacting microscopic constituents. To show this, we unravel the origin of the large eigenvalues of large covariance and correlation matrices and provide a complete understanding of the coexistence of features resembling properties of random matrices and of large "anomalous" eigenvalues. The main insight here is that, in any large system possessing non-vanishing average correlations between a finite fraction of all pairs of elements, a self-organized macroscopic state generically exists. In other words, "explanatory" factors emerge endogenously.

## Derivation of the Largest Eigenvalues

Let us first consider a large basket of $N$ assets with correlation matrix $C$ in which every non-diagonal pair of elements exhibits the same correlation coefficient $C_{ij} = \rho$ for $i \neq j$ and $C_{ii} = 1$. Its eigenvalues are

$$\lambda_1 = 1 + (N-1)\rho \quad \text{and} \quad \lambda_{i \geq 2} = 1 - \rho \tag{1.38}$$

with multiplicity $N - 1$ and with $\rho \in (0, 1)$ in order for the correlation matrix to remain positive definite. Thus, in the large size limit $N \to \infty$, even for a weak positive correlation $\rho \to 0$ (with $\rho N \gg 1$), a very large eigenvalue appears, associated with the "delocalized" (i.e., uniformly spread over all components) eigenvector $v_1 = (1/\sqrt{N})(1, 1, \cdots, 1)$, which dominates completely the correlation structure of the system. This trivial example stresses that the key point for the emergence of a large eigenvalue is not the strength of the correlations, provided that they do not vanish, but the large size $N$ of the system.

This result (1.38) still holds qualitatively when the correlation coefficients are all distinct. To see this, it is convenient to use a perturbation approach. We thus add a small random component to each correlation coefficient:

$$C_{ij} = \rho + \epsilon \cdot a_{ij} \quad \text{for} \quad i \neq j , \tag{1.39}$$

where the coefficients $a_{ij} = a_{ji}$ have zero mean, variance $\sigma^2$ and are independently distributed (there are additional constraints on the support of the distribution of the $a_{ij}$'s in order for the matrix $C_{ij}$ to remain positive definite with probability one). The determination of the eigenvalues and eigenvectors of $C_{ij}$ is performed using the perturbation theory up to the second order in $\epsilon$. We find that the largest eigenvalue satisfies

$$\mathrm{E}[\lambda_1] = (N-1)\rho + 1 + \frac{(N-1)(N-2)}{N^2} \cdot \frac{\epsilon^2 \sigma^2}{\rho} + \mathcal{O}(\epsilon^3) \tag{1.40}$$

while, at the same order, the corresponding eigenvector $v_1$ remains unchanged. The degeneracy of the eigenvalue $\lambda = 1 - \rho$ is broken and leads to a complex set of smaller eigenvalues described below.

In fact, this result (1.40) can be generalized to the non-perturbative domain of any correlation matrix with independent random coefficients $C_{ij}$, provided that they have the same mean value $\rho$ and variance $\sigma^2$. Indeed, in such a case, the expectations of the largest and second largest eigenvalues are [180]

$$\mathrm{E}[\lambda_1] = (N-1) \cdot \rho + 1 + \sigma^2/\rho + o(1) , \tag{1.41}$$

$$\mathrm{E}[\lambda_2] \leq 2\sigma\sqrt{N} + \mathcal{O}(N^{1/3} \log N) . \tag{1.42}$$

Moreover, the statistical fluctuations of these two largest eigenvalues are asymptotically (for large fluctuations $t > \mathcal{O}(\sqrt{N})$) bounded by a Gaussian term according to the following large deviation result

$$\Pr\{|\lambda_{1,2} - \mathrm{E}[\lambda_{1,2}]| \geq t\} \leq e^{-c_{1,2}t^2} , \qquad (1.43)$$

for some positive constant $c_{1,2}$ [279]. Numerical simulations of the distribution of eigenvalues of a random correlation matrix confirm indeed that the largest eigenvalue is indeed proportional to $N$, while the bulk of the eigenvalues are much smaller and are described by a modified semicircle law [357] centered on $\lambda = 1 - \rho$, in the limit of large $N$.

This result is very different from that obtained when the mean value $\rho$ vanishes. In such a case, the distribution of eigenvalues of the random matrix $C$ is given by the semicircle law [357]. However, due to the presence of the ones on the main diagonal of the correlation matrix $C$, the center of the circle is not at the origin but at the point $\lambda = 1$. Thus, the distribution of the eigenvalues of random correlation matrices with zero mean correlation coefficients is a semicircle of radius $2\sigma\sqrt{N}$ centered at $\lambda = 1$.

The result (1.41) is deeply related to the so-called *friendship theorem* in mathematical graph theory, which states that, in any finite graph such that any two vertices have exactly one common neighbor, there is one and only one vertex adjacent to all other vertices [155]. A more heuristic but equivalent statement is that, in a group of people such that any pair of persons have exactly one common friend, there is always one person (the "politician") who is the friend of everybody. Consider the matrix $C$ with its non-diagonal entries $C_{ij}$ ($i \neq j$) equal to Bernoulli random variable with parameter $\rho$, that is, $Pr[C_{ij} = 1] = \rho$ and $Pr[C_{ij} = 0] = 1 - \rho$. Then, the matrix $C_{ij} - I$, where $I$ is the unit matrix, becomes nothing but the adjacency matrix of the random graph $G(N, \rho)$ [279]. The proof of [155] of the "friendship theorem" indeed relies on the $N$-dependence of the largest eigenvalue and on the $\sqrt{N}$-dependence of the second largest eigenvalue of $C_{ij}$ as given by (1.41) and (1.42).

Figure 1.1 shows the distribution of eigenvalues of a random correlation matrix. The inset shows the largest eigenvalue lying at the predicted size $\rho N = 56.8$, while the bulk of the eigenvalues are much smaller and are described by a modified semicircle law centered on $\lambda = 1 - \rho$, in the limit of large $N$. This result, on the largest eigenvalue emerging from the collective effect of the cross-correlation between all $N(N - 1)/2$ pairs, provides a novel perspective to the observation [40, 413] that the only reasonable explanation for the simultaneous crash of 23 stock markets worldwide in October 1987 is the impact of a world market factor: according to the results (1.41) and (1.42) and the view expounded by Fig. 1.1, the simultaneous occurrence of significant correlations between the markets worldwide is bound to lead to the existence of an extremely large eigenvalue, the world market factor constructed by... a linear combination of the 23 stock markets! What this result shows is that invoking factors to explain the cross-sectional structure of stock returns is cursed by the chicken-and-egg problem: factors exist because stocks are correlated; stocks are correlated because of common factors impacting them.

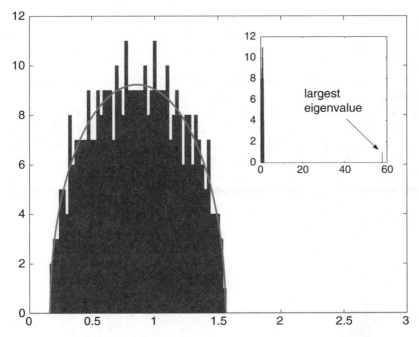

**Fig. 1.1.** Spectrum of eigenvalues of a random correlation matrix with average correlation coefficient $\rho = 0.14$ and standard deviation of the correlation coefficients $\sigma = 0.345 \sqrt{N}$: the ordinate is the number of eigenvalues in a bin with value given by the abscissa. One observes that all eigenvalues except the largest one are smaller than or equal to $\approx 1.5$. The size $N = 406$ of the matrix is the same as in previous studies [392] for the sake of comparison. The continuous curve is the theoretical translated semicircle distribution of eigenvalues describing the bulk of the distribution which passes the Kolmogorov test. The center value $\lambda = 1 - \rho$ ensures the conservation of the trace equal to $N$. There is no adjustable parameter. The inset represents the whole spectrum with the largest eigenvalue whose size is in agreement with the prediction $\rho N = 56.8$. Reproduced from [337]

## Generalization to a Segmented Market with Different Coupled Industries

Empirically [392, 288], a few other eigenvalues below the largest one have an amplitude of the order of 5–10 that deviate significantly from the bulk of the distribution. The above analysis provides a very simple mechanism for them, justifying the postulated model in [373]. The solution consists in considering, as a first approximation, the block diagonal matrix $C'$ with diagonal elements made of the matrices $A_1, \cdots, A_p$ of sizes $N_1, \cdots, N_p$ with $\sum N_i = N$, constructed according to (1.39) such that each matrix $A_i$ has the average correlation coefficient $\rho_i$. When the coefficients of the matrix $C'$ outside the matrices $A_i$ are zero, the spectrum of $C'$ is given by the union of all the spectra of the $A_i$'s, which are each dominated by a large eigenvalue $\lambda_{1,i} \simeq \rho_i \cdot N_i$.

The spectrum of $C'$ then exhibits $p$ large eigenvalues. Each block $A_i$ can be interpreted as a sector of the economy, including all the companies belonging to a same industrial branch and the eigenvector associated with each largest eigenvalue represents the main factor driving this sector of activity [343, 349]. For similar sector sizes $N_i$ and average correlation coefficients $\rho_i$, the largest eigenvalues are of the same order of magnitude. In addition, a very large unique eigenvalue is obtained by introducing some coupling constants outside the block diagonal matrices. A well-known result of the perturbation theory states that such coupling leads to a "repulsion" between the eigenvalues, which can be observed in Fig. 1.2 where $C'$ has been constructed with three block matrices $A_1$, $A_2$, and $A_3$ and non-zero off-diagonal coupling described in the figure caption. These values allow to quantitatively replicate the empirical finding of Laloux et al. in [392], where the three first eigenvalues are approximately $\lambda_1 \simeq 57$, $\lambda_2 \simeq 10$ and $\lambda_3 \simeq 8$.

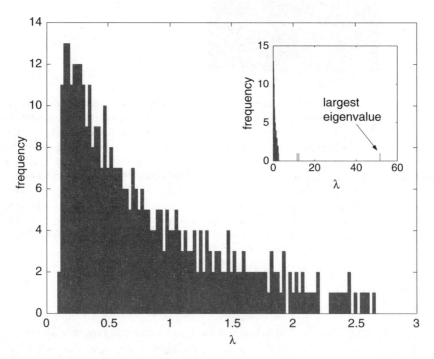

**Fig. 1.2.** Spectrum of eigenvalues estimated from the sample correlation matrix of $N = 406$ time series of length $T = 1309$. The times series have been constructed from a multivariate Gaussian distribution with a correlation matrix made of three block-diagonal matrices of sizes respectively equal to 130, 140, and 136 and mean correlation coefficients equal to 0.18 for all of them. The off-diagonal elements are all equal to 0.1. The same results hold if the off-diagonal elements are random. The inset shows the existence of three large eigenvalues, which result from the three-block structure of the correlation matrix. Reproduced from [337]

Expressions (1.40,1.41) and numerical tests for a large variety of correlation matrices show that the equally weighted eigenvector $v_1 = (1/\sqrt{N})(1, 1, \ldots, 1)$, associated with the largest eigenvalue is extremely robust and remains (on average) the same for any large system. Thus, even for time-varying correlation matrices, which is the result of heteroscedastic effects, the composition of the main factor remains almost the same. This can be seen as a generalized limit theorem reflecting the bottom-up organization of broadly correlated time series.

### 1.3.5 Large Risks in Complex Systems

These calculations show that an endogenous small positive correlation between all stock-pairs gives rise to large eigenvalues which can then be associated with "market factors." It seems that earlier researches have promoted the other way around: existing market factors (stock indices, news agencies, etc.) introduce exogenous market impact which affect different stocks similarly, thereby introducing positive correlation and thus large eigenvalues. This is clear from the general formulation of (linear) factor models such as the CAPM, APT, and Fama–French approaches in which the returns of all stocks are regressed against the same set of factors. Actually, we propose that the two chains of cause and result may be intrinsically coupled: the correlation structure between stocks is a stable attractor of a self-organized dynamics with positive and negative feedbacks in which factors exist because correlations exist, and correlations exist because factors exist. It would suggest the development of dynamical factor models, in which agents form anticipations on correlations based on their calibration of the past behavior of the regression to factors, in order to study the possible types of attractors (single or multiple equilibria) in the correlation structure of stocks. This may cast new light on the major unsolved problem stated in the introduction of this chapter concerning the relationship between return and risks: perhaps, the concept of return as the remuneration of risk which is so fundamental in financial theory should be replaced by the concept of the emergence of the risk-return duality, in which their relationship can be negative or positive, depending upon circumstances that remain to be worked out. Moreover, simulations of complex self-organizing systems show that large fluctuations and extreme variations are the rule rather than the exception.

The complex system approach, which involves seeing interconnections and relationships, *i.e.*, the whole picture as well as the component parts, is nowadays pervasive in modern control of engineering devices and business management. A central property of a complex system is the possible occurrence of coherent large-scale collective behaviors with a very rich structure, resulting from the repeated non-linear interactions among its constituents: the whole turns out to be much more than the sum of its parts. Most complex systems around us do exhibit rare and sudden transitions that occur over time intervals that are short compared with the characteristic time scales of their posterior evolution. Such extreme events express more than anything else the underlying forces usually hidden by almost perfect balance and thus provide the potential for a better scientific understanding of complex systems. These crises have fundamental societal impacts and range from large natural catastrophes, catastrophic events of environmental degradation, to the failure of engineering structures, crashes in the stock market, social unrest leading to large-scale strikes and upheaval, economic drawdowns on national and global scales, regional power blackouts, traffic gridlocks, diseases and epidemics, etc. An outstanding scientific question is how such large-scale patterns of catastrophic nature might evolve from a series of interactions on the smallest and increasingly larger scales. In complex systems, it has been found that the organization of spatial and temporal correlations do not stem, in general, from a nucleation phase diffusing across the system. It results rather from a progressive and more global cooperative process occurring over the whole system by repetitive interactions, which is partially described by the distributed correlations at the origin of a large eigenvalue as described above. An instance would be the many occurrences of simultaneous scientific and technical discoveries signaling the global nature of the maturing process. Recent developments suggest that non-traditional approaches, based on the concepts and methods of statistical and nonlinear physics coupled with ideas and tools from computation intelligence could provide novel methods in complexity to direct the numerical resolution of more realistic models and the identification of relevant signatures of large and extreme risks. To address the challenge posed by the identification and modeling of such outliers, the available theoretical tools comprise in particular bifurcation and catastrophe theories, dynamical critical phenomena and the renormalization group, nonlinear dynamical systems, and the theory of partially (spontaneously or not) broken symmetries. This field of research is presently very active and is expected to advance significantly our understanding, quantification, and control of risks.

In the mean time, both practitioners and academics need reliable metrics to characterize risks and dependences. This is the purpose of the following chapters, which expose powerful models and measures of large risks and complex dependences between time series.

# Appendix

## 1.A  Why Do Higher Moments Allow us to Assess Larger Risks?

As asserted in the main body of this chapter, the complete description of the fluctuations of an asset or a portfolio at a fixed time scale is given by the knowledge of the probability density function (pdf) of its return. The pdf encompasses all the risk dimensions associated with this asset. Unfortunately, it is impossible to classify or order the risks described by the entire pdf, except in special cases where the concept of stochastic dominance applies.

Therefore, the whole pdf cannot provide an adequate measure of risk, which should be embodied by a single variable. In order to perform a selection among a basket of assets and construct optimal portfolios, one needs measures given as real numbers, not functions, which can be ordered according to the natural ordering of real numbers on the line.

In this vein, Markowitz [347] has proposed to summarize the risk of an asset by the variance of its returns (or equivalently by the corresponding standard deviation). It is clear that this description of risks is fully satisfying only for assets with Gaussian pdfs. In any other case, the variance generally provides a very poor estimate of the real risk. Indeed, it is a well-established empirical fact that the pdfs of asset returns have fat tails (see Chap. 2), so that the Gaussian approximation underestimates significantly the large price movements frequently observed on stock markets (see Fig. 2.1). Consequently, the variance cannot be taken as a suitable measure of risks, since it only accounts for the smallest contributions to the fluctuations of the asset's returns.

The variance of the return $X$ of an asset involves its second moment $\mathrm{E}[X^2]$ and, more precisely, is equal to its second centered moment (or moment about the mean) $\mathrm{E}\left[(X - \mathrm{E}[X])^2\right]$. Thus, the weight of a given fluctuation $X$ contributing to the variance of the returns is proportional to its square. Due to the decay of the pdf of $X$ for large $X$ bounded from above by $\sim 1/|X|^{1+\mu}$ with $\mu > 2$ (see Chap. 2), the largest fluctuations do not contribute significantly to this expectation. To increase their contributions, and in this way to account for the largest fluctuations, it is natural to invoke moments of order $n$ higher

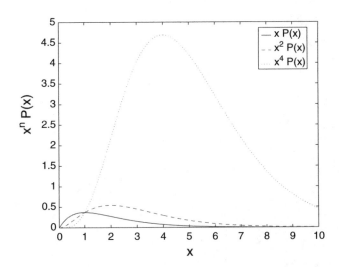

**Fig. 1.3.** This figure represents the function $x^n \cdot e^{-x}$ for $n = 1, 2$, and $4$ and shows the typical size of the fluctuations involved in the moment of order $n$. Reproduced from [333]

than 2. The larger $n$ is, the larger is the contribution of the rare and large returns in the tail of the pdf. This phenomenon is demonstrated in Fig. 1.3, where we can observe the evolution of the quantity $x^n \cdot f(x)$ for $n = 1, 2$, and $4$, where $f(x)$, in this example, denotes the density of the standard exponential distribution $e^{-x}$. The expectation $E[X^n]$ is then simply represented geometrically as equal to the area below the curve $x^n \cdot f(x)$. These curves provide an intuitive illustration of the fact that the main contributions to the moment $E[X^n]$ of order $n$ come from values of $X$ in the vicinity of the maximum of $x^n \cdot f(x)$, which increases fast with the order $n$ of the moment we consider, all the more so, the fatter is the tail of the pdf of the returns $X$. In addition, the typical size of the return assessed by the moment of order $n$ is given by $\lambda_n = E[X^p]^{1/p}$ (which coincide with the $L_p$ norm of $X$, for positive random variables). For the exponential distribution chosen to construct Fig. 1.3, the value of $x$ corresponding to the maximum of $x^n \cdot f(x)$ is exactly equal to $n$, while $\lambda_n = \frac{n}{e} + \mathcal{O}(\ln n)$. Thus, increasing the order of the moment allows one to sample larger fluctuations of the asset prices.

# 2

## Marginal Distributions of Returns

## 2.1 Motivations

As discussed in Chap. 1, the risks of a portfolio of $N$ assets are fully characterized by the (possibly time-dependent) multivariate distribution of returns, which is the joint probability of any given realization of the $N$ asset returns. For Gaussian models, this requires only the estimation of the average returns and of their covariance matrix. However, there is no doubt anymore that the Gaussian model is an inadequate description of real financial data (see for instance Fig. 2.1): the tails of the distributions are much fatter than Gaussian and the dependence between assets is not fully captured by the sole covariance matrix. The calibration and tests of multivariate models as well as their use for derivative pricing, portfolio analysis, and optimization are thus daunting tasks, characterized by the "curse of dimensionality."

The present book is constructed upon the foundation offered by the mathematical theory of copulas: as shown in Chap. 3 and used in subsequent chapters, any multivariate distribution can be uniquely decomposed into a part (the copula) capturing the intrinsic dependence between the assets and another part quantifying the risks embodied in the marginal distributions. In this representation, the information contained in a multivariate distribution of asset returns is thus decomposed in two sets: the intrinsic dependence and the marginals. Portfolio risks result from the multivariate composition of both the risks embedded in the marginals and the risks due to dependence, as well-known since Markowitz's mean-variance portfolio theory. Diversification of risks may then result from two mechanisms (working independently or in conjunction): (i) the law of large numbers (the larger the number of assets, the smaller the relative amplitude of the fluctuations of the total value relative to its mean) and (ii) anticorrelations (two assets whose prices tend to move in opposite directions give a lower risk when combined in a portfolio). The former mechanism often dominates in large portfolios while the second mechanism is at the basis of derivative hedging.

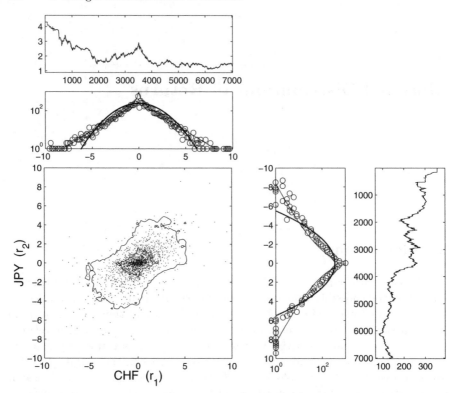

**Fig. 2.1.** Bivariate distribution of the daily annualized returns $r_i$ of the Swiss franc (CHF) in US \$ ($i = 1$) and of the Japanese yen (JPY) in US \$ ($i = 2$) for the time interval from January 1971 to October 1998. One-fourth of the data points are represented for clarity of the figure. The contour lines are obtained by smoothing the empirical bivariate density distribution and represent equilevels. The outer (respectively middle and inner) line is such that 90% (respectively 50% and 10%) of the total number of data points fall within it. It is apparent that the data is not described by an elliptically contoured pdf as it should be if the dependence was prescribed by a Gaussian (or more generally by an elliptic) distribution. Instead, the contour line takes the shape of a "bean". Also shown are the price-time series and the marginal distributions (in log-linear scales) in the panels at the top and on the side. The parabolas in thick lines correspond to the best fits by Gaussian distributions. The thin lines correspond to the best fits by stretched exponentials $\sim \exp[-(r_i/r_{0i})^{c_i}]$ with exponents $c_1 = 1.14$ for CHF and $c_2 = 0.8$ for JPY. Reproduced from [457]

In the present chapter, we review the knowledge accumulated on the characterization of marginal distributions of asset returns. This knowledge combined with adequate representations of the dependence structure between assets described in the following chapters can then be used to fully define the multivariate risks. The present chapter thus reviews the bricks of individual asset risks which can then be combined with the help of copulas to build the multivariate risk edifice. The emphasis is put on the determination of the precise shape of the tail of the distribution of returns of a given asset, which is a major issue both from a practical and from an academic point of view. Indeed, for practitioners, it is crucial to accurately estimate the high and low quantiles of the distribution of returns (profit and loss) because they are involved in almost all the modern risk management methods while from an academic perspective, many economic and financial theories rely on a specific parameterization of the distributions whose parameters are intended to represent the "macrovariables" influencing the agents.

For the purpose of practical market risk management, one typically needs to assess tail risks associated with the distribution of returns or profit and losses. Following the recommendations of the BIS,[1] one has to focus on risks associated with positions held for 10 days. Therefore, this requires to consider the distributions of 10-day returns. However, at such a large time scale, the number of (non-overlapping) historical observations dramatically decreases. Even over a century, one can only collect 2500 data points, or so, per asset. Therefore, the assessment of risks associated with high quantiles is particularly unreliable.

Recently, the use of high frequency data has allowed for an accurate estimation of the very far tails of the distributions of returns. Indeed, using samples of one to 10 million points enables one to efficiently calibrate probability distributions up to probability levels of order 99.9995%. Then, one can hope to reconstruct the distribution of returns at a larger time scale by convolution. It is the stance taken by many researchers advocating the use of Lévy processes to model the dynamics of asset prices [109, 196, and references therein]. The recent study by Eberlein and Özkan [141] shows the relevance of this approach, at least for fluctuations of moderate sizes. However, for large fluctuations, this approach is not really accurate, as shown in Fig. 2.2, which compares the probability density function (pdf) of raw 60-minute returns of the Standard & Poor's 500 index with the hypothetical pdf obtained by 60 convolution iterates of the pdf of the 1-minute returns; it is clear that the former exhibits significantly fatter tails than the latter.

This phenomenon derives naturally from the fact that asset returns cannot be merely described by *independent* random variables, as assumed when prices are modeled by Lévy processes. In fact, independence is too strong an assumption. For instance, the no free-lunch condition only implies the absence of linear time dependence since the best linear predictor of future (discounted) prices is then simply the current price. Volatility clustering, also called ARCH effect [150], is a clear manifestation of the existence of nonlinear dependences

---

[1] Bank for International Settlements. The BIS is an international organization which fosters cooperation among central banks and other agencies in pursuit of monetary and financial stability. Its banking services are provided exclusively to central banks and international organizations.

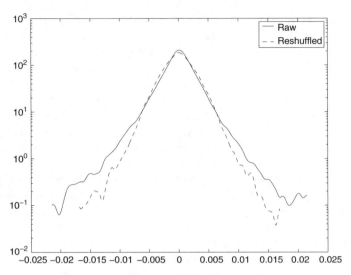

**Fig. 2.2.** Kernel density estimates of the raw 60-minute returns and the density obtained by 60 convolution iterates of the raw 1-minute returns kernel density for the Standard & Poor's 500

between returns observed at different lags. These dependences prevent the use of convolution for estimating tail risks with sufficient accuracy. Figure 2.2 illustrates the important observation that fat tails of asset return distributions owe their origin, at least in part, to the existence of volatility correlations. In the example of Fig. 2.2, a given 60-minute return is the sum of sixty 1-minute returns. If there was no dependence between these sixty 1-minute returns, the 60-minute return could be seen as the sum of 60 independent random variables; hence, its probability density could be calculated exactly by taking 60 convolutions of the probability density of the 1-minute returns. Note that this 60-fold convolution is equivalent to estimating the density of 60-minute returns in which their sixty 1-minute returns have been reshuffled randomly to remove any possible correlation. Figure 2.2 shows a faster decay of the pdf of these reshuffled 60-minute returns compared with the pdf of the true empirical 60-minute returns. Thus, assessing extreme risks at large time scales (1 or 10 days) by simple convolution of the distribution of returns at time scales of 1 or of 5 minutes leads to crude approximations and to dramatic underestimations of the amount of risk really incurred. The role of the dependence between successive returns is even more important in times of crashes: very large drawdowns (amplitudes of runs of losses) have been shown to result from anomalous transient dependences between a few successive days [249, 250]; as a consequence, they cannot be explained or modeled by the distribution calibrated for the bulk (99%) of the rest of the sample of drawdowns. These extreme events have been termed "outliers", "kings" [286] or "black swans" [470].

The only way to reliably aggregate high-frequency data is to have a consistent model at one's disposal. By *consistent* model is meant a model that accounts for the complex time structure of asset returns. (FI)-GARCH [31], $\alpha$-ARCH [132], multifractal[2] models [39, 341] or any other stochastic volatility model [232, 473] can be used for this purpose, but none of them is yet universally recognized since they do not rely on well-established founding of economic principles. As a consequence, one is exposed to model error: for instance, a simple GARCH model still underestimates the tail risks since it underestimates the long-range dependence of the volatility.

In this context, the most pragmatic approach may be to let the data speak by themselves, which is the stance taken in this chapter. For each different horizon, we discuss the possible parametric distributions that fit the data best. As we shall see, three main scales should be distinguished: small scale (a few minutes), intermediate scale (about an hour) and a large scale (1 day or more). At the smallest time scale, we will see that the tails of distributions are probably decaying *more slowly* than any power law. At the medium scale, regularly varying distributions provide a reasonable model, while at time scales of 1 day or more, rapidly varying distributions – like Weibull distributions – seem to accurately describe the tails of the distributions of asset returns, at least in the range of quantiles useful for risk management.

## 2.2 A Brief History of Return Distributions

The distribution of returns is one of the most basic characteristics of the markets and many academic studies have been devoted to it. Contrarily to the average or expected return, for which economic theory provides guidelines to assess them in relation to risk premium, firm size, or book-to-market (see Chap. 1 and [161] for instance), the functional form of the distribution of returns, and especially of extreme returns, is much less constrained and still a topic of active debate.

### 2.2.1 The Gaussian Paradigm

Generally, the central limit theorem would lead to a Gaussian distribution for sufficiently large time intervals over which the return is estimated. Taking the continuous time limit, such that any finite time interval is seen as the sum of an infinite number of increments thus leads to the paradigm of log-normal distributions of prices and equivalently of Gaussian distributions of returns. Based on the pioneering work of Bachelier [26] and later improved

---

[2] While fractal objects, processes, or measures enjoy a global scale invariance property – *i.e.*, look similar at any (time) scale – multifractals only enjoy this property locally, *i.e.*, they can be conceived as a fractal superposition of infinitely many local fractals.

**Table 2.1.** Descriptive statistics for the daily Dow Jones Industrial Average index returns (from 27, May 1896 to 31, May 2000, sample size $n = 28415$) calculated over 1 day and 1 month and for Nasdaq Composite Index returns calculated over 5 minutes and 1 hour (from 8, April 1997 to 29, May 1998, sample size $n = 22123$)

| | Mean | St. dev. | Skewness | Ex. Kurtosis | Jarque-Bera |
|---|---|---|---|---|---|
| Nasdaq (5 minutes) [†] | $1.80 \times 10^{-6}$ | $6.61 \times 10^{-4}$ | 0.0326 | 11.8535 | $1.30 \times 10^{5}$ (0.00) |
| Nasdaq (1 hour) [†] | $2.40 \times 10^{-5}$ | $3.30 \times 10^{-3}$ | 1.3396 | 23.7946 | $4.40 \times 10^{4}$ (0.00) |
| Nasdaq (5 minutes) [‡] | $-6.33 \times 10^{-9}$ | $3.85 \times 10^{-4}$ | −0.0562 | 6.9641 | $4.50 \times 10^{4}$ (0.00) |
| Nasdaq (1 hour) [‡] | $1.05 \times 10^{-6}$ | $1.90 \times 10^{-3}$ | −0.0374 | 4.5250 | $1.58 \times 10^{3}$ (0.00) |
| Dow Jones (1 day) | $8.96 \times 10^{-5}$ | $4.70 \times 10^{-3}$ | −0.6101 | 22.5443 | $6.03 \times 10^{5}$ (0.00) |
| Dow Jones (1 month) | $1.80 \times 10^{-3}$ | $2.54 \times 10^{-2}$ | −0.6998 | 5.3619 | $1.28 \times 10^{3}$ (0.00) |

[†] Raw data, [‡] data corrected for the $U$-shape of the intraday volatility due to the opening, lunch, and closing effects.

The Dow Jones exhibits a significantly negative skewness, which can probably be ascribed to the impact of the market crashes. The raw Nasdaq returns are significantly positively skewed while the returns corrected for the "lunch effect" are negatively skewed, showing that the lunch effect plays an important role in the shaping of the distribution of the intraday returns. Note also the important decrease of the kurtosis after correction of the Nasdaq returns for the lunch effect, confirming the strong impact of the lunch effect. In all cases, the excess-kurtosis are high and remain significant even after a time aggregation of one month. The numbers within parentheses represent the $p$-value of Jarque-Bera's normality test, a joint statistic using skewness and kurtosis coefficients [116]: the normality assumption is rejected for these time series. The Lagrange multiplier test proposed in [151] allows to test for heteroscedasticity. It leads to the $T \cdot R^2$ test statistic, where $T$ denotes the sample size and $R^2$ is the determination coefficient of the regression of the squared centered returns $x_t$ on a constant and on $q$ of their lags $x_{t-1}, x_{t-2}, \ldots, x_{t-q}$. Under the null hypothesis of homoscedastic time series, $T \cdot R^2$ follows a $\chi^2$-statistic with $q$ degrees of freedom. The test – performed up to lag $q = 10$ – shows that, in every case, the null hypothesis is strongly rejected at any usual significance level. Thus, the time series are heteroscedastic and exhibit volatility clustering. The BDS test [84], which allows one to detect not only volatility clustering, as in the previous test, but also departure from iid-ness due to non-linearities confirms that the null-hypothesis of iid data is strongly rejected at any usual significance level. Reproduced from [329].

by Osborne [377] and Samuelson [425], the log-normal paradigm has been the starting point of many financial theories such as Markowitz's portfolio selection method [347], Sharpe's market equilibrium model (CAPM) [429] or Black and Scholes rational option pricing theory [60]. However, for real financial data, the convergence in distribution to a Gaussian law is very slow (see for instance [72, 88]), much slower than predicted for independent returns. As shown in Table 2.1, the excess kurtosis (which is zero for a normal distribution) typically remains large even for monthly returns, testifying (i) of significant deviations from normality, (ii) of the heavy tail behavior of the distribution of returns and (iii) of significant time dependences between asset returns [88].

## 2.2.2 Mechanisms for Power Laws in Finance

Another approach rooted in economic theory, which can be invoked to derive the distribution of financial returns, consists in applying the "Gibrat principle" [441] initially introduced to account for the growth of cities and of wealth through a mechanism combining stochastic multiplicative and additive noises [55, 207, 268, 446, 454] leading to a Pareto distribution of sizes [94, 193]. Rational bubble models à la Blanchard [61] can also be cast in this mathematical framework of stochastic recurrence equations and leads to distributions with regularly varying tails, albeit with a strong constraint on the tail exponent (see [323] for the monovariate case and [331] for the multivariate case). These works suggest that an alternative and natural way to capture the heavy tail character of the distributions of returns is to use distributions with power-like tails (Pareto, generalized Pareto, Lévy stable laws) or more generally, regularly varying distributions[57],[3] the later ones encompassing all the former ones. At first glance, Fig. 2.3, which depicts the complementary sample distribution function for the 30-minute returns of the Standard & Poor's 500, seems to substantiate this thesis.

Other mechanisms involving the existence of a long memory of the volatility have been recently found to describe many of the stylized facts of monovariate financial returns. In particular, the multifractal random walk (MRW) is a process constructed with a very long memory in the volatility, such that it has a *bona fide* continuous limit with exact multifractal properties on the absolute values of the returns [27]. Appendix 2.A defines random cascade models from which the MRW derives and summarizes their main properties. For random cascade models exhibiting multifractality, it has been shown exactly that the random variables, defined in Sect. 2.A.1 as the increments $\delta_{\Delta t} X(t) = X(t) - X(t - \Delta t)$ corresponding to the log-returns calculated over the horizon $\Delta t$, are distributed in the tail according to a power law

$$\Pr[\delta_{\Delta t} X \geq x] = \mathcal{L}(x) \cdot x^{-b} , \qquad (2.1)$$

where the exponent $b$ is given by

$$b = \sup\{q, \ q > 1, \ \zeta(q) > 1\} , \qquad (2.2)$$

and $\zeta(q)$, defined in (2.A.7), is the spectrum of exponents of the moments of the absolute value of the log-returns (see Appendix 2.A). For the MRW, we have $\zeta(q) = (q - q(q-2)\lambda^2)/2$ according to (2.A.36), where $\lambda^2$ is the so-called multifractal parameter. Condition (2.2) then yields $b = 1/\lambda^2$ to leading order. Another equivalent way to arrive at the same result is to use the moments defined by (2.A.34) with (2.A.35), which can be shown to diverge ($K_q = +\infty$)

---

[3] The general representation of a regularly varying distribution $F$ is given by $1 - F(x) = \mathcal{L}(x) \cdot x^{-\alpha}$, where $\mathcal{L}(\cdot)$ is a slowly varying function, that is, a function such that $\lim_{x \to \infty} \mathcal{L}(tx)/\mathcal{L}(x) = 1$ for any finite $t$. The parameter $\alpha$ is usually called the *tail index* or the *tail exponent*.

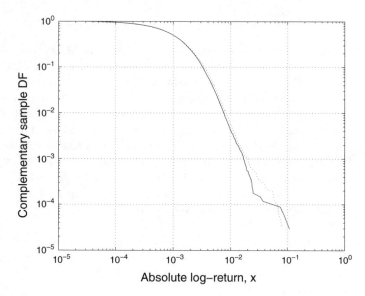

**Fig. 2.3.** Complementary sample distribution function for the Standard & Poor's 500 30-minute returns over the two decades 1980–1999. The plain (resp. dotted) line depicts the complementary distribution for the positive (the absolute value of negative) returns. Reproduced from [330]

if $\zeta(q) < 1$ [27]. The calibration[4] of $\lambda^2$ gives in general very small values in the range 0.01–0.04 leading to a tail index $b$ in the range 15–50 [366]. This has led previous workers to conclude that such a large tail exponent is unobservable with available data sets, and may well be described by other effective laws.

However, Muzy *et al.* [365] have recently shown that empirical distributions of log-returns do not give access to the unconditional prediction $b \approx 15$–50 with (2.2). This is because the value of $q$ determining the exponent $b$ according to (2.2) is itself associated with an $\alpha$ (through the Legendre transformation (2.A.20) and (2.A.21)) for which the multifractal spectrum $f(\alpha)$ defined in Sect. 2.A.2 is negative. But negative $f(\alpha)$'s are unobservable.[5] Indeed, from the definition (2.A.19) of $f(\alpha)$, only positive $f(\alpha)$'s correspond to genuine fractal dimensions and are thus observable: this is because they correspond to more than a few points of observations in the limit $\Delta t \ll T$. The key remark of Muzy *et al.* [365] is therefore that the observable exponent $b_{\text{obs}}$ for an infinite time series will be the largest positive $q$ such that $f(\alpha) \geq 0$:

$$b_{\text{obs}} = \sup\{q, \ q > 1, \ f(\alpha) > 0\} . \tag{2.3}$$

---

[4] From the correlation function of the log-volatility, from the scaling approach using the multifractal spectrum or from the generalized method of moment [321].

[5] Mandelbrot has shown that they can be interpreted in terms of singularities of large deviations.

Using the form (2.A.36) together with (2.A.20), we obtain $b_{\text{obs}} = \sqrt{2}/\lambda$. For a financial time series of finite length $L$, Muzy *et al.* [365] have shown in addition that the observed exponent is further reduced as a function of the ratio $\ln N_L / \ln N_T$ of the logarithm of the number $N_L = L/T$ of integral scales over the logarithm of the number $N_T = T/\Delta t$ of data points per integral scale. This makes a huge difference: rather than tail indices in the range 15–50, this gives observable tail indices in the range 3–5, as observed empirically.

See also [28] for extensions of the MRW to log infinitely divisible processes and [39, 341] for other applications of the multifractal process to the modeling of asset returns dynamics.

A different point of view on the underlying mechanism for the power law tails of price fluctuations has been proposed by Gabaix *et al.* [194, 195]. In essence, their proposal is that price variations are driven by fluctuations in the volume of transactions, $V$, whose cumulative distribution function $F_V$ has a regularly-varying tail with a universal exponent $\gamma \approx 1.5$. The fluctuations in the volume of transactions are argued in addition to be modulated by a deterministic market impact function, which describes the response of prices to transactions of the form $r = kV^\beta$, where $r$ is the change in the logarithm of price resulting from a transaction of volume $V$, $k$ is a constant and $\beta = 0.5$. This relationship can be derived from the assumption that agents are profit maximizers. These two ingredients $F_V(V) \sim 1 - 1/V^\gamma$ and $r = kV^\beta$ imply that large price returns $r$ have also a power law distribution with exponent $\mu = \gamma/\beta \approx 3$. Gabaix *et al.* find that their theory is consistent with the data. It is important to stress that these results are obtained by using aggregated data over a fixed time interval. Farmer and Lillo [165] argue that aggregating the data in time complicates the discussion, since the functional form of the market impact generally depends on the length of the time interval. They find that the same analysis based on individual (rather than time-aggregated) transactions does not confirm Gabaix *et al.*'s results and they suggest that the tail of price changes is driven by fluctuations in liquidity rather than in the volume of transactions. However, Plerou *et al.* [387] make the important point that individual transactions do not reflect true orders, especially the large ones, since the large orders of a large fund, say, are generally split in several transactions. Thus, the correct observable seems indeed to be the time-aggregated volume (albeit with variations in timespan), rather than individual trades. It is probable, however, that both mechanisms of fluctuations in the volume of transactions and in liquidity play a role in determining the statistics of price changes. In addition, the mechanism proposed by Gabaix *et al.* transfers the question of the origin of the power law distribution of the returns to the open question of the origin of the power law distribution of the volumes of transactions, which could reflect the power law distribution of the fund sizes, since the larger a fund is, the larger its orders can be expected to be. However, the distribution of the top 10% mutual funds [194] and of firm sizes [25, 376] are found to be regularly varying with a tail index close to 1, significantly smaller than the value 1.5 of the exponent of the distribution of the volumes of transactions. Unraveling the origin of this exponent 1.5 thus requires an understanding of the strategies of investors and how they organize, fragment, and delay their orders.

### 2.2.3 Empirical Search for Power Law Tails and Possible Alternatives

In the early 1960s, Mandelbrot [339] and Fama [157] presented evidence that distributions of returns can be well approximated by a symmetric Lévy stable law with tail index $b$ about 1.7. These estimates of the tail index have recently been supported by Mittnik *et al.* [362], and slightly different indices of the stable law ($b = 1.4$) were suggested by Mantegna and Stanley [345, 346].

On the other hand, there are numerous evidences of a larger value of the tail index $b \cong 3$ [217, 312, 320, 322, 367]. See also the various alternative parameterizations in terms of the Student distribution [62, 275], or Pearson Type-VII distributions [368], which all have an asymptotic power law tail and are regularly varying. Thus, a general conclusion of this group of authors concerning tail fatness can be formulated as follows: the tails of the distribution of returns are heavier than a Gaussian tail and heavier than an exponential tail; they certainly admit the existence of a finite variance ($b > 2$), whereas the existence of the third (skewness) and the fourth (kurtosis) moments is questionable.

These two classes of results are contradictory only on the surface, because they actually do not apply to the same quantiles of the distributions of returns. Indeed, Mantegna and Stanley [345] have shown that the distribution of returns of the Standard & Poor's 500 index can be described accurately by a Lévy stable law only within a limited range up to about 5 standard deviations, while a faster decay (approximated by an exponential or a power law with larger exponent) of the distribution is observed beyond. This almost-but-not-quite Lévy stable description could explain (at least, in part) the slow convergence of the distribution of returns to the Gaussian law under time aggregation [72, 451]; and it is precisely outside this range of up to 5 standard deviations, where the Lévy law does not apply anymore that a tail index $b \cong 3$ has been estimated. Indeed, most authors who have reported a tail index $b \cong 3$ have used some optimality criteria for choosing the sample fractions (*i.e.*, the largest values) for the estimation of the tail index. Thus, unlike the authors supporting stable laws, they have used only a fraction of the largest (positive tail) and smallest (negative tail) sample values.

It would thus seem that all has been said on the distributions of returns. However, there are still dissenting views in the literature. Indeed, the class of regularly varying distributions is not the sole one able to account for the large kurtosis and fat-tailness of the distributions of returns. Some recent works suggest alternative descriptions for the distributions of returns. For instance, Gouriéroux and Jasiak [208] claim that the distribution of returns on the French stock market decays faster than any power law. Cont *et al.* [108]

have proposed to use exponentially truncated stable distributions, Barndorff-Nielsen [37], Eberlein *et al.* [140] and Prause [393] have respectively considered normal inverse Gaussian and (generalized) hyperbolic distributions, which asymptotically decay as $x^\alpha \cdot \exp(-\beta x)$, while Laherrère and Sornette [286] suggest to fit the distributions of stock returns by the Stretched-Exponential law.[6] Of the same type are the marginal distributions of the so-called CGMY model proposed by Carr *et al.* [90]. These results, challenging the traditional hypothesis of a power-like tail, offer a new representation of the returns distributions.

In addition, real financial time series exhibit (G)ARCH effects [65, 66] leading to heteroscedasticity and to clustering of high threshold exceedances due to a long memory of the volatility. These rather complex dependent structures make difficult, if not questionable, the blind application of standard statistical tools for data analysis. In particular, the existence of significant dependence in the return volatility leads to the existence of a significant bias and an increase of the true standard deviation of the statistical estimators of tail indices. Indeed, there are now many examples showing that dependences and long memories as well as non-linearities mislead standard statistical tests (see for instance [12, 216]). Consider the Hill's and Pickands' estimators, which play an important role in the study of the tails of distributions. It is often overlooked that, for dependent time series, Hill's estimator remains only consistent but not asymptotically efficient [416]. Moreover, for financial time series with a dependence structure described by an IGARCH process, it has been shown that the standard deviation of Hill's estimator obtained by a bootstrap method can be seven to eight times larger than the standard deviation derived under the asymptotic normality assumption [267]. These figures are even worse for Pickands' estimator.

## 2.3 Constraints from Extreme Value Theory

The application of extreme value theory (EVT) to the investigation of the properties of the distributions of asset returns has grown rapidly during the last decade. Longin [312] was one of the main promoters of this method and has advocated its use for risk management purposes [313], particularly for Value-at-Risk assessment and stress testing. The conclusions drawn from the various studies of empirical distributions of log-returns, based on the extreme value theory, show that they should belong to the maximum domain of attraction of the Fréchet distribution, so that they are necessarily regularly varying laws. However, most of these studies have been performed under the

---

[6] Picoli *et al.* [385] have also presented fits comparing the relative merits of Stretched-Exponential and so-called *q*-exponentials (which are similar to Student distribution with power law tails) for the description of the frequency distributions of basketball baskets, cyclone victims, brand-name drugs by retail sales, and highway lengths.

restrictive assumption that (i) financial time series are made of independent and identically distributed returns, and (ii) the corresponding distributions of returns belong to one of only three possible maximum domains of attraction.[7] However, these assumptions are not fulfilled in general. While Smith's results [444] indicate that the dependence of the data does not constitute a major problem in the limit of large samples, so that volatility clustering of financial data does not prevent the reliability of EVT, we shall see that it can significantly bias standard statistical tools for samples of size commonly used in extreme tails studies. Moreover, the conclusions of many studies are essentially based on an *aggregation* procedure which stresses the central part of the distribution while smoothing and possibly distorting the characteristics of the tail (whose properties are obviously essential in characterizing the tail behavior).

The question then arises whether the limitations of these statistical tools could have led to erroneous conclusions about the tail behavior of the distributions of returns. In this section, presenting tests performed on synthetic time series with time dependence in the volatility with both Pareto and Stretched Exponential (SE) distributions, and on two empirical time series (the daily returns of the Dow Jones Industrial Average Index over a century ($n = 28415$ data points) and the 5-minute returns of the Nasdaq Composite index over 1 year from April 1997 to May 1998 ($n = 22123$ data points)), we exemplify the fact that the standard generalized extreme value (GEV) estimators can be quite inefficient due to the possibly slow convergence toward the asymptotic theoretical distribution and the existence of biases in the presence of dependence between data. Thus, one cannot reliably distinguish between rapidly and regularly varying classes of distributions. The generalized Pareto distribution (GPD) estimators work better, but still lack power in the presence of strong dependence. Note that the two empirical data sets used in the illustration below are justified by their similarity with (i) the data set of daily returns used in [312] particularly, and (ii) the high frequency data used in [217, 322, 367] among others.

---

[7] Extensions of the asymptotic theory of extreme values to correlated sequences have been developed by Berman [48, 49] for Gaussian sequences and Loynes [318], O'Brien [374], Leadbetter [293] and others [369] in the more general context of stationary sequences satisfying mixing conditions. See also Kotz and Nadarajah [277] for the limit distribution of extreme values of 2D correlated random variables. Recently, there is a growing interest in the extreme value theory of strongly correlated random variables in many areas of science, including applications to diffusing particles in correlated random potentials [89], to the understanding of large deviations in spin glass ground state energies [13], to front propagation and fluctuations [378], fragmentation, binary search tree problem in computer science [325, 326], to maximal height of growing surfaces [399], to the Hopfield model of brain learning [75], and so on.

### 2.3.1 Main Theoretical Results on Extreme Value Theory

Two limit theorems allow one to study the extremal properties and to determine the maximum domain of attraction (MDA) of a distribution function in two forms.

First, consider a sample of $N$ iid realizations $X_1, X_2, \ldots, X_N$ of a random variable. Let $X_N^\wedge$ denote the maximum of this sample.[8] Then, the Gnedenko theorem states that, if, after an adequate centering and normalization, the distribution of $X_N^\wedge$ converges to a *non-degenerate* distribution as $N$ goes to infinity, this limit distribution is then necessarily the generalized extreme value (GEV) distribution defined by

$$H_\xi(x) = \exp\left[-(1 + \xi \cdot x)^{-1/\xi}\right] , \qquad (2.4)$$

with $x \in [-1/\xi, \infty)$ if $\xi > 0$ and $x \in (-\infty, -1/\xi]$ if $\xi < 0$. When $\xi = 0$, $H_\xi(x)$ should be understood as

$$H_0(x) = \exp[-\exp(-x)], \quad x \in \mathbb{R} . \qquad (2.5)$$

Thus, for $N$ large enough

$$\Pr[X_N^\wedge < x] \simeq H_{\xi_N}\left(\frac{x - \mu_N}{\psi_N}\right) , \qquad (2.6)$$

for some value of the centering parameter $\mu_N$, scale factor $\psi_N$ and form parameter $\xi_N$. The form parameter $\xi$ is of paramount importance for the shape of the limiting distribution. Its sign determines the three possible limiting forms of the GEV distribution of maxima (2.4):

1. If $\xi > 0$ the limit distribution is the (shifted) Fréchet power-like distribution;
2. If $\xi = 0$, the limit distribution is the Gumbel (double-exponential) distribution;
3. If $\xi < 0$, the limit distribution has a support bounded from above.

The determination of the parameter $\xi$ is the central problem of extreme value analysis. Indeed, it allows one to determine the maximum domain of attraction of the underlying distribution and therefore its behavior in the tails. When $\xi > 0$, the underlying distribution belongs to the Fréchet maximum domain of attraction and is regularly varying (power-like tail). When $\xi = 0$, it belongs to the Gumbel maximum domain of attraction and is rapidly varying (exponential tail), while if $\xi < 0$ it belongs to the Weibull maximum domain of attraction and has a finite right endpoint, which means that there exists a finite $x_F$ such that $X \leq x_f$ with probability one.

---

[8]  Similar results hold for $X_N^\vee = \min\{X_1, \ldots, X_N\}$ since $\min\{X_1, \ldots, X_N\} = -\max\{-X_1, \ldots, -X_N\}$.

The usefulness of formula (2.6) for risk assessment purposes seems obvious as it provides a universal estimation of the Value-at-Risk. If $X$ denotes the profit and loss, $X_N^\wedge$ represents the largest among $N$ losses. The Value-at-Risk at confidence level $\alpha$, denoted by $\mathrm{VaR}_\alpha$, is given by the unique solution of:

$$F_X\left(\mathrm{VaR}_\alpha\right) = \Pr\left[X < \mathrm{VaR}_\alpha\right] = \alpha , \qquad (2.7)$$

provided that $F_X$ is increasing.[9] For $N$ iid observations of the profits and losses, we have

$$\Pr\left[X_N^\wedge < \mathrm{VaR}_\alpha\right] = \Pr\left[X < \mathrm{VaR}_\alpha\right]^N = \alpha^N , \qquad (2.8)$$

so that $\mathrm{VaR}_\alpha$ is (asymptotically) solution of

$$H_{\xi_N}\left(\frac{\mathrm{VaR}_\alpha - \mu_N}{\psi_N}\right) = \alpha^N , \qquad (2.9)$$

which with (2.4) yields

$$\mathrm{VaR}_\alpha \simeq \mu_N + \frac{\psi_N}{\xi_N}\left[(-N\ln\alpha)^{-\xi_N} - 1\right]. \qquad (2.10)$$

When the observations are not *iid*, one can generally replace $N$ by $\theta \cdot N$, where $\theta \in [0,1]$ is the so-called *extremal index* [146, 293, 313], related to the size of the clusters of extremes which may appear when the data exhibit temporal dependence. Indeed, generally speaking, one can write [146, p.419]:

$$\Pr\left[X_N^\wedge < \mathrm{VaR}_\alpha\right] \simeq \Pr\left[X < \mathrm{VaR}_\alpha\right]^{\theta \cdot N} = \alpha^{\theta \cdot N} , \qquad (2.11)$$

so that

$$\mathrm{VaR}_\alpha \simeq \mu + \frac{\psi}{\xi}\left[(-\theta \cdot N\ln\alpha)^{-\xi} - 1\right]. \qquad (2.12)$$

The second limit theorem is called after Gnedenko-Pickands-Balkema-de Haan (GPBH) and its formulation is as follows [146, pp. 152–168] (see also [451, Chap. 1] for an intuitive exposition). In order to state the GPBH theorem, let us define the right endpoint $x_F$ of a distribution function $F(x)$ as $x_F = \sup\{x : F(x) < 1\}$. Let us call the function

$$\Pr\{X - u \ge x \mid X > u\} \equiv \bar{F}_u(x) \qquad (2.13)$$

the excess distribution function. Then, this (survival) distribution function $\bar{F}_u(x)$ belongs to the maximum domain of attraction of $H_\xi(x)$ defined by (2.4) if and only if there exists a positive scale-function $s(u)$, depending on the threshold $u$, such that

---

[9] For a more general definition of VaR, see (3.85) page 125.

$$\lim_{u \to x_F} \sup_{0 \le x \le x_F - u} |\bar{F}_u(x) - \bar{G}(x \mid \xi, s(u))| = 0 \;, \tag{2.14}$$

where

$$G(x \mid \xi, s) = 1 + \ln H_\xi \left(\frac{x}{s}\right) = 1 - \left(1 + \xi \cdot \frac{x}{s}\right)^{-1/\xi} \tag{2.15}$$

is called the generalized Pareto distribution (GPD). By taking the limit $\xi \to 0$, expression (2.15) leads to the exponential distribution. The support of the distribution function (2.15) is defined as follows:

$$\begin{cases} 0 \le x < \infty, & \text{if } \xi \ge 0 \\ 0 \le x \le -s/\xi, & \text{if } \xi < 0 \,. \end{cases} \tag{2.16}$$

Thus, the GPD has a finite support for $\xi < 0$.

Again, this theorem has important practical implications for risk management, since it provides a general assessment of the expected-shortfall of a position $X$ associated with a given distribution of profits and losses. The expected-shortfall, at confidence level $\alpha$, is given by:

$$\mathrm{ES}_\alpha = \mathrm{E}\left[X | X \ge \mathrm{VaR}_\alpha\right] \,, \tag{2.17}$$

which can be evaluated with the help of relation (2.15):

$$\mathrm{ES}_\alpha = \mathrm{VaR}_\alpha + \frac{s}{1 - \xi} \,, \tag{2.18}$$

with $\xi < 1$, in order for the expectation to exist.[10]

As a note of caution, it should be underlined that the existence of a non-degenerate limit distribution of properly centered and normalized maxima $X_N^\wedge$ or peaks over threshold $X - u|X > u$ is a rather strong requirement. There are a lot of distribution functions which do not satisfy this condition, $e.g.$, infinitely alternating functions between a power-like and an exponential behavior.

## 2.3.2 Estimation of the Form Parameter and Slow Convergence to Limit Generalized Extreme Value (GEV) and Generalized Pareto (GPD) Distributions

There exist two main ways of estimating the form parameter $\xi$. First, if there is a sample of maxima (taken from subsamples of sufficiently large size), then one can fit to this sample the GEV distribution, thus estimating the parameters by the maximum likelihood method, for instance. Alternatively, one can prefer the distribution of exceedances over a large threshold given by the GPD (2.15), whose tail index can be estimated with Pickands' estimator or by maximum

---

[10] Recall that $\xi$ is the inverse of the tail exponent.

likelihood, as previously. Hill's estimator cannot be used in the present case since it assumes $\xi > 0$, while the essence of extreme value analysis is, as we said, to test for all the classes of limit distributions without excluding any possibility, and not only to determine the quantitative value of an exponent. Each of these methods has its advantages and drawbacks, especially when one has to study dependent data, as we show below.

Given a sample of size $N$, one can consider the $q$-maxima drawn from $q$ subsamples of size $p$ (such that $p \cdot q = N$) to estimate the parameters $(\mu, \psi, \xi)$ in (2.6) by maximum likelihood. This procedure yields consistent and asymptotically efficient Gaussian estimators, provided that $\xi > -1/2$ [444]. The properties of the estimators still hold approximately for dependent data, provided that the interdependence remains weak. However, it is difficult to choose the optimal value $q$ of the number of subsamples as it depends both on the size $N$ of the entire sample and on the underlying distribution: the maxima drawn from an exponential distribution are known to converge very quickly to Gumbel's distribution [220], while for the Gaussian law, convergence is particularly slow [219].

The second possibility is to estimate the parameter $\xi$ from the distribution of exceedances (*i.e.*, from the GPD). For this, one can use either the maximum likelihood estimator or Pickands' estimator. Maximum Likelihood estimators are well known to be asymptotically the most efficient ones (at least for $\xi > -1/2$ and for independent data) but, in this particular case, Pickands' estimator works reasonably well. Given an ordered sample $x_1 \leq x_2 \leq \cdots \leq x_N$ of size $N$, Pickands' estimator is given by

$$\hat{\xi}_{k,N} = \frac{1}{\ln 2} \ln \frac{x_k - x_{2k}}{x_{2k} - x_{4k}} \ . \tag{2.19}$$

For independent and identically distributed data, this estimator is consistent provided that $k$ is chosen so that $k \longrightarrow \infty$ and $k/N \longrightarrow 0$ as $N \longrightarrow \infty$. Moreover, $\hat{\xi}_{k,N}$ is asymptotically normal with variance

$$\sigma(\hat{\xi}_{k,N})^2 \cdot k \longrightarrow \frac{\xi^2(2^{2\xi+1} + 1)}{(2(2^\xi - 1)\ln 2)^2}, \quad \text{as } N \longrightarrow \infty \ . \tag{2.20}$$

In the presence of dependence between data, one can expect an increase of the standard deviation, as reported by Kearns and Pagan [267]. For time dependence of the GARCH class, they have indeed demonstrated a significant increase of the standard deviation of the tail index estimator, such as Hill's estimator, by a factor more than seven with respect to their asymptotic properties for iid samples. This leads to very inaccurate index estimates for time series with this kind of temporal dependence. Another problem lies in the determination of the optimal threshold $u$ of the GPD, which is in fact related to the optimal determination of the subsamples size $p$ in the case of the estimation of the parameters of the distribution of maximum.

In order to compare the performance of the various estimators of the tail index $\xi$ for iid data, Malevergne *et al.* [329] have considered several numerically generated samples respectively drawn from (i) an asymptotic power law distribution with tail index $b = 3$, (ii) a SE distribution, $i.e.$, such that

$$\ln \Pr [X \leq x] \propto -x^c, \quad \text{as } x \longrightarrow \infty, \tag{2.21}$$

with fractional exponent $c = 0.7$ and (iii) a SE with fractional exponent $c = 0.3$. Considering 1000 replications of each of these three samples (made of $10,000$ data each), they show that the estimates of $\xi$ obtained from the distribution of maxima (2.6) are compatible (at the 95% confidence level) with the theoretical value for the first two distributions (Pareto and SE with $c = 0.7$) as soon as the size $p$ of the subsamples, from which the maxima are drawn, is larger than 10. For the SE with fractional exponent $c = 0.3$, an average value $\xi$ larger than 0.2 is obtained even for large subsample sizes ($p = 200$). This value is reported to be significantly different from the theoretical value $\xi = 0.0$. These results clearly show that the distribution of the maximum drawn from a SE distribution with $c = 0.7$ converges quickly toward the theoretical asymptotic GEV distribution, while for $c = 0.3$ the convergence is very slow. A fast convergence for $c = 0.7$ is not surprising since, for this value of the fractional index $c$, the SE distribution remains close to the exponential distribution, which is known to converge very quickly to the GEV distribution [220]. For $c = 0.3$, the SE distribution behaves, over a wide range, like the power law (see page 59 hereafter for a theoretical formalization with an exact embedding of the power law into the SE family). Thus, it is not surprising to obtain an estimate of $\xi$ which remains significantly positive for SE distributions with small exponents $c$'s.

Overall, the results reported in [329] are slightly better for the maximum likelihood estimates obtained from the GPD. Indeed, the bias observed for the SE with $c = 0.3$ seems smaller for large quantiles than the smallest biases reached by the GEV method. Thus, it appears that the distribution of exceedance converges faster to its asymptotic distribution than the distribution of maximum. However, while in line with the theoretical values, the standard deviations are found to be almost always larger than in the previous case, which testifies of the higher variability of this estimator. Thus, for sample of sizes of $10,000$ or so – a typical size for most financial samples – the GEV and GPD maximum likelihood estimates should be handled with care and their results interpreted with caution due to possibly important bias and statistical fluctuations. If a small value of $\xi$ seems to allow one to reliably conclude in favor of a rapidly varying distribution, a positive estimate does not appear informative, and in particular does not allow one to reject the rapidly varying behavior of a distribution. Pickands' estimator does not perform better, in so far as it is also unable to distinguish between a regularly varying distribution and a SE with a low fractional exponent [329].

As another example illustrating the very slow convergence to the limit distributions of the extreme value theory mentioned above, *even with very large samples*, let us consider a simulated sample of iid random variables (we thus fulfill the most basic assumption of extreme values theory, i.e, iid-ness) with Weibull distribution defined by

$$F_u(x) = 1 - \exp\left[-\left(\frac{x}{d}\right)^c\right],\qquad(2.22)$$

with parameter set $(c > 0, d > 0)$, for $x \geq 0$. This distribution belongs to the class of Stretched-Exponential distributions when the exponent $c$ is smaller than one, namely when the distribution decays more slowly than an exponential distribution (but still faster than any power law). We consider two values for the exponent of the Weibull distribution: $c = 0.7$ and $c = 0.3$, with $d = 1$. Theoretically, using for instance the GPD of exceedances should give estimated values of $\xi$ close to zero in the limit of large $N$, since the SE distribution belongs to the basin of attraction of the Gumbel distribution. In order to use the GPD, we construct the conditional Weibull distribution under the condition $X > U_k, k = 1, \ldots, 15$, where the thresholds $U_k$ are chosen as: $U_1 = 0.1$; $U_2 = 0.3$; $U_3 = 1$; $U_4 = 3$; $U_5 = 10$; $U_6 = 30$; $U_7 = 100$; $U_8 = 300$; $U_9 = 1000$; $U_{10} = 3000$; $U_{11} = 10^4$; $U_{12} = 3 \cdot 10^4$; $U_{13} = 10^5$; $U_{14} = 3 \cdot 10^5$ and $U_{15} = 10.^6$

For each simulation, the size of the sample above a given threshold $U_k$ is set equal to 50,000 in order to get small standard deviations. The maximum-likelihood estimates of the GPD form parameter $\xi$ are shown in Fig. 2.4 as a function of the index $k$ of $U_k$. For $c = 0.7$, the threshold $U_7$ gives an estimate $\xi = 0.0123$ with standard deviation equal to 0.0045, i.e., the estimate for $\xi$ differs significantly from zero (recall that $\xi = 0$ is the theoretical limit value).

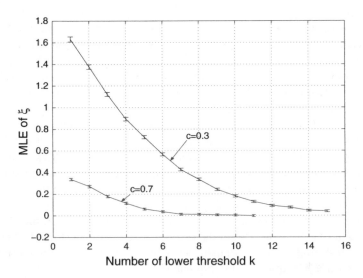

**Fig. 2.4.** Maximum likelihood estimates of the GPD form parameter $\xi$ in (2.15) as a function of the index $k$ of the thresholds $U_k$ defined in the text for stretched-exponential samples of size 50,000 and their 95% confidence interval. Reproduced from [329]

Stronger deviations from the correct value $\xi = 0$ are found for the smaller thresholds $U_1, ..., U_6$ while the discrepancy abates for larger thresholds $U_k$'s for $k > 7$. These results occur notwithstanding the huge size of the implied data set; indeed, the probability $\Pr(X > U_7)$ for $c = 0.7$ is about $10^{-9}$, so that in order to obtain a data set of conditional samples from an unconditional data set of the size studied here (50,000 realizations above $U_7$), the size of such an unconditional sample should be approximately $10^9$ times larger than the number of "peaks over threshold." It is practically impossible to have such a sample. For $c = 0.3$, the convergence to the theoretical value zero is much slower and the discrepancy with the correct value $\xi = 0$ remains even for the largest financial data sets: for a single asset, the largest data sets, drawn from high frequency data, are no larger than or of the order of one million points;[11] the situation does not improve for data sets one or two orders of magnitudes larger as considered in [211], obtained by aggregating thousands of stocks.[12] Thus, although the GPD form parameter should be theoretically zero in the limit of a large sample for the Weibull distribution, this limit cannot be reached for any available sample sizes. This is another clear illustration that a rapidly varying distribution, like the SE distribution, can be mistaken for a regularly varying distribution for any practical applications.

### 2.3.3 Can Long Memory Processes Lead to Misleading Measures of Extreme Properties?

As we already mentioned, Kearns and Pagan [267] have reported how misleading could be Hill's and Pickands' estimators in the presence of dependence in data. Focusing on IGARCH processes, they show that the estimated standard deviations of these estimators increase significantly with respect to the theoretical standard deviations derived under the iid assumption. They also find an important bias. Generalizing these results, the study by Malevergne et al. [329] shows that the presence of simple Markovian time dependences is sufficient to draw erroneous conclusions from GEV or GPD maximum likelihood estimates and Pickands estimates as well. Considering Markovian processes with different stationary distributions including a regularly varying distribution with the tail index $b = 3$ and two SEs with fractional exponents $c = 0.3$ and $c = 0.7$, they report the presence of a significant downward bias (with respect to the iid case) in almost every situation for the GPD estimates: the stronger the dependence (measured by the correlation time varying from 20 to 100), the more important is the bias. At the same time, the empirical values of the standard deviations remain comparable with those obtained for iid

---

[11] One year of data sampled at the 1-minute time scale gives approximately $1.2 \cdot 10^5$ data points.

[12] In this case, another issue arises concerning the fact that the aggregation of returns from different assets may distort the information and the very structure of the tails of the probability density functions (pdf), if they exhibit some intrinsic variability [351].

data. The downward bias can be ascribed to the dependence between data. Indeed, positive dependence yields important clustering of extremes and accumulation of realizations around some values, which – for small samples – could (misleadingly) appear as the consequence of the compactness of the support of the underlying distribution. In other words, for finite samples, the dependence prevents the full exploration of the tails and creates clusters that mimic a thinner tail (even if the clusters are all occurring at large values since the range of exploration of the tail controls the value of $\xi$).

The situation is different for the GEV estimates which exhibit biases which can be either upward or downward (with respect to the iid case). For the GEV estimates, two effects are competing. On the one hand, the dependence creates a downward bias, as explained above, while, on the other hand, the lack of convergence of the distribution of maxima toward its GEV asymptotic distribution results in an upward bias, as observed on iid data (see the previous section). This last phenomenon is strengthened by the existence of time dependence which leads to decrease the "effective" sample size (the actual size divided by the correlation time of the time series) and thus slows down the convergence rate toward the asymptotic distribution even more. Interestingly, both the GEV and GPD estimators for the Pareto distribution may be utterly wrong in presence of long-range dependence for any cluster sizes.

The same kind of results are reported for Pickands' estimator. However, the estimated standard deviations reported in [329] remain of the same order as the theoretical ones, contrarily to results reported by [267] for IGARCH processes. Nonetheless, in both studies, a very significant bias, either positive or negative, is found, which can lead to misclassify a SE distribution for a regularly varying distribution. Thus, in presence of dependence, Pickands' estimator becomes unreliable.

To summarize, the determination of the maximum domain of attraction with usual estimators does not appear to be a very efficient way to study the extreme properties of financial time series. Many studies on the tail behavior of the distributions of asset returns have focused on these methods (see the influential study [312] for instance) and may thus have led to spurious conclusions. In particular, the fact that rapidly varying distribution functions may be mistaken for regularly varying distribution functions casts doubts on the strength of the seeming consensus according to which the distributions of returns are regularly varying. It also casts doubts on the reliability of EVT for risk assessment. If an accurate estimation of the shape parameter $\xi$ is so difficult to reach, how can one hope to obtain trustful estimates of the Value-at-Risk or expected-shortfall by use of EVT?

### 2.3.4 GEV and GPD Estimators of the Distributions of Returns of the Dow Jones and Nasdaq Indices

As an illustration, let us apply the GEV and GDP estimators to the daily returns of the Dow Jones Industrial Average Index over the last century and

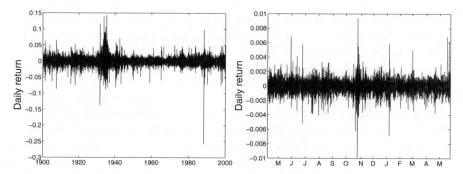

**Fig. 2.5.** Daily returns of the Dow Jones Industrial Average Index from 1900 to 2000 (*left panel*) and 5-minute returns of the Nasdaq Composite index over 1 year from April 1997 to May 1998 (*right panel*)

to the 5-minute returns of the Nasdaq Composite index over 1 year from April 1997 to May 1998. These two time series are depicted on Fig. 2.5.

For the intraday Nasdaq data, there are two caveats that must be addressed before any estimation can be made. First, in order to remove the effect of overnight price jumps, the intraday returns have to be determined separately for each of 289 days contained in the Nasdaq data. Then, the union of all these 289 return data sets provide a better global return data set. Second, the volatility of intraday data is known to exhibit a U-shape, also called "lunch effect", that is, an abnormally high volatility at the beginning and the end of the trading day compared with a low volatility at the approximate time of lunch. Such an effect is present in this data set and it is desirable to correct it. Such a correction has been performed by renormalizing the 5-minute returns at a given instant of the trading day by the corresponding average absolute return at the same instant (when the average is performed over the 289 days). We shall refer to this time series as the corrected Nasdaq returns in contrast with the raw (incorrect) Nasdaq returns and we shall examine both data sets for comparison.

The daily returns of the Dow Jones also exhibit some non-stationarity. Indeed, one can observe a clear excess volatility roughly covering the time of the bubble ending in the October 1929 crash followed by the Great Depression. To investigate the influence of this non-stationarity, the statistical study presented below has been performed twice: first with the entire sample, and then after having removed the period from 1927 to 1936 from the sample. The results are somewhat different, but on the whole, the conclusions about the nature of the tail are the same.

Although the distributions of positive and negative returns are known to be very similar (see for instance [256]), we have chosen to treat them separately. For the Dow Jones, this gives us 14949 positive and 13464 negative data points while, for the Nasdaq index, we have 11241 positive and 10751 negative data points.

Given these precautionary measures, the analysis of the previous section has been applied to the the Dow Jones and Nasdaq (raw and corrected) returns. In order to estimate the standard deviations of Pickands' estimator for the GPD derived from the upper quantiles of these distributions, and of the Maximum Likelihood estimators of the distribution of the maximum and of the GPD, we have randomly generated 1000 subsamples, each subsample being constituted of 10,000 data points in the positive or negative parts of the samples respectively (with replacement). It should be noted that the Maximum Likelihood estimates themselves were derived from the full samples. The results are given in Tables 2.2 and 2.3.

These results confirm the difficulties in obtaining a clear conclusion concerning the nature of the tail behavior of the distributions of returns. In particular, it seems impossible to exclude a rapidly varying behavior of their tails. Even the estimations obtained with the maximum likelihood of the GPD tail index do not allow one to reject clearly the hypothesis that the tails of the empirical distributions of returns are rapidly varying, in particular for large quantile values. For the Nasdaq data set, accounting for the lunch effect does not yield any significant change in the estimations.

## 2.4 Fitting Distributions of Returns with Parametric Densities

Since it is particularly difficult to conclude with enough certainty on the regularly or rapidly varying behavior of the tails of distributions of asset returns by using the nonparametric methods of the extreme value theory, it may be more appropriate to consider a parametric approach. However, in order to avoid – or at least to lower – the risk of misspecification inherent in any parametric approach, it is mandatory to use models as versatile as possible. In particular, it is necessary to consider models which encompass both regularly and rapidly varying distributions. Many examples of such models have been described in the literature, such as the generalized $t$-distribution of McDonald and Newey [353] or of the $q$-exponential and $q$-Weibull distributions [385].

In the remaining of this section, relying on the results presented in [330], we introduce two versatile families to characterize the behavior of the tails of asset return distributions. The implications of the choice of these parametric families for the assessment of tail risk will be discussed at the end of this chapter.

### 2.4.1 Definition of Two Parametric Families

#### A General 3-Parameters Family of Distributions

We consider a general 3-parameters family of distributions and its particular restrictions corresponding to some fixed value(s) of one (two) parameters. This family is defined by its density function given by:

**Table 2.2.** Mean values and standard deviations of the maximum likelihood estimates of the parameter $\xi$ for the distribution of the maximum (cf. (2.6)) when data are grouped in samples of size $20, 40, 200,$ and $400$ and for the generalized pareto distribution (2.15) for thresholds $u$ corresponding to quantiles $90\%, 95\%, 99\%,$ and $99.5\%$

| (a) | | | | | Dow Jones | | | | |
|-----|-----|-----|-----|-----|-----|-----|-----|-----|-----|
| | Positive Tail | | | | | Negative Tail | | | |
| | GEV | | | | | GEV | | | |
| cluster | 20 | 40 | 200 | 400 | cluster | 20 | 40 | 200 | 400 |
| $\xi$ | 0.273 | 0.280 | 0.304 | 0.322 | $\xi$ | 0.262 | 0.295 | 0.358 | 0.349 |
| Emp Std | 0.029 | 0.039 | 0.085 | 0.115 | Emp Std | 0.030 | 0.045 | 0.103 | 0.143 |
| | GPD | | | | | GPD | | | |
| quantile | 0.9 | 0.95 | 0.99 | 0.995 | quantile | 0.9 | 0.95 | 0.99 | 0.995 |
| $\xi$ | 0.248 | 0.247 | 0.174 | 0.349 | $\xi$ | 0.214 | 0.204 | 0.250 | 0.345 |
| Emp Std | 0.036 | 0.053 | 0.112 | 0.194 | Emp Std | 0.041 | 0.062 | 0.156 | 0.223 |
| Theor Std | 0.032 | 0.046 | 0.096 | 0.156 | Theor Std | 0.033 | 0.046 | 0.108 | 0.164 |

| (b) | | | | | Nasdaq (raw data) | | | | |
|-----|-----|-----|-----|-----|-----|-----|-----|-----|-----|
| | GEV | | | | | GEV | | | |
| cluster | 20 | 40 | 200 | 400 | cluster | 20 | 40 | 200 | 400 |
| $\xi$ | 0.209 | 0.193 | 0.388 | 0.516 | $\xi$ | 0.191 | 0.175 | 0.292 | 0.307 |
| Emp Std | 0.031 | 0.115 | 0.090 | 0.114 | Emp Std | 0.030 | 0.038 | 0.094 | 0.162 |
| | GPD | | | | | GPD | | | |
| quantile | 0.9 | 0.95 | 0.99 | 0.995 | quantile | 0.9 | 0.95 | 0.99 | 0.995 |
| $\xi$ | 0.200 | 0.289 | 0.389 | 0.470 | $\xi$ | 0.143 | 0.202 | 0.229 | 0.242 |
| Emp Std | 0.040 | 0.058 | 0.120 | 0.305 | Emp Std | 0.040 | 0.057 | 0.143 | 0.205 |
| Theor Std | 0.036 | 0.054 | 0.131 | 0.196 | Theor Std | 0.035 | 0.052 | 0.118 | 0.169 |

| (c) | | | | | Nasdaq (corrected data) | | | | |
|-----|-----|-----|-----|-----|-----|-----|-----|-----|-----|
| | GEV | | | | | GEV | | | |
| cluster | 20 | 40 | 200 | 400 | cluster | 20 | 40 | 200 | 400 |
| $\xi$ | 0.090 | 0.175 | 0.266 | 0.405 | $\xi$ | 0.099 | 0.132 | 0.138 | 0.266 |
| Emp Std | 0.029 | 0.039 | 0.085 | 0.187 | Emp Std | 0.030 | 0.041 | 0.079 | 0.197 |
| | GPD | | | | | GPD | | | |
| quantile | 0.9 | 0.95 | 0.99 | 0.995 | quantile | 0.9 | 0.95 | 0.99 | 0.995 |
| $\xi$ | 0.209 | 0.229 | 0.307 | 0.344 | $\xi$ | 0.165 | 0.160 | 0.210 | 0.054 |
| Emp Std | 0.039 | 0.052 | 0.111 | 0.192 | Emp Std | 0.039 | 0.052 | 0.150 | 0.209 |
| Theor Std | 0.036 | 0.052 | 0.123 | 0.180 | Theor Std | 0.036 | 0.050 | 0.116 | 0.143 |

Panel (a) gives the results for the Dow Jones index, panel (b) for the raw Nasdaq index, and in panel (c) for the Nasdaq index corrected for the "lunch effect." Reproduced from [329]

**Table 2.3.** Pickand's estimates (2.19) of the parameter $\xi$ for the generalized Pareto distribution (2.15) for thresholds $u$ corresponding to quantiles $90\%, 95\%, 99\%$ and $99.5\%$ and two different values of the ratio $N/k$ respectively equal to 4 and 10

| (a) | | | | | Dow Jones | | | | |
|---|---|---|---|---|---|---|---|---|---|
| | Negative Tail | | | | | Positive Tail | | | |
| quantile | 0.9 | 0.95 | 0.99 | 0.995 | quantile | 0.9 | 0.95 | 0.99 | 0.995 |
| $N/k$ | 4 | | | | $N/k$ | 4 | | | |
| $\xi$ | 0.2314 | 0.2944 | –0.1115 | 0.3314 | $\xi$ | 0.2419 | 0.4051 | –0.3752 | 0.5516 |
| emp. Std | 0.1073 | 0.1550 | 0.3897 | 0.6712 | emp. Std | 0.0915 | 0.1274 | 0.3474 | 0.5416 |
| th. Std | 0.1176 | 0.1680 | 0.3563 | 0.5344 | th. Std | 0.1178 | 0.1712 | 0.3497 | 0.5562 |
| | | | | | | | | | |
| $N/k$ | 10 | | | | $N/k$ | 10 | | | |
| mean | 0.3119 | 0.0890 | –0.3452 | 0.9413 | $\xi$ | 0.3462 | 0.3215 | 0.9111 | –0.3873 |
| emp. Std | 0.1523 | 0.2219 | 0.8294 | 1.1352 | emp. Std | 0.1766 | 0.1929 | 0.6983 | 1.6038 |
| th. Std | 0.1883 | 0.2577 | 0.5537 | 0.9549 | th. Std | 0.1894 | 0.2668 | 0.6706 | 0.7816 |

| (b) | | | | | Nasdaq (raw data) | | | | |
|---|---|---|---|---|---|---|---|---|---|
| $N/k$ | 4 | | | | $N/k$ | 4 | | | |
| $\xi$ | 0.0493 | 0.0539 | –0.0095 | 0.4559 | $\xi$ | 0.0238 | 0.1511 | 0.1745 | 1.1052 |
| emp. Std | 0.1129 | 0.1928 | 0.4393 | 0.6205 | emp. Std | 0.1003 | 0.1599 | 0.4980 | 0.6180 |
| th. Std | 0.1147 | 0.1623 | 0.3601 | 0.5462 | th. Std | 0.1143 | 0.1644 | 0.3688 | 0.6272 |
| | | | | | | | | | |
| $N/k$ | 10 | | | | $N/k$ | 10 | | | |
| $\xi$ | 0.2623 | 0.1583 | –0.8781 | 0.8855 | $\xi$ | 0.2885 | 0.1435 | 1.3734 | –0.8395 |
| emp. Std | 0.1940 | 0.3085 | 0.9126 | 1.5711 | emp. Std | 0.2166 | 0.3220 | 0.7359 | 1.5087 |
| th. Std | 0.1868 | 0.2602 | 0.5543 | 0.9430 | th. Std | 0.1876 | 0.2596 | 0.7479 | 0.7824 |

| (c) | | | | | Nasdaq (Corrected data) | | | | |
|---|---|---|---|---|---|---|---|---|---|
| $N/k$ | 4 | | | | $N/k$ | 4 | | | |
| $\xi$ | 0.2179 | 0.0265 | 0.3977 | 0.1073 | $\xi$ | 0.2545 | –0.0402 | –0.0912 | 1.3915 |
| emp. Std | 0.1211 | 0.1491 | 0.4585 | 0.7206 | emp. Std | 0.1082 | 0.1643 | 0.4317 | 0.6220 |
| th. Std | 0.1174 | 0.1617 | 0.3822 | 0.5167 | th. Std | 0.1180 | 0.1605 | 0.3570 | 0.6720 |
| | | | | | | | | | |
| $N/k$ | 10 | | | | $N/k$ | 10 | | | |
| $\xi$ | –0.0878 | 0.4619 | 0.0329 | 0.3742 | $\xi$ | 0.0877 | 0.3907 | 1.4680 | 0.1098 |
| emp. Std | 0.1882 | 0.2728 | 0.7561 | 1.1948 | emp. Std | 0.1935 | 0.2495 | 0.8045 | 1.2345 |
| th. Std | 0.1786 | 0.2734 | 0.5722 | 0.8512 | th. Std | 0.1822 | 0.2699 | 0.7655 | 0.8172 |

Panel (a) gives the results for the Dow Jones, panel (b) for the raw Nasdaq data and panel (c) for the Nasdaq corrected for the "lunch effect." Reproduced from [329]

$$f_u(x|b,c,d) = \begin{cases} A(b,c,d,u) \; x^{-(b+1)} \exp\left[-\left(\frac{x}{d}\right)^c\right] & \text{if } x \geqslant u > 0 \\ 0 & \text{if } x < u . \end{cases} \qquad (2.23)$$

Here, $b, c$, and $d$ are unknown parameters, $u$ is a known lower threshold that will be varied for the purposes of analysis and $A(b,c,d,u)$ is a normalizing constant given by the expression:

$$A(b,c,d,u) = \frac{d^b \, c}{\Gamma(-b/c,(u/d)^c)} , \qquad (2.24)$$

where $\Gamma(a,x)$ denotes the (non-normalized) incomplete Gamma function:

$$\Gamma(a,x) = \int_x^\infty t^{a-1} e^{-t} \, dt . \qquad (2.25)$$

The parameter $b$ ranges from minus infinity to infinity while $c$ and $d$ range from zero to infinity. In the particular case where $c = 0$, the parameter $b$ also needs to be positive to ensure the normalization of the probability density function. The family (2.23) includes several well-known pdfs often used in different applications. We enumerate them.

1. The Pareto distribution:

$$F_u(x) = 1 - (u/x)^b , \qquad (2.26)$$

   which corresponds to the set of parameters $(b > 0, c = 0)$ with $A(b,c,d,u) = b \cdot u^b$. Several works have attempted to derive or justify the existence of a power tail of the distribution of returns from agent-based models [91], from optimal trading of large funds with sizes distributed according to the Zipf law, as recalled in Sect. 2.2.2, or from $ad\ hoc$ stochastic processes [55, 445].

2. The Weibull distribution:

$$F_u(x) = 1 - \exp\left[-\left(\frac{x}{d}\right)^c + \left(\frac{u}{d}\right)^c\right] , \qquad (2.27)$$

   with parameter set $(b = -c, c > 0, d > 0)$ and normalization constant $A(b,c,d,u) = \frac{c}{d^c} \exp\left[\left(\frac{u}{d}\right)^c\right]$. Recall that this distribution is said to be a Stretched-Exponential distribution when the exponent $c$ is smaller than one, namely when the distribution decays more slowly than an exponential distribution. Stationary distributions exhibiting this kind of tails arise, for instance, from the so called $\alpha$-ARCH processes introduced in [132].
   From a theoretical viewpoint, this class of distributions is motivated in part by the fact that the large deviations of multiplicative processes are generically distributed with Stretched-Exponential distributions [191]. Stretched-Exponential distributions are also parsimonious examples of the important subset of subexponentials, that is, of the general class of distributions decaying slower than an exponential [487]. This class of subexponentials share several important properties of heavy-tailed distributions

[146], not shared by exponentials or distributions decreasing faster than exponentials: for instance, they have "fat tails" in the sense of the asymptotic probability weight of the maximum compared with the sum of large samples [167] (see also [451], Chaps. 1 and 6).

Notwithstanding their fat-tailness, SE distributions have all their moments finite,[13] in contrast with regularly varying distributions for which moments of order equal to or larger than the tail index $b$ are not defined. This property may provide a substantial advantage to exploit in generalizations of the mean-variance portfolio theory using higher-order moments (see for instance [6, 162, 241, 259, 333, 421, 453] among many others). In addition, the existence of all moments is an important property allowing for an efficient estimation of any high-order moment, since it ensures that the estimators are asymptotically Gaussian. In particular, for Stretched-Exponentially distributed random variables, the variance, skewness and kurtosis can be accurately estimated, contrarily to random variables with regularly varying distribution with tail index in the range 3–5 [356].

3. The Exponential distribution:

$$F_u(x) = 1 - \exp\left(-\frac{x}{d} + \frac{u}{d}\right) , \qquad (2.28)$$

with parameter set ($b = -1$, $c = 1$, $d > 0$) and normalization constant $A(b, c, d, u) = \frac{1}{d}\exp\left(-\frac{u}{d}\right)$. For sufficiently high quantiles, the exponential behavior can, for instance, derive from the hyperbolic model introduced by Eberlein *et al.* [140] or from a simple model where stock price dynamics is governed by a diffusion with stochastic volatility. Dragulescu and Yakovenko [136] have found an excellent fit of the Dow Jones index for time lags from 1 to 250 trading days with a model exhibiting an asymptotic exponential tail of the distribution of log-returns.

4. The incomplete Gamma distribution:

$$F_u(x) = 1 - \frac{\Gamma(-b, x/d)}{\Gamma(-b, u/d)} \qquad (2.29)$$

with parameter set ($b$, $c = 1$, $d > 0$) and normalization $A(b, c, d, u) = \frac{d^b}{\Gamma(-b, u/d)}$. Such an asymptotic tail behavior can, for instance, be observed for the generalized hyperbolic models, whose description can be found in [393].

The Pareto distribution (PD) and Exponential distribution (ED) are one-parameter families, whereas the Weibull/Stretched-exponential (SE) and the incomplete Gamma distribution (IG) are two-parameter families. The comprehensive distribution (CD) given by (2.23) contains three unknown parameters.

---

[13] However, they do not admit an exponential moment, which leads to problems in the reconstruction of the distribution from the knowledge of their moments [465].

Links between these different models reveal themselves under specific asymptotic conditions. Very interesting is the behavior of the SE model when $c \to 0$ and $u > 0$. In this limit, and provided that

$$c \cdot \left(\frac{u}{d}\right)^c \to \beta, \quad \text{as } c \to 0 \,, \tag{2.30}$$

where $\beta$ is a positive constant, the SE model tends to the Pareto model. Indeed, we can write

$$\frac{c}{d^c} \cdot x^{c-1} \cdot \exp\left(-\frac{x^c - u^c}{d^c}\right) = c\left(\frac{u}{d}\right)^c \cdot \frac{x^{c-1}}{u^c} \exp\left[-\left(\frac{u}{d}\right)^c \cdot \left(\left(\frac{x}{u}\right)^c - 1\right)\right],$$

$$\simeq \beta \cdot x^{-1} \exp\left[-c\left(\frac{u}{d}\right)^c \cdot \ln\frac{x}{u}\right], \quad \text{as } c \to 0$$

$$\simeq \beta \cdot x^{-1} \exp\left[-\beta \cdot \ln\frac{x}{u}\right] \,,$$

$$\simeq \beta \frac{u^\beta}{x^{\beta+1}} \,, \tag{2.31}$$

which is the pdf of the Pareto model with tail index $\beta$. The condition (2.30) comes naturally from the properties of the maximum likelihood estimator of the scale parameter $d$ given by (2.B.53) in Appendix 2.B. It implies that, as $c \to 0$, the characteristic scale $d$ of the SE model must also go to zero with $c$ to ensure the convergence of the SE model toward the Pareto model.

The Pareto model with exponent $\beta$ can therefore be approximated with any desired accuracy on any finite interval $[u, U]$, $U > u > 0$, by the SE model with parameters $(c, d)$ satisfying $c\left(\frac{u}{d}\right)^c = \beta$ (cf. (2.30), where the arrow is replaced by an equality). Although the value $c = 0$ does not give, strictly speaking, a SE distribution, the limit $c \longrightarrow 0$ provides any desired approximation to the Pareto distribution, uniformly on any finite interval $[u, U]$. This deep relationship between the SE and PD models allows us to understand why it can be very difficult to decide, on a statistical basis, which of these models fits the data best.

Another interesting behavior is obtained in the limit $b \to +\infty$, where the Pareto model tends to the exponential model [72]. Indeed, provided that the scale parameter $u$ of the power law is simultaneously scaled as $u^b = (b/\alpha)^b$, we can write the tail of the cumulative distribution function of the PD as $u^b/(u+x)^b$ which is indeed of the form $u^b/x^b$ for large $x$. Then,

$$\frac{u^b}{(u+x)^b} = \left(1 + \alpha\frac{x}{b}\right)^{-b} \to \exp(-\alpha x) \quad \text{for} \quad b \to +\infty \,. \tag{2.32}$$

This shows that the exponential model can be approximated with any desired accuracy on intervals $[u, u + A]$ by the PD model with parameters $(\beta, u)$ satisfying $u^b = (b/\alpha)^b$, for any positive constant A. Although the value $b \to +\infty$ does not give, strictly speaking, an exponential distribution, the limit $u \propto b \longrightarrow +\infty$ provides any desired approximation to the exponential distribution, uniformly on any finite interval $[u, u + A]$. This limit is thus less general than the SE $\rightarrow$ PD limit since it is valid only asymptotically for $u \longrightarrow +\infty$, while $u$ can be finite in the SE $\rightarrow$ PD limit.

## The Log-Weibull Family of Distributions

Another interesting family is the two-parameter log-Weibull family:

$$F_u(x) = 1 - \exp\left[-b\left(\ln(x/u)\right)^c\right] , \quad \text{for } x \geq u . \tag{2.33}$$

whose density is

$$f_u(x|b,c,d) = \begin{cases} \frac{b \cdot c}{x} \left(\ln \frac{x}{u}\right)^{c-1} \exp\left[-b\left(\ln \frac{x}{u}\right)^c\right], & \text{if } x \geqslant u > 0 \\ 0, & \text{if } x < u . \end{cases} \tag{2.34}$$

This family of pdf interpolates smoothly between the SE and the Pareto classes. It recovers the Pareto family for $c = 1$, in which case the parameter $b$ is the tail exponent. For $c$ larger than one, the tail of the log-Weibull is thinner than any Pareto distribution but heavier than any Stretched-Exponential.[14] In particular, when $c$ equals two, the log-normal distribution is retrieved (above threshold $u$). For $c$ smaller than one, the tails of the log-Weibull distributions are even heavier than any regularly varying distribution. It is interesting to note that in this case the log-Weibull distributions do not belong to the domain of attraction of a law of the maximum. Therefore, the standard extreme values theory cannot apply to such distributions. If it would appear that the log-Weibull distributions with an index $c < 1$ provides a reasonable description of the tails of distributions of returns, this would mean that risk management methods based upon EVT are particularly unreliable (see below).

### 2.4.2 Parameter Estimation Using Maximum Likelihood and Anderson-Darling Distance

It is instructive to fit the two data sets used in Sect. 2.3.4 – *i.e.* the Dow Jones daily returns and the Nasdaq 5-minute returns – in addition to a sample of returns of the Standard & Poor's 500[15] over the two decades 1980–1999 by the distributions enumerated above (2.23), (2.26–2.29) and (2.34). We will show that no single parametric representation among any of the cited pdfs fits the *whole range* of the data sets. Positive and negative returns will be analyzed separately, the later being converted to the positive semi-axis. The analysis

---

[14] A generalization of the log-Weibull distributions to the following three-parameter family also contains the SE family in some formal limit. Consider indeed $1 - F(x) = \exp(-b(\ln(1 + x/D))^c)$ for $x > 0$, which has the same tail as expression (2.33). Taking $D \rightarrow +\infty$ together with $b = (D/d)^c$ with $d$ finite yields $1 - F(x) = \exp(-(x/d))^c)$.

[15] The returns on the Standard & Poor's 500 are calculated at five different time scales: 1 minute, 5 minutes, 30 minutes, an hour and 1 day.

uses a *movable* lower threshold $u$, restricting by this threshold the study to the observations satisfying the condition $x > u$.

In addition to estimating the parameters involved in each representation (2.23, 2.26–2.29, 2.34) by maximum likelihood[16] for each particular threshold $u$, it is important to characterize the goodness-of-fit. There are many measures of goodness-of-fit; a natural class consists in the distances between the estimated distribution and the sample distribution. Many distances can be used: mean-squared error, Kullback-Leibler distance,[17] Kolmogorov distance, Sherman distance (as in [312]) or Anderson-Darling distance, to cite a few. The parameters of each pdf can also be determined according to the criterion of minimizing the distance between the estimated distribution and the sample distribution. The chosen distance is thus useful both for characterizing and for estimating the parametric pdf. In this case, once an estimation of the parameters of a particular distribution family has been obtained according to the selected distance, the quantification of the statistical significance of the fit requires to derive the statistics associated with the chosen distance. These statistics are known for most of the examples cited above, in the limit of large sample.

In addition to the maximum likelihood method (which is associated as we said with the Kullback-Leibler distance), it is instructive to use the Anderson-Darling distance to estimate the parameters and perform the tests of goodness-of-fit. The Anderson-Darling distance between a theoretical distribution function $F(x)$ and its empirical analog $F_N(x)$, estimated from a sample of $N$ realizations, is defined by

$$\text{ADS} = N \cdot \int \frac{[F_N(x) - F(x)]^2}{F(x)(1 - F(x))} \, dF(x) \tag{2.35}$$

and evaluated as

$$= -N - 2 \sum_1^N \left\{ w_k \log(F(x_k)) + (1 - w_k) \log(1 - F(x_k)) \right\} , \tag{2.36}$$

where $w_k = 2k/(2N + 1)$, $k = 1 \ldots N$ and $x_1 \leqslant \ldots \leqslant x_N$ is its ordered sample. If the sample is drawn from a population with distribution function $F(x)$, the Anderson-Darling statistic (ADS) has a standard AD-distribution *free of the theoretical distribution function $F(x)$* [11], similarly to the $\chi^2$ for the $\chi^2$-statistic, or the Kolmogorov distribution for the Kolmogorov statistic. It should be noted that the ADS weights $[F_N(x) - F(x)]^2$ in (2.35) by $N/F(x)(1 - F(x))$ which is nothing but the inverse of its variance. Thus, the AD distance emphasizes more the tails of the distribution than, say, the Kolmogorov distance which is determined by the *maximum absolute* deviation of

---

[16] The estimators and their asymptotic properties are summarized in Appendix 2.B.

[17] This distance (or *divergence*) is the natural distance associated with the maximum likelihood estimation since it is for the maximum likelihood values that the distance between the true model and the assumed model reaches its minimum.

$F_N(x)$ from $F(x)$ or the mean-squared error, which is mostly controlled by the middle range of the distribution.

Since we have to insert the estimated parameters into the ADS, this statistic does not obey any more the standard AD-distribution: the ADS decreases because the use of the fitting parameters ensures a better fit to the sample distribution (we will come back later, with more details, on this topic in Chap. 5). However, we can still use the standard quantiles of the AD-distribution as *upper* boundaries of the ADS. If the observed ADS is larger than the standard quantile with a high significance level $(1 - \varepsilon)$, we can then conclude that the null hypothesis $F(x)$ is rejected with a significance level larger than $(1 - \varepsilon)$. If one wishes to estimate the real significance level of the ADS in the case where it does not exceed the standard quantile of a high significance level, one is forced to use some other method, such as the bootstrap method.

In the following, the estimates minimizing the Anderson-Darling distance will be referred to as AD-estimates. The maximum likelihood estimates (ML-estimates) are asymptotically more efficient than AD-estimates for independent data and under the condition that the null hypothesis (given by one of the four distributions (2.26–2.29), for instance) corresponds to the true data-generating model. When this is not the case, the AD-estimates can provide a *better practical tool* for approximating sample distributions compared with the ML-estimates. These estimates will be reported for the thresholds $u(q_k)$ determined by the probability levels $q_1 = 0$, $q_2 = 0.1$, $q_3 = 0.2$, $q_4 = 0.3$, $q_5 = 0.4$, $q_6 = 0.5$, $q_7 = 0.6$, $q_8 = 0.7$, $q_9 = 0.8$, $q_{10} = 0.9$, $q_{11} = 0.925$, $q_{12} = 0.95$, $q_{13} = 0.96$, $q_{14} = 0.97$, $q_{15} = 0.98$, $q_{16} = 0.99$, $q_{17} = 0.9925$, $q_{18} = 0.995$, $q_{19} = 0.999$, $q_{20} = 0.9995$ and $q_{21} = 0.9999$.

Despite the fact that threshold $u(q_k)$ varies from sample to sample, it always corresponds to the same fixed probability level $q_k$ which allows one to compare the goodness-of-fit for samples of different sizes. In the statistics presented below, only subsamples with at least 100 data points or so are considered, in order to allow for a sufficiently accurate assessment of the quantile under consideration.

### 2.4.3 Empirical Results on the Goodness-of-Fits

The Anderson-Darling statistics (ADS) for four parametric distributions (Weibull or Stretched-Exponential, Exponential, Pareto and Log-Weibull) are shown in Table 2.4 for two quantile ranges, the first top half of the table corresponding to the 90% lowest thresholds while the second bottom half corresponds to the 10% highest ones. For the lowest thresholds, the ADS rejects all distributions at the 95% confidence level, except the SE for the negative tail of the Standard & Poor's 500 for the 60-minute returns and for the Nasdaq. Thus, none of the considered distributions is adequate to model the data over such large ranges. For the 10% highest quantiles, the exponential model is rejected at the 95% confidence level except for the negative tails of the Dow Jones (daily returns) and the Nasdaq. The Log-Weibull and the SE

**Table 2.4.** Mean Anderson-Darling distances in the range of thresholds $u(q_1)-u(q_9)$ (90% lowest thresholds) and in the range $u(q_i) \geq u(q_{10})$ (10% highest thresholds)

Mean AD-statistics for $u_1-u_9$

| | S&P 500 1 min | | | | S&P 500 5 min | | | | S&P 500 30 min | | | |
| --- | --- | --- | --- | --- | --- | --- | --- | --- | --- | --- | --- | --- |
| | Pos. tail | | Neg. tail | | Pos. tail | | Neg. tail | | Pos. tail | | Neg. tail | |
| Weibull | 292.85 | (100%) | 299.46 | (100%) | 36.62 | (100%) | 41.04 | (100%) | 7.36 | (100%) | 4.84 | (100%) |
| Exponential | 771.70 | (100%) | 718.56 | (100%) | 86.79 | (100%) | 108.17 | (100%) | 17.47 | (100%) | 16.36 | (100%) |
| Pareto | 23998.94 | (100%) | 23337.60 | (100%) | 6834.06 | (100%) | 6563.26 | (100%) | 1847.40 | (100%) | 1298.47 | (100%) |
| Log-Weibull | 1559.70 | (100%) | 1470.11 | (100%) | 360.18 | (100%) | 331.45 | (100%) | 60.03 | (100%) | 67.22 | (100%) |

Mean AD-statistics for $u_i \geq u_{10}$

| | S&P 500 1 min | | | | S&P 500 5 min | | | | S&P 500 30 min | | | |
| --- | --- | --- | --- | --- | --- | --- | --- | --- | --- | --- | --- | --- |
| | Pos. tail | | Neg. tail | | Pos. tail | | Neg. tail | | Pos. tail | | Neg. tail | |
| Weibull | 6.80 | (100%) | 5.80 | (100%) | 1.81 | (88%) | 1.93 | (90%) | 0.67 | (42%) | 0.79 | (51%) |
| Exponential | 143.97 | (100%) | 136.66 | (100%) | 28.12 | (100%) | 30.88 | (100%) | 8.19 | (100%) | 9.75 | (100%) |
| Pareto | 19.97 | (100%) | 19.24 | (100%) | 8.10 | (100%) | 7.61 | (100%) | 1.63 | (85%) | 1.77 | (88%) |
| Log-Weibull | 3.60 | (99%) | 4.10 | (99%) | 1.20 | (73%) | 1.55 | (84%) | 0.64 | (39%) | 0.42 | (17%) |

Mean AD-statistics for $u_1-u_9$

| | S&P 500 60 min | | | | Nasdaq | | | | Dow Jones | | | |
| --- | --- | --- | --- | --- | --- | --- | --- | --- | --- | --- | --- | --- |
| | Pos. tail | | Neg. tail | | Pos. tail | | Neg. tail | | Pos. tail | | Neg. tail | |
| Weibull | 3.58 | (99%) | 2.36 | (94%) | 1.37 | (80%) | 0.85 | (55%) | 4.96 | (100%) | 3.86 | (99%) |
| Exponential | 8.12 | (100%) | 12.20 | (100%) | 5.41 | (100%) | 3.33 | (98%) | 16.48 | (100%) | 10.30 | (100%) |
| Pareto | 1001.68 | (100%) | 702.47 | (100%) | 475.00 | (100%) | 441.40 | (100%) | 691.30 | (100%) | 607.30 | (100%) |
| Log-Weibull | 34.44 | (100%) | 36.55 | (100%) | 35.90 | (100%) | 30.92 | (100%) | 32.30 | (100%) | 28.27 | (100%) |

Mean AD-statistics for $u_i \geq u_{10}$

| | S&P 500 60 min | | | | Nasdaq | | | | Dow Jones | | | |
| --- | --- | --- | --- | --- | --- | --- | --- | --- | --- | --- | --- | --- |
| | Pos. tail | | Neg. tail | | Pos. tail | | Neg. tail | | Pos. tail | | Neg. tail | |
| Weibull | 0.66 | (41%) | 0.68 | (42%) | 0.67 | (42%) | 0.50 | (29%) | 0.38 | (13%) | 0.35 | (10%) |
| Exponential | 4.99 | (100%) | 4.89 | (100%) | 3.06 | (97%) | 1.97 | (90%) | 3.06 | (97%) | 1.89 | (89%) |
| Pareto | 1.12 | (70%) | 1.28 | (76%) | 1.30 | (78%) | 1.33 | (78%) | 0.78 | (50%) | 1.26 | (75%) |
| Log-Weibull | 0.48 | (23%) | 0.57 | (32%) | 0.46 | (29%) | 0.49 | (30%) | 0.38 | (13%) | 0.69 | (43%) |

The figures within parenthesis characterize the goodness of fit: they represent the significance levels with which the considered model can be rejected. Note that these significance levels are only lower bounds since one or two parameters are fitted. Reproduced from [330] distributions are the best since they are only rejected at the 1-minute time scale for the Standard & Poor's 500. The Pareto distribution provides a reliable description for time scales larger than or equal to 30 minutes. However, it remains less accurate than the log-Weibull and the SE distributions, on average. Overall, it can be noted that the Nasdaq and the 60-minute returns of the Standard & Poor's 500 behave very similarly. Let us now analyze each distribution in more detail.

**Pareto Distribution**

Figures 2.3 and 2.6 show the complementary sample distribution functions $1 - F_N(x)$ for the Standard & Poor's 500 index at the 30-minute time scale and for the daily Dow Jones Industrial Average index, respectively. In Fig. 2.6, the mismatch between the Pareto distribution and the data can be seen with the naked eye: even in the tails, one observes a continuous downward curvature in the double logarithmic diagram, instead of a straight line as would be the case if the distribution ultimately behaved like a Pareto law. To formalize this impression, we calculate the ML and AD estimators for each threshold $u$. For the Pareto law, the ML estimator is well known to agree with Hill's estimator. Indeed, denoting $x_1 \geqslant \ldots \geqslant x_{N_u}$ the ordered subsample of values exceeding $u$ where $N_u$ is the size of this subsample, the Hill maximum likelihood estimate of the parameter $b$ is [233]

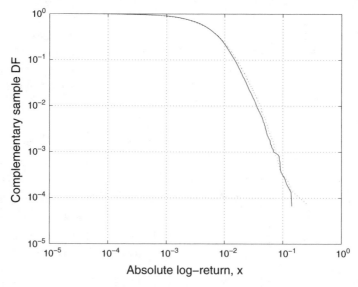

**Fig. 2.6.** Complementary sample distribution function for the daily returns of the Dow Jones index over the time period from 1900–2000. The plain (resp. dotted) line shows the complementary distribution for the positive (resp. the absolute value of negative) returns. Reproduced from [330]

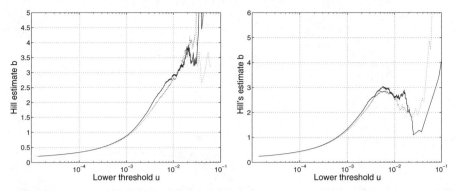

**Fig. 2.7.** Hill estimate $\hat{b}_u$ as a function of the threshold $u$ for the Dow Jones (*left panel*) and for the Standard & Poor's 1-minute returns(*right panel*)

$$\hat{b}_u = \left[ \frac{1}{N_u} \sum_{k=1}^{N_u} \log\left(\frac{x_k}{u}\right) \right]^{-1} . \tag{2.37}$$

Its standard deviation can be asymptotically estimated as

$$\mathrm{Std}(b_u) = \hat{b}_u / \sqrt{N_u} , \tag{2.38}$$

under the assumption of iid data, but very severely underestimate the true standard deviation when samples exhibit dependence, as reported by Kearns and Pagan [267] (see the previous section of this chapter).

Figure 2.7 shows the Hill estimates $\hat{b}_u$ as a function of $u$ for the Dow Jones and for the Standard & Poor's 500 1-minute returns. Instead of an approximately constant exponent (as would be the case for true Pareto samples), the tail index estimator, for the Dow Jones, increases until $u \cong 0.04$, beyond which it seems to slow its growth and oscillates around a value $\approx 3-4$ up to the threshold $u \cong .08$. It should be noted that the interval $[0, 0.04]$ contains 99.12% of the sample whereas the interval $[0.04, 0.08]$ contains only 0.64% of the sample. The behavior of $\hat{b}_u$ is very similar for the Nasdaq (not shown). The behavior of $\hat{b}_u$ for the Standard & Poor's 500 shown on the right panel of Fig. 2.7 is somewhat different: Hill's estimate $\hat{b}_u$ slows its growth at $u \cong 0.006$, corresponding to the 95% quantile, then decays until $u \cong 0.05$ (99.99% quantile) and then strongly increases again. Are these slowdowns of the growth of $\hat{b}_u$ genuine signatures of a possible constant well-defined asymptotic value that would qualify a regularly varying function?

To answer this question, let us have a look at Fig. 2.8 which shows the Hill estimator $\hat{b}_u$ for all data sets (positive and negative branches of the distribution of returns for the Dow Jones, the Nasdaq and the Standard & Poor's 500 (SP)) as a function of the index $n = 1, 2, \ldots, 18$ of the quantiles or standard significance levels $q_1, \ldots, q_{18}$. Similar results are obtained with the AD estimates. The three branches of the distribution of returns for the Dow Jones

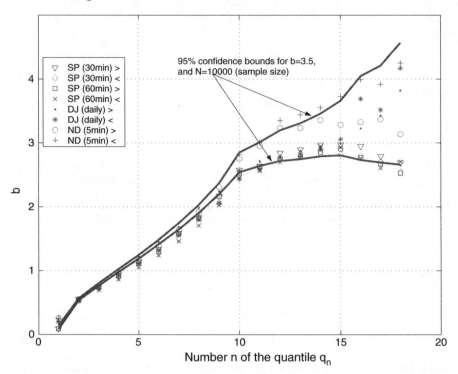

**Fig. 2.8.** Hill estimator $\hat{b}_u$ for all sets (positive and negative branches of the distribution of returns for the Dow Jonesc (DJ), Nasdaq (ND) and Standard & Poor's 500 (SP)) as a function of the index $n = 1, \ldots, 18$ of the 18 quantiles or standard significance levels $q_1, \ldots, q_{18}$ given in Table 6.3. The two thick lines (in red) show the 95% confidence bounds obtained from synthetic time series of 10000 data points generated with a Student distribution with exponent $b = 3.5$. Reproduced from [330]

and the negative tail of the Nasdaq suggest a continuous growth of the Hill estimator $\hat{b}_u$ as a function of $n = 1, \ldots, 18$. However, it turns out that this apparent growth may be explained solely on the basis of statistical fluctuations and slow convergence to a moderate $b$-value. Indeed, the two thick lines show the 95% confidence bounds obtained from synthetic time series of 10000 data points generated with a Student distribution with exponent $b = 3.5$. It is clear that the growth of the upper bound can explain the observed behavior of the $b$-value obtained for the Dow Jones and Nasdaq data. It would thus be incorrect to extrapolate this apparent growth of the $b$-value. However, conversely, we cannot conclude with certainty that the growth of the $b$-value has been exhausted and that we have access to the asymptotic value. Indeed, large values of tail indices are for instance predicted by traditional GARCH models giving b~ 10–20 [153, 463].

## Weibull Distributions

We now present the results of the fits of the same data with the SE distribution (2.27). The corresponding Anderson-Darling statistics (ADS) are shown in Table 2.4. The ML-estimates and AD-estimates of the form parameter $c$ are represented in Table 2.5. Table 2.4 shows that, for the highest quantiles, the ADS for the SE is the smallest of all ADS, suggesting that the SE is the best model of all. Moreover, for the lowest quantiles, it is the sole model not systematically rejected at the 95% level.

The $c$-estimates are found to decrease when increasing the order $q$ of the threshold $u(q)$ beyond which the estimations are performed. In addition, several $c$-estimates are found very close to zero. However, this does not automatically imply that the SE model is not the correct model for the data even for these highest quantiles. Indeed, numerical simulations show that, even for synthetic samples drawn from genuine SE distributions with exponent $c$ smaller than 0.5 and whose size is comparable with that of our data, in about one case out of three (depending on the exact value of $c$) the estimated value of $c$ is zero. This *a priori* surprising result comes from condition (2.B.57) in Appendix 2.B which is not fulfilled with certainty even for samples drawn for SE distributions.

Notwithstanding this cautionary remark, note that the $c$-estimate of the positive tail of the Nasdaq data equals zero for all quantiles higher than $q_{14} = 0.97\%$. In fact, in every case, the estimated $c$ is not significantly different from zero – at the 95% significance level – for quantiles higher than $q_{12}$–$q_{14}$, except for quantile $q_{21}$ of the negative tail of the Standard & Poor's 500, but this value is probably doubtful. In addition, the values of the estimated scale parameter $d$, not reported here, are found very small, particularly for the Nasdaq – beyond $q_{12} = 95\%$ – and the S&P 500 – beyond $q_{10} = 90\%$. In contrast, the Dow Jones keeps significant scale factors until $q_{16}$–$q_{17}$.

These evidences taken all together provide a clear indication on the existence of a change of behavior of the true pdf of these distributions: while the bulks of the distributions seem rather well approximated by a SE model, a distribution with a tail fatter than that of the SE model is required for the highest quantiles. Actually, the fact that both $c$ and $d$ are extremely small may be interpreted according to the asymptotic correspondence given by (2.30) and (2.31) as the existence of a possible power law tail.

At this stage, we can state the following conservative statement: the *true* distribution of returns is probably bracketed by a power law, as a lower bound and a SE as an upper bound. It is therefore particularly interesting to focus on distributions such as log-Weibull distributions which interpolate between these two classes in order to obtain – hopefully – a better description of the data.

**Table 2.5.** Maximum likelihood (MLE) and Anderson-Darling (ADE) estimates of the form parameter $c$ of the Weibull (Stretched-Exponential) distribution

| | S&P 500 (1 min) | | | | Nasdaq | | | | Dow Jones | | | |
|---|---|---|---|---|---|---|---|---|---|---|---|---|
| | Pos. tail | | Neg. tail | | Pos. tail | | Neg. tail | | Pos. tail | | Neg. tail | |
| | MLE | ADE | MLE | ADE | MLE | ADE | MLE | ADE | MLE | ADE | MLE | ADE |
| $q_1$ | 1.065 (0.001) | 1.175 | 1.051 (0.001) | 1.158 | 1.007 (0.008) | 1.053 | 0.987 (0.008) | 1.017 | 1.040 (0.007) | 1.104 | 0.975 (0.007) | 1.026 |
| $q_2$ | 0.927 (0.002) | 1.049 | 0.915 (0.002) | 1.035 | 0.983 (0.011) | 1.051 | 0.953 (0.011) | 0.993 | 0.973 (0.010) | 1.075 | 0.910 (0.010) | 0.989 |
| $q_3$ | 0.8754 (0.002) | 1.0196 | 0.8634 (0.002) | 1.0027 | 0.944 (0.014) | 1.031 | 0.912 (0.014) | 0.955 | 0.931 (0.013) | 1.064 | 0.856 (0.012) | 0.948 |
| $q_4$ | 0.813 (0.002) | 0.970 | 0.799 (0.002) | 0.947 | 0.896 (0.018) | 0.995 | 0.876 (0.018) | 0.916 | 0.878 (0.015) | 1.038 | 0.821 (0.015) | 0.933 |
| $q_5$ | 0.763 (0.003) | 0.952 | 0.752 (0.003) | 0.932 | 0.857 (0.021) | 0.978 | 0.861 (0.021) | 0.912 | 0.792 (0.019) | 0.955 | 0.767 (0.018) | 0.889 |
| $q_6$ | 0.733 (0.003) | 0.985 | 0.727 (0.003) | 0.971 | 0.790 (0.026) | 0.916 | 0.833 (0.026) | 0.891 | 0.708 (0.023) | 0.873 | 0.698 (0.022) | 0.819 |
| $q_7$ | 0.593 (0.004) | 0.799 | 0.590 (0.004) | 0.791 | 0.732 (0.033) | 0.882 | 0.796 (0.033) | 0.859 | 0.622 (0.028) | 0.788 | 0.612 (0.028) | 0.713 |
| $q_8$ | 0.504 (0.005) | 0.740 | 0.502 (0.005) | 0.730 | 0.661 (0.042) | 0.846 | 0.756 (0.042) | 0.834 | 0.480 (0.035) | 0.586 | 0.531 (0.035) | 0.597 |
| $q_9$ | 0.337 (0.007) | 0.537 | 0.342 (0.007) | 0.531 | 0.509 (0.058) | 0.676 | 0.715 (0.059) | 0.865 | 0.394 (0.047) | 0.461 | 0.478 (0.047) | 0.527 |
| $q_{10}$ | 0.152 (0.010) | 0.394 | 0.159 (0.010) | 0.387 | 0.359 (0.092) | 0.631 | 0.522 (0.099) | 0.688 | 0.304 (0.074) | 0.346 | 0.403 (0.076) | 0.387 |
| $q_{11}$ | 0.079 (0.012) | 0.327 | 0.091 (0.012) | 0.339 | 0.252 (0.110) | 0.515 | 0.481 (0.120) | 0.697 | 0.231 (0.087) | 0.158 | 0.379 (0.091) | 0.337 |
| $q_{12}$ | $< 10^{-8}$ | 0.151 | $< 10^{-8}$ | 0.169 | 0.039 (0.138) | 0.177 | 0.273 (0.155) | 0.275 | 0.269 (0.111) | 0.207 | 0.357 (0.119) | 0.288 |
| $q_{13}$ | $< 10^{-8}$ | 0.0793 | $< 10^{-8}$ | 0.084 | 0.057 (0.155) | 0.233 | 0.255 (0.177) | 0.274 | 0.253 (0.127) | 0.147 | 0.428 (0.136) | 0.465 |
| $q_{14}$ | $< 10^{-8}$ | 0.008 | $< 10^{-8}$ | 0.020 | $< 10^{-8}$ | 0 | 0.215 (0.209) | 0.194 | 0.290 (0.150) | 0.174 | 0.448 (0.164) | 0.641 |
| $q_{15}$ | $< 10^{-8}$ | 0.008 | $< 10^{-8}$ | 0.008 | $< 10^{-8}$ | 0 | 0.103 (0.260) | 0 | 0.379 (0.192) | 0.407 | 0.451 (0.210) | 0.863 |
| $q_{16}$ | $< 10^{-8}$ | 0.008 | $< 10^{-8}$ | 0.008 | $9.6 \times 10^{-8}$ | 0 | 0.064 (0.390) | 0 | 0.398 (0.290) | 0.382 | 0.022 (0.319) | 0.110 |
| $q_{17}$ | $< 10^{-8}$ | 0.008 | $< 10^{-8}$ | 0.008 | $< 10^{-8}$ | 0 | 0.158 (0.452) | 0.224 | 0.307 (0.346) | 0.255 | 0.178 (0.367) | 0.703 |
| $q_{18}$ | $< 10^{-8}$ | 0.008 | $< 10^{-8}$ | 0.008 | $< 10^{-8}$ | 0 | $< 10^{-8}$ | 0 | $2 \times 10^{-8}$ | 0 | $< 10^{-8}$ | 0 |
| $q_{19}$ | 0.035 (0.082) | 0.007 | 0.009 (0.032) | 0.007 | — | — | — | — | — | — | — | — |
| $q_{20}$ | 0.111 (0.119) | 0.075 | 0.316 (0.117) | 0.007 | — | — | — | — | — | — | — | — |
| $q_{21}$ | $< 10^{-8}$ | 0.008 | 0.827 (0.393) | 0.900 | — | — | — | — | — | — | — | — |

Reproduced from [330]

## Log-Weibull Distributions

The parameters $b$ and $c$ of the log-Weibull distribution defined by (2.33) are estimated with both the maximum likelihood and Anderson-Darling methods for the 18 standard significance levels $q_1, \ldots, q_{18}$ (given on page 62) for the Dow Jones and Nasdaq data and up to $q_{21}$ for the Standard & Poor's 500 data. The results for the Dow Jones and the Standard & Poor's 500 are given in Table 2.6. For both positive and negative tails of the Dow Jones, the results are very stable for all quantiles lower than $q_{10}$: $c = 1.09 \pm 0.02$ and $b = 2.71 \pm 0.07$. These results reject the Pareto distribution degeneracy $c = 1$ at the 95% confidence level. Only for the quantiles higher than or equal to $q_{16}$, an estimated value $c$ compatible with the Pareto distribution is found. Moreover both for the positive and negative Dow Jones tails, one finds that $c \approx 0.92$ and $b \approx 3.6-3.8$, suggesting either a possible change of regime or a sensitivity to "outliers" or a lack of robustness due to a too small sample size. For the positive Nasdaq tail, the exponent $c$ is found compatible with $c = 1$ (the Pareto value), at the 95% significance level, above $q_{11}$ while $b$ remains almost stable at $b \simeq 3.2$. For the negative Nasdaq tail, we find that $c$ decreases almost systematically from 1.1 for $q_{10}$ to 1 for $q_{18}$ for both estimators while $b$ regularly increases from about 3.1 to about 4.2. The Anderson-Darling distances are significantly better than for the SE and this statistics cannot be used to conclude neither in favor of nor against the log-Weibull class.

The situation is different for the Standard & Poor's 500 (1-min). For the positive tail, the parameter $c$ remains significantly smaller than 1 from $q_{14} = 97\%$ to $q_{21}$ except for $q_{19}$ and $q_{20}$. Therefore, it seems that for very small time scales, the tails of the distribution of returns might be even fatter than a power law. As stressed in Sect. 2.4.1, when $c$ is less than one, the log-Weibull distribution does not belong to the domain of attraction of a law of the maximum. As a consequence, EVT cannot provide reliable results when applied to such data, neither from a theoretical point of view nor from a practical stance (*e.g.* extreme risk assessment). The conclusions are the same for the 5-minute time scale. For the 30-minute and 60-minute time scales, $c$ remains systematically less than one for the highest quantiles but this difference ceases to be significant. In the negative tail, the situation is overall the same.

### 2.4.4 Comparison of the Descriptive Power of the Different Families

The previous sections have shown that none of the considered distributions (2.26–2.29) and (2.34) fit the data over the entire range, which is not a surprise. For the highest quantiles, several models seem to be able to represent the data, including the Pareto model, the SE model and the log-Weibull model discussed above. The last two models seem to be the most reasonable models among the models compatible with the data. For all the samples, their Anderson-Darling statistics remain so close to each other for the highest quantiles that the descriptive power of these two models cannot be distinguished.

**Table 2.6.** Maximum likelihood (MLE) and Anderson-Darling (ADE) estimates of the parameters $b$ and $c$ of the log-Weibull distribution defined by (2.33)

| | Dow Jones (1 day) Positive tail | | | | Dow Jones (1 day) Negative tail | | | |
|---|---|---|---|---|---|---|---|---|
| | MLE | | ADE | | MLE | | ADE | |
| | c | b | c | b | c | b | c | b |
| $q_1$ | 5.262 (0.005) | 0.000 (0.000) | 5.55 | 0.000 | 5.085 (0.005) | 0.000 (0.000) | 5.320 | 0.000 |
| $q_2$ | 2.140 (0.009) | 0.241 (0.002) | 2.25 | 0.220 | 2.125 (0.009) | 0.211 (0.002) | 2.240 | 0.191 |
| $q_3$ | 1.790 (0.010) | 0.531 (0.005) | 1.87 | 0.510 | 1.751 (0.010) | 0.495 (0.005) | 1.800 | 0.481 |
| $q_4$ | 1.616 (0.012) | 0.830 (0.008) | 1.65 | 0.820 | 1.593 (0.012) | 0.744 (0.008) | 1.630 | 0.735 |
| $q_5$ | 1.447 (0.012) | 1.165 (0.012) | 1.47 | 1.160 | 1.459 (0.013) | 1.022 (0.011) | 1.480 | 1.015 |
| $q_6$ | 1.339 (0.012) | 1.472 (0.017) | 1.36 | 1.473 | 1.353 (0.013) | 1.311 (0.016) | 1.370 | 1.311 |
| $q_7$ | 1.259 (0.013) | 1.768 (0.023) | 1.28 | 1.773 | 1.269 (0.014) | 1.609 (0.022) | 1.270 | 1.610 |
| $q_8$ | 1.173 (0.013) | 2.097 (0.031) | 1.17 | 2.096 | 1.188 (0.015) | 1.885 (0.030) | 1.190 | 1.887 |
| $q_9$ | 1.125 (0.015) | 2.362 (0.043) | 1.12 | 2.358 | 1.158 (0.017) | 2.178 (0.042) | 1.150 | 2.174 |
| $q_{10}$ | 1.090 (0.020) | 2.705 (0.070) | 1.08 | 2.695 | 1.087 (0.022) | 2.545 (0.069) | 1.090 | 2.545 |
| $q_{11}$ | 1.035 (0.022) | 2.771 (0.083) | 1.03 | 2.762 | 1.074 (0.024) | 2.688 (0.085) | 1.070 | 2.681 |
| $q_{12}$ | 1.047 (0.027) | 2.867 (0.105) | 1.04 | 2.857 | 1.068 (0.029) | 2.880 (0.111) | 1.050 | 2.857 |
| $q_{13}$ | 1.046 (0.030) | 2.960 (0.121) | 1.03 | 2.933 | 1.067 (0.032) | 2.900 (0.125) | 1.080 | 2.924 |
| $q_{14}$ | 1.044 (0.034) | 3.000 (0.142) | 1.03 | 2.976 | 1.132 (0.038) | 3.171 (0.158) | 1.120 | 3.155 |
| $q_{15}$ | 1.090 (0.043) | 3.174 (0.184) | 1.09 | 3.165 | 1.163 (0.047) | 3.439 (0.209) | 1.180 | 3.472 |
| $q_{16}$ | 1.085 (0.059) | 3.424 (0.280) | 1.09 | 3.425 | 1.025 (0.056) | 3.745 (0.322) | 1.010 | 3.731 |
| $q_{17}$ | 1.093 (0.066) | 3.666 (0.345) | 1.09 | 3.650 | 1.108 (0.069) | 3.822 (0.380) | 1.120 | 3.891 |
| $q_{18}$ | 0.935 (0.071) | 3.556 (0.411) | 0.902 | 3.484 | 0.921 (0.071) | 3.804 (0.461) | 0.933 | 3.846 |
| | S&P 500 (1 min) Positive tail | | | | S&P 500 (1 min) Negative tail | | | |
| | MLE | | ADE | | MLE | | ADE | |
| | c | b | c | b | c | b | c | b |
| $q_1$ | 3.261 (0.003) | 0.029 (0.000) | 3.298 | 0.027 | 3.232 (0.003) | 0.030 (0.000) | 3.264 | 0.028 |
| $q_2$ | 1.875 (0.002) | 0.433 (0.001) | 1.878 | 0.410 | 1.884 (0.002) | 0.420 (0.001) | 1.881 | 0.399 |
| $q_3$ | 1.645 (0.002) | 0.723 (0.001) | 1.642 | 0.690 | 1.647 (0.002) | 0.707 (0.001) | 1.641 | 0.676 |
| $q_4$ | 1.471 (0.002) | 1.017 (0.001) | 1.477 | 0.970 | 1.465 (0.002) | 1.000 (0.001) | 1.470 | 0.954 |
| $q_5$ | 1.414 (0.002) | 1.277 (0.002) | 1.405 | 1.233 | 1.411 (0.002) | 1.251 (0.002) | 1.401 | 1.208 |
| $q_6$ | 1.382 (0.002) | 1.512 (0.002) | 1.387 | 1.477 | 1.383 (0.002) | 1.477 (0.002) | 1.389 | 1.443 |
| $q_7$ | 1.233 (0.002) | 1.862 (0.003) | 1.234 | 1.811 | 1.232 (0.002) | 1.823 (0.003) | 1.239 | 1.776 |
| $q_8$ | 1.187 (0.002) | 2.155 (0.005) | 1.192 | 2.116 | 1.192 (0.002) | 2.117 (0.005) | 1.196 | 2.079 |
| $q_9$ | 1.112 (0.002) | 2.508 (0.007) | 1.111 | 2.470 | 1.113 (0.002) | 2.455 (0.007) | 1.112 | 2.415 |
| $q_{10}$ | 1.069 (0.003) | 2.876 (0.011) | 1.078 | 2.896 | 1.062 (0.003) | 2.818 (0.011) | 1.074 | 2.831 |
| $q_{11}$ | 1.048 (0.003) | 2.961 (0.014) | 1.066 | 3.016 | 1.055 (0.003) | 2.927 (0.014) | 1.069 | 2.972 |
| $q_{12}$ | 1.016 (0.004) | 3.048 (0.018) | 1.033 | 3.123 | 1.015 (0.004) | 3.006 (0.017) | 1.034 | 3.076 |
| $q_{13}$ | 1.002 (0.004) | 3.063 (0.020) | 1.021 | 3.151 | 1.001 (0.004) | 3.033 (0.020) | 1.020 | 3.115 |
| $q_{14}$ | 0.981 (0.005) | 3.054 (0.023) | 1.003 | 3.153 | 0.990 (0.005) | 3.033 (0.023) | 1.012 | 3.134 |
| $q_{15}$ | 0.961 (0.006) | 3.015 (0.027) | 0.985 | 3.133 | 0.978 (0.006) | 3.004 (0.027) | 1.003 | 3.132 |
| $q_{16}$ | 0.941 (0.008) | 2.867 (0.036) | 0.961 | 2.980 | 0.937 (0.008) | 2.871 (0.037) | 0.957 | 2.987 |
| $q_{17}$ | 0.937 (0.010) | 2.798 (0.040) | 0.951 | 2.899 | 0.927 (0.010) | 2.780 (0.041) | 0.947 | 2.887 |
| $q_{18}$ | 0.902 (0.011) | 2.649 (0.046) | 0.902 | 2.677 | 0.925 (0.012) | 2.644 (0.046) | 0.940 | 2.726 |
| $q_{19}$ | 0.994 (0.028) | 2.256 (0.084) | 0.971 | 2.201 | 0.962 (0.027) | 2.134 (0.080) | 0.923 | 2.063 |
| $q_{20}$ | 0.999 (0.039) | 2.245 (0.118) | 0.967 | 2.139 | 1.011 (0.040) | 2.037 (0.107) | 0.933 | 1.879 |
| $q_{21}$ | 0.949 (0.083) | 2.686 (0.330) | 0.957 | 2.801 | 1.288 (0.115) | 3.387 (0.455) | 1.234 | 3.272 |

The numbers in parenthesis give the standard deviations of the estimates. Reproduced from [330]

One can go further and ask which of these models are sufficient to describe the data compared with the comprehensive distribution (2.23) encompassing all of them. Here, the four distributions (2.26–2.29) are compared with the comprehensive distribution (2.23) using Wilks' theorem [485] on maximum likelihood ratios, which allows to compare nested hypotheses. It will be shown that the Pareto and the SE models are the most parsimonious. We then turn to a direct comparison of the best two-parameter models (the SE and log-Weibull models) with the best one-parameter model (the Pareto model), which will require an extension of Wilks' theorem derived in Appendix 2.D. This extension allows us to directly test the SE model against the Pareto model.

## Comparison Between the Four Parametric Families (2.26–2.29) and the Comprehensive Distribution (2.23)

According to Wilks' theorem, the doubled log-likelihood ratio $\Lambda$:

$$\Lambda = 2 \, \log \frac{\max \mathcal{L}(CD, X, \Theta)}{\max \mathcal{L}(z, X, \theta)} \, , \tag{2.39}$$

has asymptotically (as the size $N$ of the sample $X$ tends to infinity) the $\chi^2$-distribution. Here $\mathcal{L}$ denotes the likelihood function, $\theta$ and $\Theta$ are parametric spaces corresponding to hypotheses $z$ and $CD$ (comprehensive distribution defined in (2.23)) correspondingly (hypothesis $z$ is one of the four hypotheses (2.26–2.29) that are particular cases of the $CD$ under some parameter restrictions recalled in Sect. 2.4.1). The statement of the theorem is valid under the condition that the *sample $X$ obeys the hypothesis $z$ for some particular value of its parameter belonging to the space $\theta$*. The number of degrees of freedom of the $\chi^2$-distribution is equal to the difference of the dimensions of the two spaces $\Theta$ and $\theta$. We have $\dim(\Theta) = 3, \dim(\theta) = 2$ for the SE and for the incomplete Gamma distributions while $\dim(\theta) = 1$ for the Pareto and the Exponential distributions. This leads to one degree of freedom of the $\chi^2$-distribution for the two former cases and two degrees of freedom of the $\chi^2$-distribution for the later models. The maximum of the likelihood in the numerator of (2.39) is taken over the space $\Theta$, whereas the maximum of the likelihood in the denominator of (2.39) is taken over the space $\theta$. Since we have always $\theta \subset \Theta$, the likelihood ratio is always larger than 1, and the log-likelihood ratio is non-negative. If the observed value of $\Lambda$ does not exceed some high-confidence level (say, 99% confidence level) of the $\chi^2$, we then reject the hypothesis CD in favor of the hypothesis $z$, considering the space $\Theta$ redundant. Otherwise, we accept the hypothesis CD, considering the space $\theta$ insufficient.

The double log-likelihood ratios (2.39) are shown for the positive and negative branches of the distribution of returns in Fig. 2.9 for the Nasdaq Composite index. Similar results (not shown) are obtained for the Dow Jones and the Standard & Poor's 500 (1, 5, 30 and 60 minutes) indices.

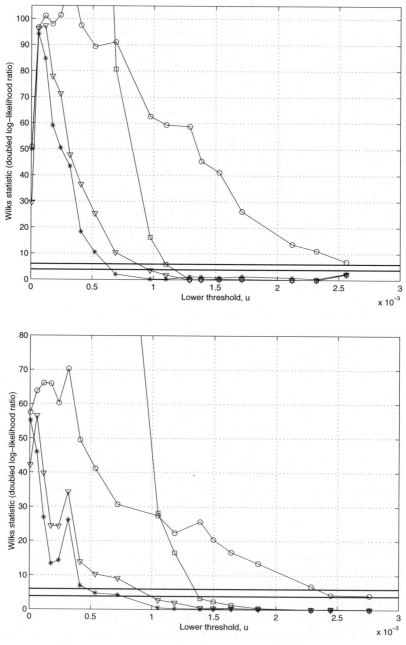

**Fig. 2.9.** Wilks statistic for the comprehensive distribution versus the four parametric distributions: Pareto ($\square$), Stretched-Exponential ($*$), Exponential, ($\circ$) and incomplete Gamma ($\nabla$) for the Nasdaq 5-minute returns. The upper (lower) panel refers to the positive (negative) returns. The horizontal lines represent the critical values at the 95% confidence level of the test for the $\chi^2$-distribution with one (lower line) and two (upper line) degrees of freedom. Reproduced from [330]

For the Nasdaq data, Figure 2.9 clearly shows that the Exponential distribution is completely insufficient: for all lower thresholds, the Wilks log-likelihood ratio exceeds the critical value corresponding to the 95% level of the $\chi_1^2$ function. The Pareto distribution is insufficient for thresholds corresponding to quantiles less than $q_{11} = 92.5\%$ and becomes comparable with the comprehensive distribution beyond. It is natural that the families with two parameters, the incomplete Gamma and the SE, have higher goodness-of-fit than the one-parameter Exponential and Pareto distributions. The incomplete Gamma distribution is comparable with the comprehensive distribution beyond quantile $q_{10} = 90\%$, whereas the SE is somewhat better beyond quantile $q_8 = 70\%$. For the tails representing 7.5% of the data, all parametric families except for the Exponential distribution fit the sample distribution with almost the same efficiency according to this test.

The results obtained for the Dow Jones data are similar. The SE is comparable with the comprehensive distribution starting with $q_8 = 70\%$. On the whole, one can say that the SE distribution performs better than the three other parametric families.

The situation is somewhat different for the Standard & Poor's 500 index. For the positive tail, none of the four distributions is really sufficient in order to accurately describe the data. The comprehensive distribution is overall the best. In the negative tail, we retrieve a behavior more similar to that observed in the two previous cases, except for the Exponential distribution which also appears to be better than the comprehensive distribution. However, it should be noted that the comprehensive distribution is only rejected in the very far tail. The four models (2.26–2.29) are better than the comprehensive distribution only for the two highest quantiles ($q_{20}$ and $q_{21}$) of the negative tail. In contrast, the Pareto, SE and incomplete Gamma models are better than the comprehensive distribution over the 10 highest quantiles (or so) for the Nasdaq and the Dow Jones.

We should stress again that each log-likelihood ratio, so-to say "acts on its own ground" that is, the corresponding $\chi^2$-distribution is valid *under the assumption of the validity of each particular hypothesis whose likelihood stands in the numerator of the double log-likelihood (2.39)*. It would be desirable to compare all combinations of pairs of hypotheses directly, in addition to comparing each of them with the comprehensive distribution. Unfortunately, the Wilks theorem cannot be used in the case of pair-wise comparison because the problem is no more that of comparing nested hypothesis (*i.e.*, one hypothesis is a particular case of the comprehensive model). As a consequence, the previous results on the comparison of the relative merits of each of the four distributions using the generalized log-likelihood ratio should be interpreted with care, in particular, in a case of contradictory conclusions. Fortunately, the main conclusion of the comparison (an advantage of the SE distribution over the three other distributions) does not contradict the earlier results discussed above.

## Pair-Wise Comparison of the Pareto Model
## with the Stretched-Exponential and Log-Weibull Models

Let us compare formally the descriptive power of the SE distribution and the log-Weibull distribution (the two best two-parameter models qualified until now) with that of the Pareto distribution (the best one-parameter model). For the comparison of the log-Weibull model versus the Pareto model, Wilks' theorem can still be applied since the log-Weibull distribution encompasses the Pareto distribution. *A contrario*, the comparison of the SE versus the Pareto distribution should in principle require that we use the methods for testing non-nested hypotheses [209], such as the Wald encompassing test or the Bayes factors [266]. Indeed, the Pareto model and the (SE) model are not, strictly speaking, nested. However, as shown in Sect. 2.4.1, the Pareto distribution is a limited case of the SE distribution, as the fractional exponent $c$ goes to zero. Changing the parametric representation of the (SE) model into

$$f(x|b,c) = b\, u^{-c}\, x^{c-1} \exp\left[-\frac{b}{c}\left(\left(\frac{x}{u}\right)^c - 1\right)\right], \quad x > u, \tag{2.40}$$

*i.e.*, setting $b = c \cdot \left(\frac{u}{d}\right)^c$, where the parameter $d$ refers to the former (SE) representation (2.27), Appendix 2.D shows that the doubled log-likelihood ratio

$$W = 2\log\frac{\max_{b,c}\mathcal{L}_{SE}}{\max_b \mathcal{L}_{PD}} \tag{2.41}$$

still follows Wilks' statistic, namely is asymptotically distributed according to a $\chi^2$-distribution, with one degree of freedom in the present case. Thus, even in this case of non-nested hypotheses, Wilks' statistic still allows us to test the null hypothesis $H_0$ according to which the Pareto model is sufficient to describe the data.

Concerning the comparison between the Pareto model and the SE one, the null hypothesis is found to be more often rejected for the Dow Jones than for the Nasdaq and the Standard & Poor's 500 [330]. Indeed, beyond the quantile $q_{12} = 95\%$, the Pareto model cannot be rejected in favor of the SE model at the 95% confidence level for the Nasdaq and the Standard & Poor's 500 data. For the Dow Jones, one must consider quantiles higher than $q_{16} = 99\%$ – at least for the negative tail – in order not to reject $H_0$ at the 95% significance level. These results are in qualitative agreement with what we could expect from the action of the central limit theorem: the power law regime (if it really exists) is pushed back to higher quantiles due to time aggregation (recall that the Dow Jones data is at the daily scale while the Nasdaq data is at the 5-minute time scale).

It is, however, more difficult to rationalize the fact reported in [330] that the SE model is not rejected (at the 99% confidence level) for the two highest quantiles ($q_{20} = 99.95\%$ and $q_{21} = 99.99\%$) of the negative tail of the 1 minute returns of the Standard & Poor's 500 and for the quantiles $q_{19} = 99.9\%$ and $q_{20} = 99.95\%$ for its positive tail. This might be ascribed to a lack of power of the test, but recall that we have restricted our investigation to empirical quantiles with more than a hundred points (or so). Therefore, invoking a lack of power is not very convincing. In addition, for these high quantiles, the fractional exponent $c$ in the SE model becomes significantly different from zero (see Table 2.5). It could be an empirical illustration of the existence of a cut-off beyond which the power law regime is replaced by an exponential (or stretched-exponential) decay of the distribution function as suggested by Mantegna and Stanley [344] and by the recent model [493] based upon a pure jump Lévy process, whose jump arrival rate obeys a power law dampened by an exponential function. To strengthen this idea, it can be noted that the exponential distribution is found sufficient to describe the distributions of the 1 minute returns of the Standard & Poor's 500, while it is always rejected (with respect to the comprehensive distribution) for the Nasdaq and the Dow Jones. Thus, this non-rejection could really be the genuine signature of a cut-off beyond which the decay of the distribution is faster than any power law. However, this conclusion is only drawn from the one hundred most extreme data points and, therefore, should be considered with caution. Larger samples should be considered to obtain a confirmation of this intuition. Unfortunately, samples with more than 10 million (non zero) data points (for a single asset) are not yet accessible.

Based upon the study of [330], Wilks' test for the Pareto distribution versus the log-Weibull distribution shows that, for quantiles above $q_{12}$, the Pareto distribution cannot be rejected in favor of the log-Weibull for the Dow Jones, the Nasdaq and the Standard & Poor's 500 30-minute returns. This parallels the lack of rejection of the Pareto distribution against the SE beyond the significance level $q_{12}$. The picture is different for the 1-minute returns of the Standard & Poor's 500. The Pareto model is almost always rejected. The most interesting point is the following: in the negative tail, the Pareto model is always strongly rejected except for the highest quantiles. Comparing with Table 2.6, one clearly sees that between $q_{15}$ and $q_{18}$ the exponent $c$ is significantly (at the 95% significance level) *less than one*, indicating a tail fatter than any power law. On the contrary, for $q_{21}$, the exponent $c$ is found significantly *larger than one*, indicating a change of regime and again an ultimate decay of the tail of the distribution faster than any power law.

In summary, the null hypothesis that the true distribution is the Pareto distribution is strongly rejected until quantiles 90–95% or so. Thus, within this range, the Stretched-Exponential and log-Weibull models seem the best and the Pareto model is insufficient to describe the data. But, for the very highest quantiles (above 95%–98%), one cannot reject any more the hypothesis that the Pareto model is sufficient compared with the SE and log-Weibull models. These two parameter models can then be seen as a redundant parameterization for the extremes compared with the Pareto distribution, except for the returns calculated at the smallest time scales.

## 2.5 Discussion and Conclusions

### 2.5.1 Summary

This chapter has revisited the generally accepted fact that the tails of the distributions of returns present a power-like behavior. Often, the conviction of the existence of a power-like tail is based on the Gnedenko theorem stating the existence of only three possible types of limit distributions of normalized maxima (a finite maximum value, an exponential tail, and a power-like tail) together with the exclusion of the first two types by empirical evidence. The power-like character of the tails of the distribution of log-returns follows then simply from the power-like distribution of maxima. However, in this chain of arguments, the conditions needed for the fulfillment of the corresponding mathematical theorems are often omitted and not discussed properly. In addition, widely used arguments in favor of power law tails invoke the *self-similarity* of the data but are often *assumptions* rather than experimental evidence or consequences of economic and financial laws.

Sharpening and generalizing the results obtained by Kearns and Pagan [267], Sect. 2.3.3 has recalled that standard statistical estimators of heavy tails are much less efficient than often assumed and cannot in general clearly distinguish between a power law tail and a SE tail (even in the absence of long-range dependence in the volatility). So, in view of the stalemate reached with the nonparametric approaches and in particular with the standard extreme value estimators, resorting to a parametric approach appears essential. The parametric approach is useful to decide which class of extreme value distributions – rapidly versus regularly varying – accounts best for the empirical distributions of returns at different time scales. However, here again, the problem is not as straightforward as its appears. Indeed, in order to apply statistical methods to the study of empirical distributions of returns and to derive their resulting implication for risk management, it is necessary to keep in mind the existence of necessary conditions that the empirical data must obey for the conclusions of the statistical study to be valid. Maybe the most important condition in order to speak meaningfully about distribution functions is the stationarity of the data, a difficult issue that we have barely touched upon here. In particular, the importance of regime switching is now well established [14, 397] and its possible role should be assessed and accounted for.

### 2.5.2 Is There a Best Model of Tails?

The results that standard statistical estimators of heavy tails are much less efficient than often assumed and cannot in general clearly distinguish between a power law tail and a SE tail, can be rationalized by the fact that, into a certain limit, the Stretched-Exponential pdf tends to the Pareto distribution (see (2.30–2.31) and Appendix 2.B). Thus, the Pareto (or power law) distribution can be approximated with any desired accuracy on an arbitrary interval by a suitable adjustment of the pair $(c, d)$ of the parameters of the Stretched Exponential pdf. The parametric tests presented above indicate that the class of SE and log-Weibull distributions provide a significantly better fit to empirical returns than the Pareto, the exponential or the incomplete Gamma distributions. All these tests are consistent with the conclusion that these two models provide the best effective apparent and parsimonious models to account for the empirical data on the largest possible range of returns.

However, this does not mean that the Stretched Exponential or the log-Weibull model is the correct description of the tails of empirical distributions of returns. Again, as already mentioned, the strength of these models come from the fact that they encompass the Pareto model in the tail and offers a better description in the bulk of the distribution. To see where the problem arises, Table 2.7 summarizes the best ML-estimates for the SE parameters $c$ (form parameter) and $d$ (scale parameter) restricted to the quantiles beyond $q_{12} = 95\%$, which offers a good compromise between a sufficiently large sample size and a restricted tail range leading to an accurate approximation in this range.

One can see that $c$ is very small (and all the more so for the scale parameter $d$) for the tail of the distribution of positive returns of the Nasdaq data, suggesting a convergence to a power law tail. The exponents $c$ for the negative returns of the Nasdaq data and for both positive and negative returns of the Dow Jones data are an order of magnitude larger but the statistical tests show that they are not incompatible with an asymptotic power tail either. Indeed, Sect. 2.4.4 has shown that, for the very highest quantiles (above 95–98%), one cannot reject the hypothesis that the Pareto model is sufficient compared with the SE model. The values of $c$ and $d$ are even smaller for the Standard & Poor's 500 data both at the 1-minute and 5-minute time scales.

**Table 2.7.** Best parameters $c$ and $d$ of the Stretched-Exponential model and best parameter $b$ of the Pareto model estimated beyond quantile $q_{12} = 95\%$ for the Dow Jones (DJ), the Nasdaq (ND) and the Standard & Poor's 500 (SP) indices. The apparent Pareto exponent $c(u(q_{12})/d)^c$ (see expression (2.30)) is also shown

| Sample | $c$ | $d$ | $c(u(q_{12})/d)^c$ | $b$ |
|---|---|---|---|---|
| DJ pos. returns | 0.274 (0.111) | $4.81 \times 10^{-6}$ | 2.68 | 2.79 (0.10) |
| DJ neg. returns | 0.362 (0.119) | $1.02 \times 10^{-4}$ | 2.57 | 2.77 (0.11) |
| ND pos. returns | 0.039 (0.138) | $4.54 \times 10^{-52}$ | 3.03 | 3.23 (0.14) |
| ND neg. returns | 0.273 (0.155) | $1.90 \times 10^{-7}$ | 3.10 | 3.35 (0.15) |
| SP pos. returns (1min) | – | – | 3.01 | 3.02 (0.02) |
| SP neg returns (1min) | – | – | 2.97 | 2.97 (0.02) |
| SP pos. returns (5min) | 0.033 (0.031) | $3.06 \times 10^{-59}$ | 2.95 | 2.95 (0.03) |
| SP neg. returns (5min) | 0.033 (0.031) | $3.26 \times 10^{-56}$ | 2.87 | 2.86 (0.03) |

Note also that the exponents $c$ are larger for the daily Dow Jones data than for the 5-minute Nasdaq data and the 1-minute and 5-minute Standard & Poor's 500 data, in agreement with an expected (slow) convergence to the Gaussian law according to the central limit theory.[18] However, a $t$-test does not allow one to reject the hypothesis that the exponents $c$ remain the same for the positive and negative tails of the Dow Jones data. This confirms previous results, for instance [319, 256] according to which the extreme tails can be considered as symmetric, at least for the Dow Jones data. In contrast, there is a very strong asymmetry for the 5-minute sampled Nasdaq and the Standard & Poor's 500 data.

These are the evidences in favor of the existence of an asymptotic power law tail. Balancing this view, many of the tests have shown that the power law model is not as powerful compared with the SE and log-Weibull models, even arbitrarily far in the tail (as far as the available data allows us to probe). In addition, for the smallest time scales, the tail of the distribution of return is, over a large range, well-described by a log-Weibull distribution with an exponent $c$ less than one, *i.e.*, is fatter than any power law. A change of regime is ultimately observed and the very extreme tail decays faster than any power law. Both a SE or a log-Weibull model with exponent $c > 1$ provide a reasonable description.

Attempting to wrap up the different results obtained by the battery of tests presented here, we can offer the following conservative conclusion: it seems that the tails of the distributions examined here are decaying faster than any (reasonable) power law but slower than any Stretched-Exponentials. Maybe log-normal distributions could offer a better effective description of the distribution of returns,[19] as suggested in [436].

In sum, in the most practical case, the Pareto distribution is sufficient above quantiles $q_{12} = 95\%$ but is not stable enough to ascertain with strong confidence an asymptotic power law nature of the pdf.

### 2.5.3 Implications for Risk Assessment

The correct description of the distribution of returns has important implications for the assessment of large risks not yet sampled by historical time series. Indeed, the whole purpose of a characterization of the functional form of the distribution of returns is to extrapolate currently available historical

---

[18] See [453] and Figs. 3.6–3.9 pp. 81–82 of [451] where it is shown that SE distributions are approximately stable in family and the effect of aggregation can be seen to slowly increase the exponent $c$. See also [137] which studies specifically this convergence to a Gaussian law as a function of the time scale.

[19] Let us stress that we are speaking of a log-normal distribution of returns, not of price! Indeed, the standard Black and Scholes model of a log-normal distribution of prices is equivalent to a Gaussian distribution of returns. Thus, a log-normal distribution of returns is much more fat-tailed, and in fact bracketed by power law tails and Stretched-Exponential tails.

time series beyond the range provided by the empirical reconstruction of the distributions. For risk management, the determination of the tail of the distribution is crucial. Indeed, many risk measures, such as the Value-at-Risk or the expected-shortfall, are based on the properties of the tail of the distributions of returns. In order to assess risk at probability levels of 95% or so, nonparametric methods have merits. However, in order to estimate risks at high probability level such as 99% or larger, nonparametric estimations fail by lack of data and parametric models become unavoidable. This shift in strategy has a cost and replaces sampling errors by model errors. The considered distribution can be too thin-tailed as when using normal laws, and risk will be underestimated, or it can be too fat-tailed and risk will be overestimated as with Lévy law and possibly with regularly varying distributions. In each case, large amounts of money are at stake and can be lost due to a too conservative or too optimistic risk measurement.

In order to bypass these problems, many authors [34, 313, 355, among others] have proposed to estimate the extreme quantiles of the distributions in a semiparametric way, which allows one (i) to avoid the model errors and (ii) to limit the sampling errors with respect to nonparametric methods and thus to keep a reasonable accuracy in the estimation procedure. For this aim, it has been suggested to use the extreme value theory.[20] However, as emphasized in Sect. 2.3.3, estimates of the parameters of such (GEV or GPD) distributions can be very unreliable in the presence of dependence, so that these methods finally appear to be not very accurate and one cannot avoid a parametric approach for the estimation of the highest quantiles.

The above analysis suggests that the Paretian paradigm leads to an overestimation of the probability of large events and therefore leads to the adoption of too conservative positions. Generalizing to larger time scales, the overly pessimistic view of large risks deriving from the Paretian paradigm should be all the more revised, due to the action of the central limit theorem. The above comparison between several models, which turn out to be almost undistinguishable such as the Stretched-Exponential, the Pareto and the log-Weibull distributions, offers the important possibility of developing scenarios that can test the sensitivity of risk assessment to errors in the determination of parameters and even more interesting with respect to the choice of models, often referred to as model errors.

Finally, an additional note of caution is in order. This chapter has focused on the marginal distributions of returns calculated at fixed time scales and thus neglects the possible occurrence of runs of dependencies, such as in cumulative drawdowns. In the presence of dependencies between returns, and especially if the dependence is nonstationary and increases in time of stress, the characterization of the marginal distributions of returns is not sufficient. As an example, Johansen and Sornette [249] (see also Chap. 3 of [450])

---

[20] See, for instance, http://www.gloriamundi.org for an overview of the extensive application of EVT methods for VaR and expected-shortfall estimation.

have recently shown that the recurrence time of very large drawdowns cannot be predicted from the sole knowledge of the distribution of returns and that transient dependence effects occurring in times of stress make very large drawdowns more frequent, qualifying them as abnormal "outliers" (other names are "kings" or "black swans").

# Appendix

## 2.A Definition and Main Properties of Multifractal Processes

The traditional description of the dynamics of asset prices initiated by Bachelier [26] was based upon the Brownian motion and then the geometric Brownian motion [377, 425]. But it is now widely recognized that these descriptions suffer from two major discrepancies. As shown in this chapter, the stationary distribution of asset returns is far from the Gaussian law (it exhibits fat tails) and, in addition, the volatility of asset returns has long range dependence (or long memory), which is characterized by the alternation of periods of small price changes and periods of large price changes.

In the mathematical literature, stochastic processes are said to exhibit a long memory when their autocovariance function decays hyperbolically [85, 213, 222]. Fractionally integrated processes like ARFIMA[21] and FIGARCH[22] [31] processes are discrete time processes that enjoy this property. The first class is not suitable for the modeling of financial assets returns in so far as it yields long memory in the returns themselves. In contrast, the second class leads to long memory properties in the squared returns, which is more appropriate. However, an important question remains open concerning FIGARCH processes: is this kind of representation time consistent? That is, given that the daily returns of an asset can be modeled by a FIGARCH process, can we still model the monthly returns of this asset by a FIGARCH process? In other words, is the class of FIGARCH processes closed under time aggregation?

If time-consistency is not obeyed, the comparison of the discrete-time model with empirical data at all time-scales simultaneously imposes strong additional restrictions on the model. It is thus highly desirable that a suitable discrete-time model be time-consistent. Note that continuous-time models are time-consistent by construction, justifying the emphasis on continuous-time stochastic processes with long memory.

This Appendix presents useful results on a family of continuous-time stochastic processes which enjoy the property of long memory, the so-called multifractal process, born from the generalization of the seminal works by Mandelbrot on the notions of *self-similarity* and *fractality*.

---

[21] Fractionally integrated autoregressive moving average.
[22] Fractionally integrated GARCH.

## 2.A.1 Self-similar Processes, Multiplicative Cascades and Multifractal Processes

Before presenting two examples of multifractal processes with suitable properties for the modeling of financial asset prices, it is useful to describe their underpinning. First, let us recall that given the filtered space $(\Omega, \{\mathcal{F}_t\}_{t \geq 0}, \mathbb{P})$, the stochastic process $\{X(t)\}$ (with $X(0) = 0$) is self-similar with exponent $H > 0$ if, by definition, for all $\lambda, k, t_1, \ldots, t_k \geq 0$

$$\{X(\lambda t_1), \ldots, X(\lambda t_k)\} \overset{law}{=} \{\lambda^H \cdot X(t_1), \ldots, \lambda^H \cdot X(t_k)\}. \tag{2.A.1}$$

The most famous self-similar stochastic process is obviously the Brownian motion whose exponent $H = 1/2$. It belongs to the family of self-similar processes with stationary Gaussian increments, namely the Fractional Brownian Motions whose exponent $H$ range is $]0, 1[$; when $0 < H < 1/2$, the autocorrelation of the increments is negative (antipersistence) while it is positive (persistence) when $1/2 < H < 1$. In the later case, the Fractional Brownian Motion exhibits long memory.

Let us consider the law of the increments $\delta_l X(t) = X(t) - X(t-l)$. Assuming the stationarity of these increments, the law of $\delta_l X(t)$ is the same as the law of $X(l)$ (since $X(0) = 0$). Thus, if $X(t)$ is self-similar with an exponent $H$, it is easy to prove that the $q$-order moment of $\delta_l X(t)$ and $\delta_L X(t)$, denoted by $M(q, l)$ and $M(q, L)$ respectively, are related by

$$M(q, l) = \left( \frac{l}{L} \right)^{qH} M(q, L). \tag{2.A.2}$$

This is called a "monofractal" process characterized by a linear dependence of the moment exponent $\zeta(q) = qH$ as a function of the moment order $q$.

A multifractal process is obtained by using a weaker form of self-similarity. Instead of the simple scaling rule

$$X(\lambda t) \overset{law}{=} \lambda^H \cdot X(t) \tag{2.A.3}$$

induced by (2.A.1), multifractal processes enjoy the more general property

$$X(\lambda t) \overset{law}{=} K(\lambda) \cdot X(t) \tag{2.A.4}$$

where $X$ and $K$ are independent random variables. This generalized scaling rule induces strong restrictions on the distribution of the stochastic process. For instance, it is straightforward to show that for all $\lambda, k, t_1, \ldots, t_k > 0$

$$\frac{X(\lambda t_1)}{X(t_1)} \overset{law}{=} \frac{X(\lambda t_2)}{X(t_2)} \overset{law}{=} \cdots \overset{law}{=} \frac{X(\lambda t_k)}{X(t_k)}, \tag{2.A.5}$$

and that $K(\mu \cdot \nu) \overset{law}{=} K'(\mu) \cdot K''(\nu)$, where $K'$ and $K''$ are two independent copies of $K$. This last relation implies (provided that the expectations are finite)

$$E\left[|K(\mu \cdot \nu)|^q\right] = E\left[|K(\mu)|^q\right] \cdot E\left[|K(\nu)|^q\right] , \tag{2.A.6}$$

which immediately yields

$$E\left[|K(\lambda)|^q\right] = \lambda^{\zeta(q)}, \quad \forall \lambda > 0 , \tag{2.A.7}$$

for some real-valued function $\zeta(\cdot)$ such that $\zeta(0) = 0$. Considering the relation between the $q$-order moments of $\delta_l X(t)$ and $\delta_L X(t)$, (2.A.2) generalizes as follows

$$M(q, l) = \left(\frac{l}{L}\right)^{\zeta(q)} M(q, L) . \tag{2.A.8}$$

The function $\zeta(q)$ defines the *multifractal spectrum* of the process.

Processes enjoying this scaling property can be derived from so-called multiplicative cascades [17, 190, 340, 341]. It is convenient to present multiplicative cascades with discrete scales $l_n = 2^{-n} L$. A multiplicative cascade for the increments $\delta X$ is defined by relating the local variation of the process $\delta_{l_n} X$ at scale $l_n$ to the variation at scale $L$ according to

$$\delta_{l_n} X(t) = \left(\prod_{i=1}^{n} W_i\right) \delta_L X(t) , \tag{2.A.9}$$

where $W_i$ are i.i.d. random positive factors. Realizations of such processes can be constructed using orthonormal wavelet bases [16]. If one defines the *magnitude* $\omega(t, l)$ at time $t$ and scale $l$ as [17],

$$\omega(t, l) = \frac{1}{2} \ln(|\delta_l X(t)|^2) , \tag{2.A.10}$$

then the cascade (2.A.9) becomes a simple random walk as a function of the logarithm of scales, at a fixed time $t$:

$$\omega(t, l_{n+1}) = \omega(t, l_n) + \ln(W_{n+1}) . \tag{2.A.11}$$

Assuming that $W$ follows a log-normal law with parameters $(\mu, \lambda^2)$, the magnitude $\omega$ admits a density at scale $l_n$, $Q_{l_n}(\omega)$, which satisfies the simple equation

$$Q_{l_n}(\omega) = \left(\varphi(\mu, \lambda^2)^{*n} * Q_L\right)(\omega) \tag{2.A.12}$$

where $*$ is the convolution product and $\varphi(\mu, \lambda^2)$ denotes the Gaussian density function with mean $\mu$ and variance $\lambda^2$. Going back to the original variable $\delta X$, the previous equation provides us with the expression of the density function of $\delta X$ at scale $l_n$

$$P_{l_n}(x) = \int G_{l_n, L}(u) e^{-u} P_L(e^{-u} x) du , \tag{2.A.13}$$

where

$$G_{l_n,L} = \varphi(\mu, \lambda^2)^{*n} = \varphi(n\mu, n\lambda^2) \ .$$

Conversely, a process that satisfies (2.A.13) with a normal kernel $G$ can be written as

$$\delta_l X(t) \overset{law}{=} W \cdot \delta_{2l} X(t) \tag{2.A.14}$$

where $W$ is distributed according to a log-normal law with mean $\mu$ and variance $\lambda^2$. By iterating this equation $n$ times, one thus recovers the cascade (2.A.9). Therefore, the cascade picture across scales constitutes a paradigm of multifractal self-similar processes. The log-normal cascade model on the dyadic tree associated to the orthonormal wavelet representation leads to a magnitude correlation function given by

$$C_\omega(\tau, l) = \mathrm{Cov}(\omega(t,l), \omega(t+\tau, l)) \propto -\lambda^2 \ln(\tau/T) \ , \quad \text{for } l < \tau < T \ , \tag{2.A.15}$$

which is proportional to the logarithm of the lag $\tau$. The parameter $T$ is called the "integral time scale" and is such that $C_\omega(\tau, l)$ is exactly 0 for $\tau > T$.

## 2.A.2 The Multifractal Spectrum

We have introduced in (2.A.8) the $q$-order moment $M(q,l)$ of the increment $\delta_l X(t)$ as scale $l$, which follows the scaling law

$$M(q,l) \sim l^{\zeta(q)} \ , \tag{2.A.16}$$

and allows us to explore the multifractal properties of the multifractal processes.

Let us define the Hölder exponent $\alpha(t_0)$ at time $t_0$ as

$$\delta_l X(t_0) \sim_{l \to 0} l^{\alpha(t_0)} \ . \tag{2.A.17}$$

The multifractal spectrum $f(\alpha)$ is the fractal (Haussdorf) dimension of the iso-Hölder exponent sets:

$$f(\alpha) = \mathrm{Dim}\{t, \ \alpha(t) = \alpha\} \ . \tag{2.A.18}$$

Roughly speaking, this means that, at scale $l \ll T$, the number of times where $\delta_l X(t) \sim l^\alpha$ is

$$\mathcal{N}(t, \alpha) \sim l^{-f(\alpha)} \ . \tag{2.A.19}$$

The multifractal formalism obtains that $f(\alpha)$ and $\zeta(q)$ are Legendre transform of each other:

$$f(\alpha) = 1 + \min_q (q\alpha - \zeta(q)) \ , \tag{2.A.20}$$
$$\zeta(q) = 1 + \min_\alpha (q\alpha - f(\alpha)) \ . \tag{2.A.21}$$

Therefore, in the multifractal formalism, $q$ is nothing but the value of the derivative of $f(\alpha)$ and conversely $\alpha$ is the value of the derivative of $\zeta(q)$:

$$q(\alpha^*) = \frac{\partial f}{\partial \alpha}\big|_{\alpha^*} \ . \tag{2.A.22}$$

### 2.A.3 The Multifractal Model of Asset Returns of Mandelbrot *et al.*

Mandelbrot *et al.* [341] have proposed a very simple way to obtain a multifractal process with suitable properties for the modeling of asset returns. In its simplest form, it is based upon the subordination of a Brownian motion by a multifractal process. Indeed, considering the price process $\{P(t)\}_{t\geq 0}$, the logarithm of the price

$$X(t) = \ln P(t) - \ln P(0) \qquad (2.A.23)$$

is assumed to be defined by

$$X(t) = B[\theta(t)] , \qquad (2.A.24)$$

where $B(t)$ denotes the standard Brownian motion assumed independent of the stochastic process $\theta(t)$ which is a multifractal process with continuous, nondecreasing paths and stationary increments.

It is easy to check that $X(t)$ enjoys the multifractal property and it is straightforward to show that its multifractal spectrum $\zeta_X$ is related to the multifractal spectrum $\zeta_\theta$ of $\theta$ by

$$\zeta_X(q) = \zeta_\theta\left(\frac{q}{2}\right). \qquad (2.A.25)$$

In addition, as long as $B(t)$ is a Brownian motion without drift, the stochastic process $\{X(t)\}_{t>0}$ is a martingale with respect to its natural filtration provided that $\mathrm{E}\left[\theta^{1/2}\right] < \infty$. Besides, if $\mathrm{E}[\theta]$ is finite, the autocovariance function of the price return process $\delta_l X(t)$ vanishes for all lag larger than $l$. The covariance of the absolute values of the price returns (raised to the power $2q$) satisfies

$$\mathrm{Cov}\left(|\delta_l X(t)|^{2q}, |\delta_l X(t+\tau)|^{2q}\right) = \kappa(q) \cdot \mathrm{Cov}\left(|\delta_l \theta(t)|^q, |\delta_l \theta(t+\tau)|^q\right) ,$$
$$(2.A.26)$$

with $\kappa(q) = \left(\mathrm{E}\left[|B(1)|^{2q}\right]\right)^2$, so that the volatility of assets returns exhibits long memory if the volatility of $\theta$ itself exhibits long memory, *i.e.*, if periods of intense trading activity alternates with periods of weak activity.

### 2.A.4 The Multifractal Random Walk (MRW)

An alternative approach for modeling the dynamics of asset returns in terms of multifractal processes has been introduced by Bacry *et al.* [27, 28] with the aim of constructing a stationary process with continuous scale invariance inspired from the standard hierarchical models presented in Sect. 2.A.1.

## Definition

The MRW is a stochastic volatility model which has exact multifractal properties, is invariant under continuous dilations, and possesses stationary increments. It is constructed so as to mimic the crucial logarithmic dependence (2.A.15) of the magnitude correlation function, at the basis of multifractality in cascade processes.

The MRW is constructed as the continuous limit for $\Delta t \to 0$ of the discretized version $X_{\Delta t}$ (using a time discretization step $\Delta t$) defined by adding up $t/\Delta t$ random variables:

$$X_{\Delta t}(t) = \sum_{k=1}^{t/\Delta t} \delta X_{\Delta t}[k] \ . \tag{2.A.27}$$

The process $\{\delta X_{\Delta t}[k]\}_k$ is a noise whose variance is stochastic, $i.e.$,

$$\delta X_{\Delta t}[k] = \epsilon_{\Delta t}[k] e^{\omega_{\Delta t}[k]} \ , \tag{2.A.28}$$

where $\omega_{\Delta t}[k]$ is the logarithm of the stochastic variance. $\epsilon_{\Delta t}$ is a Gaussian white noise independent of $\omega$ and of variance $\sigma^2 \Delta t$.[23]

Following the cascade model, $\omega_{\Delta t}$ is a Gaussian stationary process whose covariance reads

$$\mathrm{Cov}(\omega_{\Delta t}[k], \omega_{\Delta t}[l]) = \lambda^2 \ln \rho_{\Delta t}[|k - l|] \ , \tag{2.A.29}$$

where $\rho_{\Delta t}$ is chosen in order to mimic the correlation structure (2.A.15) observed in cascade models:

$$\rho_{\Delta t}[k] = \begin{cases} \frac{T}{(|k|+1)\Delta t} & \text{for } |k| \le T/\Delta t - 1 \\ 1 & \text{otherwise} \end{cases} \tag{2.A.30}$$

In order for the variance of $X_{\Delta t}(t)$ to converge when $\Delta t \to 0$, one must choose the mean of the process $\omega_{\Delta t}$ such that [27]

$$E\left(\omega_{\Delta t}[k]\right) = -\mathrm{Var}\left(\omega_{\Delta t}[k]\right) = -\lambda^2 \ln(T/\Delta t) \ , \tag{2.A.31}$$

for which $\mathrm{Var}(X_{\Delta t}(t)) = \sigma^2 \ t$ .

## Multifractal Spectrum

Since, by construction, the increments of the model are stationary, the pdf of $X_{\Delta t}(t+l) - X_{\Delta t}(t)$ does not depend on $t$ and is the same as that of $X_{\Delta t}(l)$. In

---

[23] Introducing an asymmetric dependence between $\omega$ and the noise $\epsilon$ allows one to account for the Leverage effect [170] while preserving the scale invariance properties of the MRW, but forbids the existence of a limit as $\Delta t \longrightarrow 0$ [388].

[27], it was proven that the moments of $X(l) \equiv X_{\Delta t \to 0+}(l)$ can be expressed as

$$E(X(l)^{2p}) = \frac{\sigma^{2p}(2p)!}{2^p p!} \int_0^l du_1 ... \int_0^l du_p \prod_{i<j} \rho(u_i - u_j)^{4\lambda^2} ,\qquad (2.A.32)$$

where $\rho$ is defined by

$$\rho(t) = \begin{cases} T/|t| & \text{for } |t| \leq T \\ 1 & \text{otherwise} \end{cases} .\qquad (2.A.33)$$

Using this expression in the multiple integrals in (2.A.32), a straightforward scaling argument leads to

$$M(2p, l) = K_{2p} \left(\frac{l}{T}\right)^{p - 2p(p-1)\lambda^2} ,\qquad (2.A.34)$$

where

$$K_{2p} = T^p \sigma^{2p}(2p-1)!! \int_0^1 du_1 ... \int_0^1 du_p \prod_{i<j} |u_i - u_j|^{-4\lambda^2} .\qquad (2.A.35)$$

$K_{2p}$ is nothing but the moment of order $2p$ of the random variable $X(T)$ or equivalently of $\delta_T X(t)$. Expression (2.A.35) leads to $\zeta_{2p} = p - 2p(p-1)\lambda^2$, and by analytical continuation, the corresponding full $\zeta_q$ spectrum is thus the parabola

$$\zeta_q = (q - q(q-2)\lambda^2)/2 .\qquad (2.A.36)$$

### Approximate form in Terms of a Long Memory Kernel in the Discrete Time Approximation

Consider the returns at scale $\Delta t$, defined by $r_{\Delta t}(t) \equiv \ln[p(t)/p(t - \Delta t)]$. Then, mapping the increments $\delta X_{\Delta t}[k]$ defined in (2.A.28) onto $r_{\Delta t}(t)$ makes the price $p(t)$ a multifractal random walk in the continuous limit $\Delta t \to 0$. The discrete return $r_{\Delta t}(t)$ can thus be written as

$$r_{\Delta t}(t) = \epsilon(t) \cdot \sigma_{\Delta t}(t) = \epsilon(t) \cdot e^{\omega_{\Delta t}(t)} ,\qquad (2.A.37)$$

where $\epsilon(t)$ is a standardized Gaussian white noise independent of $\omega_{\Delta t}(t)$ and $\omega_{\Delta t}(t)$ is a nearly Gaussian process (exactly Gaussian for $\Delta t \to 0$) with mean and covariance:

$$\mu_{\Delta t} = \frac{1}{2} \ln(\sigma^2 \Delta t) - C_{\Delta t}(0) \qquad (2.A.38)$$

$$C_{\Delta t}(\tau) = \text{Cov}[\omega_{\Delta t}(t), \omega_{\Delta t}(t + \tau)],$$

$$= \begin{cases} \lambda^2 \ln \left(\frac{T}{|\tau| + e^{-3/2}\Delta t}\right) & \text{if } |\tau| < T - e^{-3/2}\Delta t , \\ 0 & \text{if } |\tau| \geq T - e^{-3/2}\Delta t \end{cases} \qquad (2.A.39)$$

$\sigma^2 \Delta t$ is the variance of the returns at scale $\Delta t$ and $T$ is the "integral" (correlation) time scale. Typical values for $T$ and $\lambda^2$ are respectively 1 year and 0.02.

The MRW model can be expressed in a more familiar form, in which the log-volatility $\omega_{\Delta t}(t)$ obeys an autoregressive equation whose solution reads [456]

$$\omega_{\Delta t}(t) = \mu_{\Delta t} + \int_{-\infty}^{t} dW(\tau) \, K_{\Delta t}(t - \tau) \, , \qquad (2.A.40)$$

where $W(t)$ denotes a standard Wiener process and the memory kernel $K_{\Delta t}(\cdot)$ is a causal function, ensuring that the system is not anticipative. The process $W(t)$ can be seen as the cumulative information flow. Thus $\omega(t)$ represents the response of the price to incoming information up to the date $t$. At time $t$, the distribution of $\omega_{\Delta t}(t)$ is Gaussian with mean $\mu_{\Delta t}$ and variance $V_{\Delta t} = \int_0^{\infty} d\tau \, K_{\Delta t}^2(\tau) = \lambda^2 \ln\left(\frac{Te^{3/2}}{\Delta t}\right)$. Its covariance, which entirely specifies the random process, is given by

$$C_{\Delta t}(\tau) = \int_0^{\infty} dt \, K_{\Delta t}(t) K_{\Delta t}(t + |\tau|) \, . \qquad (2.A.41)$$

Performing a Fourier transform, we obtain

$$\hat{K}_{\Delta t}(f)^2 = \hat{C}_{\Delta t}(f) = 2\lambda^2 \, f^{-1} \left[ \int_0^{Tf} \frac{\sin(t)}{t} dt + O\left(f \Delta t \ln(f \Delta t)\right) \right] \, , \qquad (2.A.42)$$

which shows that, for $\tau$ small enough,

$$K_{\Delta t}(\tau) \sim K_0 \sqrt{\frac{\lambda^2 T}{\tau}} \qquad \text{for} \quad \Delta t \ll \tau \ll T \, . \qquad (2.A.43)$$

This slow inverse square root power law decay (2.A.43) of the memory kernel in (2.A.40) ensures the long-range dependence and multifractality of the stochastic volatility process (2.A.37). Note that (2.A.40) for the log-volatility $\omega_{\Delta t}(t)$ takes a form similar to but simpler than the ARFIMA models often used to account for the very slow decay of the sample ACF of the log-volatility of assets returns [30].

## 2.B A Survey of the Properties of Maximum Likelihood Estimators

This appendix summarizes the expressions of the maximum likelihood estimators derived from the four distributions (2.26–2.29). In the following, we consider an iid sample $X_1, \ldots, X_T$ drawn from one of the distributions under consideration, namely the Pareto, the Weibull, the Exponential, the incomplete Gamma and the log-Weibull distributions.

### 2.B.1 The Pareto Distribution

According to expression (2.26), the Pareto distribution is given by

$$F_u(x) = 1 - \left(\frac{u}{x}\right)^b, \quad x \geq u \tag{2.B.44}$$

and its density is

$$f_u(x|b) = b\frac{u^b}{x^{b+1}} . \tag{2.B.45}$$

Let us denote by

$$L_T^{PD}(\hat{b}) = \max_b \sum_{i=1}^T \ln f_u(X_i|b) \tag{2.B.46}$$

the maximum of the log-likelihood function derived under hypothesis (PD). $\hat{b}$ is the maximum likelihood estimator of the tail index $b$ under the PD hypothesis.

The maximum of the likelihood function is solution of

$$\frac{1}{b} + \ln u - \frac{1}{T}\sum_{i=1}^T \ln X_i = 0 , \tag{2.B.47}$$

which yields

$$\hat{b} = \left[\frac{1}{T}\sum_{i=1}^T \ln X_i - \ln u\right]^{-1}, \quad \text{and} \quad \frac{1}{T} L_T^{PD}(\hat{b}) = \ln\frac{\hat{b}}{u} - \left(1 + \frac{1}{\hat{b}}\right). \tag{2.B.48}$$

Moreover, one easily shows that $\hat{b}$ is asymptotically normally distributed:

$$\sqrt{T}(\hat{b} - b) \sim \mathcal{N}(0, b) . \tag{2.B.49}$$

### 2.B.2 The Weibull Distribution

The Weibull distribution is given by (2.27) and its density is

$$f_u(x|c, d) = \frac{c}{d^c} \cdot e^{\left(\frac{u}{d}\right)^c} x^{c-1} \cdot \exp\left[-\left(\frac{x}{d}\right)^c\right], \quad x \geq u . \tag{2.B.50}$$

The maximum of the log-likelihood function is

$$L_T^{SE}(\hat{c}, \hat{d}) = \max_{c,d} \sum_{i=1}^T \ln f_u(X_i|c, d) \tag{2.B.51}$$

Thus, the maximum likelihood estimators $(\hat{c}, \hat{d})$ are solution of

$$\frac{1}{c} = \frac{\frac{1}{T}\sum_{i=1}^{T}\left(\frac{X_i}{u}\right)^c \ln\frac{X_i}{u}}{\frac{1}{T}\sum_{i=1}^{T}\left(\frac{X_i}{u}\right)^c - 1} - \frac{1}{T}\sum_{i=1}^{T}\ln\frac{X_i}{u} , \tag{2.B.52}$$

$$d^c = \frac{u^c}{T}\sum_{i=1}^{T}\left(\frac{X_i}{u}\right)^c - 1 . \tag{2.B.53}$$

Equation (2.B.52) depends on $c$ only and must be solved numerically. Then, the resulting value of $c$ can be put in (2.B.53) to get $d$. The maximum of the log-likelihood function is

$$\frac{1}{T}L_T^{SE}(\hat{c}, \hat{d}) = \ln\frac{\hat{c}}{\hat{d}^{\hat{c}}} + \frac{\hat{c}-1}{T}\sum_{i=1}^{T}\ln X_i - 1 . \tag{2.B.54}$$

Since $c > 0$, the vector $\sqrt{T}(\hat{c} - c, \hat{d} - d)$ is asymptotically normal, with a covariance matrix whose expression is given in Appendix 2.C.

It should be noted that the maximum likelihood (2.B.52–2.B.53) do not admit a solution with positive $c$ for all possible samples $(X_1, \ldots, X_T)$. Indeed, the function

$$h(c) = \frac{1}{c} - \frac{\frac{1}{T}\sum_{i=1}^{T}\left(\frac{X_i}{u}\right)^c \ln\frac{X_i}{u}}{\frac{1}{T}\sum_{i=1}^{T}\left(\frac{X_i}{u}\right)^c - 1} + \frac{1}{T}\sum_{i=1}^{T}\ln\frac{X_i}{u} , \tag{2.B.55}$$

which is the total derivative of $L_T^{SE}(c, \hat{d}(c))$, is a decreasing function of $c$. This means, as one can expect, that the likelihood function is concave. Thus, a necessary and sufficient condition for (2.B.52) to admit a solution is that $h(0)$ is positive. After some calculations, we find

$$h(0) = \frac{2\left(\frac{1}{T}\sum\ln\frac{X_i}{u}\right)^2 - \frac{1}{T}\sum\ln^2\frac{X_i}{u}}{\frac{2}{T}\sum\ln\frac{X_i}{u}} , \tag{2.B.56}$$

which is positive if and only if

$$2\left(\frac{1}{T}\sum\ln\frac{X_i}{u}\right)^2 - \frac{1}{T}\sum\ln^2\frac{X_i}{u} > 0 . \tag{2.B.57}$$

A finite sample may not automatically obey this condition even if it has been generated by the SE distribution. However, the probability of occurrence of a sample leading to a negative maximum likelihood estimate of $c$ tends to zero (under the SE Hypothesis with a positive $c$) as

$$\Phi\left(-\frac{c\sqrt{T}}{\sigma}\right) \simeq \frac{\sigma}{\sqrt{2\pi T}\,c}e^{-\frac{c^2 T}{2\sigma^2}} , \tag{2.B.58}$$

*i.e.* exponentially with respect to $T$. Here, $\sigma^2$ is the variance of the limit Gaussian distribution of the maximum likelihood $c$-estimator that can be derived explicitly. If $h(0)$ is negative, $L_T^{SE}$ reaches its maximum at $c = 0$ and in such a case

$$\frac{1}{T} L_T^{SE}(c = 0) = -\ln\left(\frac{1}{T}\sum \ln\frac{X_i}{u}\right) - \frac{1}{T}\sum \ln X_i - 1 \ . \tag{2.B.59}$$

In contrast, if the maximum likelihood estimation based on the SE assumption is applied to samples distributed differently from the SE, negative $c$-estimate can then be obtained with some positive probability not tending to zero with $T \longrightarrow \infty$. If the sample is distributed according to the Pareto distribution, for instance, then the maximum-likelihood $c$-estimate converges in probability to a Gaussian random variable with zero mean, and thus the probability for negative $c$-estimates converges to 0.5.

### 2.B.3 The Exponential Distribution

The Exponential distribution function is given by (2.28), and its density is

$$f_u(x|d) = \frac{\exp\left[\frac{u}{d}\right]}{d} \exp\left[-\frac{x}{d}\right], \quad x \geq u \ . \tag{2.B.60}$$

The maximum of the log-likelihood function is reached at

$$\hat{d} = \frac{1}{T} \sum_{i=1}^{T} X_i - u \ , \tag{2.B.61}$$

and is given by

$$\frac{1}{T} L_T^{ED}(\hat{d}) = -(1 + \ln \hat{d}) \ . \tag{2.B.62}$$

The random variable $\sqrt{T}(\hat{d} - d)$ is asymptotically normally distributed with zero mean and variance $d^2/T$.

### 2.B.4 The Incomplete Gamma Distribution

The expression of the incomplete Gamma distribution function is given by (2.29) and its density is

$$f_u(x|b, d) = \frac{d^b}{\Gamma\left(-b, \frac{u}{d}\right)} \cdot x^{-(b+1)} \exp\left[-\left(\frac{x}{d}\right)\right], \quad x \geq u \ . \tag{2.B.63}$$

Let us introduce the partial derivative of the logarithm of the incomplete Gamma function:

$$\Psi(a, x) = \frac{\partial}{\partial a} \ln \Gamma(a, x) = \frac{1}{\Gamma(a, x)} \int_x^\infty dt \, \ln t \, t^{a-1} \, e^{-t} \, . \qquad (2.B.64)$$

The maximum of the log-likelihood function is reached at the point $(\hat{b}, \hat{d})$ solution of

$$\frac{1}{T} \sum_{i=1}^T \ln \frac{X_i}{d} = \Psi\left(-b, \frac{u}{d}\right), \qquad (2.B.65)$$

$$\frac{1}{T} \sum_{i=1}^T \frac{X_i}{d} = \frac{1}{\Gamma\left(-b, \frac{u}{d}\right)} \left(\frac{u}{d}\right)^{-b} e^{-\frac{u}{d}} - b \, , \qquad (2.B.66)$$

and is equal to

$$\frac{1}{T} L_T^{IG}(\hat{b}, \hat{d}) = -\ln \hat{d} - \ln \Gamma\left(-b, \frac{u}{d}\right)$$
$$+ (b+1) \cdot \Psi\left(-b, \frac{u}{d}\right) + b - \frac{1}{\Gamma\left(-b, \frac{u}{d}\right)} \left(\frac{u}{d}\right)^{-b} e^{-\frac{u}{d}} \quad (2.B.67)$$

## 2.B.5 The Log-Weibull Distribution

The Log-Weibull distribution is given by (2.34) and its density is

$$f_u(x|b, c) = \frac{b \cdot c}{x} \cdot \left(\ln \frac{x}{u}\right)^{c-1} \cdot \exp\left[-b\left(\ln \frac{x}{d}\right)^c\right], \quad x \geq u \, . \qquad (2.B.68)$$

The maximum of the log-likelihood function is

$$L_T^{SE}(\hat{b}, \hat{c}) = \max_{b,c} \sum_{i=1}^T \ln f_u(X_i|b, c) \, . \qquad (2.B.69)$$

Thus, the maximum likelihood estimators $(\hat{b}, \hat{b})$ are solution of

$$b^{-1} = \frac{1}{T} \sum_{i=1}^T \left(\ln \frac{X_i}{u}\right)^c , \qquad (2.B.70)$$

$$\frac{1}{c} = \frac{\frac{1}{T} \sum_{i=1}^T \left(\ln \frac{X_i}{u}\right)^c \ln \left(\ln \frac{X_i}{u}\right)}{\frac{1}{T} \sum_{i=1}^T \left(\ln \frac{X_i}{u}\right)^c} - \frac{1}{T} \sum_{i=1}^T \ln \left(\ln \frac{X_i}{u}\right) . \qquad (2.B.71)$$

The solution of these equations is unique and it can be shown that the vector $\sqrt{T}(\hat{b} - b, \hat{c} - c)$ is asymptotically Gaussian with a covariance which can be deduced from the matrix (2.C.88) given in Appendix 2.C.

## 2.C Asymptotic Variance–Covariance
## of Maximum Likelihood Estimators of the SE Parameters

We consider the Stretched-Exponential (SE) parametric family with complementary distribution function

$$\bar{F} = 1 - F(x) = \exp\left[-\left(\frac{x}{d}\right)^c + \left(\frac{u}{d}\right)^c\right] \quad x \geqslant u, \tag{2.C.72}$$

where $c, d$ are unknown parameters and $u$ is a known lower threshold.

Let us take a new parameterization of the SE distribution, more appropriate for the derivation of asymptotic variances. It should be noted that this change of parameters does not affect the asymptotic variance of the form parameter $c$. In the new parameterization, the complementary distribution function has the form:

$$\bar{F}(x) = \exp\left[-v\left(\left(\frac{x}{u}\right)^c - 1\right)\right], \quad x \geqslant u. \tag{2.C.73}$$

Here, the parameter $v$ involves both the unknown parameters $c, d$ and the known threshold $u$:

$$v = \left(\frac{u}{d}\right)^c. \tag{2.C.74}$$

The log-likelihood $L$ for sample $(X_1, \ldots, X_T)$ has the form:

$$L = N\ln v + N\ln c + (c-1)\sum_{i=1}^{N}\ln\frac{X_i}{u} - v\sum_{i=1}^{N}\left[\left(\frac{X_i}{u}\right)^c - 1\right]. \tag{2.C.75}$$

Now, we derive the Fisher matrix $\Phi$:

$$\Phi = \begin{pmatrix} E\left[-\partial_v^2 L\right] & E\left[-\partial_{v,c}^2 L\right] \\ E\left[-\partial_{c,v}^2 L\right] & E\left[-\partial_c^2 L\right] \end{pmatrix}, \tag{2.C.76}$$

and find

$$\frac{\partial^2 L}{\partial v^2} = -\frac{N}{v^2}, \tag{2.C.77}$$

$$\frac{\partial^2 L}{\partial v \partial c} = -N \cdot \frac{1}{N}\sum_{i=1}^{N}\left(\frac{X_i}{u}\right)^c \ln\frac{X_i}{u} \xrightarrow{N\to\infty} -NE\left[\left(\frac{X}{u}\right)^c \ln\frac{X}{u}\right] \tag{2.C.78}$$

$$\frac{\partial^2 L}{\partial c^2} = -\frac{N}{c^2} - Nv \cdot \frac{1}{N}\sum_{i=1}^{N}\left(\frac{X_i}{u}\right)^c \ln^2\frac{X_i}{u}$$

$$\xrightarrow{N\to\infty} -\frac{N}{c^2} - Nv \cdot E\left[\left(\frac{X}{u}\right)^c \ln^2\frac{X}{u}\right]. \tag{2.C.79}$$

After some calculations, we obtain:

$$E\left[\left(\frac{X}{u}\right)^c \ln\left(\frac{X}{u}\right)\right] = \frac{1 + E_1(v)}{c \cdot v}, \tag{2.C.80}$$

where $E_1(v)$ is the integral exponential function:

$$E_1(v) = \int_v^\infty \frac{e^{-t}}{t}\, dt. \tag{2.C.81}$$

Similarly we find:

$$\mathrm{E}\left[\left(\frac{X}{u}\right)^c \ln^2 \frac{X}{u}\right] = \frac{2e^v}{v \cdot c^2}[E_1(v) + E_2(v) - \ln(v)E_1(v)] , \qquad (2.\mathrm{C}.82)$$

where $E_2(v)$ is the partial derivative of the incomplete Gamma function:

$$E_2(v) = \int_v^\infty \frac{\ln(t)}{t} \cdot e^{-t} dt = \frac{\partial}{\partial a} \int_v^\infty t^{a-1} e^{-t} dt \bigg|_{a=0} = \frac{\partial}{\partial a} \Gamma(a, x) \bigg|_{a=0} . \qquad (2.\mathrm{C}.83)$$

The Fisher matrix (multiplied by $N$) then reads:

$$N\Phi = \begin{pmatrix} \frac{1}{v^2} & \frac{1+e^v E_1(v)}{c \cdot v} \\ \frac{1+e^v E_1(v)}{c \cdot v} & \frac{1}{c^2}(1 + 2e^v [E_1(v) + E_2(v) - \ln(v)E_1(v)]) \end{pmatrix}. \qquad (2.\mathrm{C}.84)$$

The covariance matrix $B$ of the ML-estimates $(\tilde{v}, \tilde{c})$ is equal to the inverse of the Fisher matrix. Thus, inverting the Fisher matrix $\Phi$ in (2.C.84) provides the desired covariance matrix:

$$B = \begin{pmatrix} \frac{v^2}{NH(v)}[1 + 2e^v E_1(v) + 2e^v E_2(v) - \ln(v)e^v E_1(v)] & -\frac{cv}{NH(v)}[1 + e^v E_1(v)] \\ -\frac{cv}{NH(v)}[1 + e^v E_1(v)] & \frac{c^2}{NH(v)} \end{pmatrix}, \qquad (2.\mathrm{C}.85)$$

where $H(v)$ has the form:

$$H(v) = 2e^v E_2(v) - 2\ln(v)e^v E_1(v) - (e^v E_1(v))^2 . \qquad (2.\mathrm{C}.86)$$

We also present here the covariance matrix of the limit distribution of ML-estimates for the SE distribution on the whole semi-axis $(0, \infty)$:

$$1 - F(x) = \exp(-g \cdot x^c), \quad x \geqslant 0 . \qquad (2.\mathrm{C}.87)$$

After some calculations following the same steps as above, we find the covariance matrix $B$ of the limit Gaussian distribution of ML-estimates $(\tilde{g}, \tilde{c})$:

$$B = \frac{6}{N\pi^2} \begin{pmatrix} g^2 \left[\frac{\pi^2}{6} + (\gamma + \ln(g) - 1)^2\right] & g \cdot c \, [\gamma + \ln(g) - 1] \\ g \cdot c \, [\gamma + \ln(g) - 1] & c^2 \end{pmatrix} , \qquad (2.\mathrm{C}.88)$$

where $\gamma$ is the Euler number: $\gamma \simeq 0.577\ 215\ldots$

## 2.D Testing the Pareto Model versus the Stretched-Exponential Model

This Appendix derives the statistic that allows one to test the SE hypothesis $f_1(x|c, b)$ versus the Pareto hypothesis $f_0(x|\beta)$ on a semi-infinite interval $(u, \infty)$, $u > 0$. The following parameterization is used:

$$f_1(x|c,b) = b\, u^{-c} x^{c-1} \exp\left[-\frac{b}{c}\left(\left(\frac{x}{u}\right)^c - 1\right)\right]; \quad x \geq u \qquad (2.D.89)$$

for the Stretched-Exponential distribution and

$$f_0(x|\beta) = \beta\, \frac{u^\beta}{x^{1+\beta}}; \quad x \geq u \qquad (2.D.90)$$

for the Pareto distribution.

**Theorem**: Assuming that the sample $X_1, \ldots, X_N$ is generated from the Pareto distribution (2.D.90), and taking the supremum of the log-likelihoods $L_0$ and $L_1$ of the Pareto and (SE) models respectively over the domains $(\beta > 0)$ for $L_0$ and $(b > 0, c > 0)$ for $L_1$, then Wilks' log-likelihood ratio $W$:

$$W_N = 2\left[\sup_{b,c} L_1 - \sup_\beta L_0\right], \qquad (2.D.91)$$

is distributed according to the $\chi^2$-distribution with one degree of freedom, in the limit $N \to \infty$.

**Proof**

The log-likelihood $L_0$ reads

$$L_0 = -\sum_{i=1}^N \log X_i + N \log(\beta) - \beta \sum_{i=1}^N \log \frac{X_i}{u}. \qquad (2.D.92)$$

The supremum over $\beta$ of $L_0$ given by (2.D.92) is reached at

$$\hat{\beta}_N = \left[\frac{1}{N}\sum_{i=1}^N \log \frac{X_i}{u}\right]^{-1}, \qquad (2.D.93)$$

and is equal to

$$\sup_\beta L_0 = -N\left(1 + \log u + \frac{1}{\hat{\beta}_N} - \log \hat{\beta}_N\right). \qquad (2.D.94)$$

The log-likelihood $L_1$ is

$$L_1 = -N\left\{\log u - (c-1)\frac{1}{N}\sum_{i=1}^N \log \frac{X_i}{u} - \log b + \frac{b}{c}\frac{1}{N}\sum_{i=1}^N\left[\left(\frac{X_i}{u}\right)^c - 1\right]\right\}. \qquad (2.D.95)$$

The supremum over $b$ of $L_1$ given by (2.D.95) is reached at

$$\hat{b}_N = c\left(\frac{1}{N}\sum_{i=1}^N\left[\left(\frac{X_i}{u}\right)^c - 1\right]\right)^{-1} \qquad (2.D.96)$$

and is equal to

$$\sup_b L_1 = -N \left( 1 + \log u - (c-1) \frac{1}{N} \sum_{i=1}^N \log \frac{X_i}{u} - \log \hat{b}_N \right) . \tag{2.D.97}$$

Taking the derivative of expression (2.D.97) with respect to $c$, we obtain the maximum likelihood equation for the SE parameter $c$

$$\frac{1}{\hat{c}_N} = \frac{\frac{1}{N} \sum_{i=1}^N \left( \frac{X_i}{u} \right)^{\hat{c}_N} \log \frac{X_i}{u}}{\frac{1}{N} \sum_{i=1}^N \left( \frac{X_i}{u} \right)^{\hat{c}_N} - 1} - \frac{1}{N} \sum_{i=1}^N \log \frac{X_i}{u} . \tag{2.D.98}$$

If the sample $X_1, \ldots, X_N$ is generated by the Pareto distribution (2.D.90), then by the strong law of large numbers, we have with probability 1 as $N \rightarrow +\infty$

$$\frac{1}{N} \sum_{i=1}^N \log \frac{X_i}{u} \longrightarrow \mathrm{E}_0 \left[ \log \frac{X}{u} \right] = \frac{1}{\beta}, \tag{2.D.99}$$

$$\frac{1}{N} \sum_{i=1}^N \left[ \left( \frac{X_i}{u} \right)^c - 1 \right] \longrightarrow \mathrm{E}_0 \left[ \left( \frac{X}{u} \right)^c - 1 \right] = \frac{c}{\beta - c}, \tag{2.D.100}$$

$$\frac{1}{N} \sum_{i=1}^N \left( \frac{X_i}{u} \right)^c \log \frac{X_i}{u} \longrightarrow \mathrm{E}_0 \left[ \left( \frac{X}{u} \right)^c \log \frac{X}{u} \right] = \frac{\beta}{(\beta - c)^2} , \tag{2.D.101}$$

where $\mathrm{E}_0[\cdot]$ denotes the expectation with respect to $f_0(\cdot|\beta)$.

Inserting these limit values into (2.D.98), the only limit solution of this equation is $c = 0$. Thus, the solution of (2.D.98) for finite $N$, denoted as $\hat{c}_N$, converges to zero with probability 1 to zero as $N \rightarrow +\infty$.

Expanding $(X_i/u)^c$ in power series in the neighborhood of $c = 0$ gives

$$\left( \frac{X_i}{u} \right)^c \cong 1 + c \cdot \log \left( \frac{X_i}{u} \right) + \frac{c^2}{2} \cdot \log^2 \left( \frac{X_i}{u} \right) + \frac{c^3}{6} \cdot \log^3 \left( \frac{X_i}{u} \right) + \cdots , \tag{2.D.102}$$

which yields

$$\frac{1}{N} \sum \left( \frac{X_i}{u} \right)^c \cong 1 + c \cdot S_1 + \frac{c^2}{2} \cdot S_2 + \frac{c^3}{6} S_3 , \tag{2.D.103}$$

$$\frac{1}{N} \sum \left( \frac{X_i}{u} \right)^c \log \left( \frac{X_i}{u} \right) \cong S_1 + c \cdot S_2 + \frac{c^2}{2} S_3 , \tag{2.D.104}$$

where

$$S_1 = \frac{1}{N} \sum_{i=1}^{N} \log \left( \frac{X_i}{u} \right),$$

(2.D.105)

$$S_2 = \frac{1}{N} \sum_{i=1}^{N} \log^2 \left( \frac{X_i}{u} \right),$$

(2.D.106)

$$S_3 = \frac{1}{N} \sum_{i=1}^{N} \log^3 \left( \frac{X_i}{u} \right).$$

(2.D.107)

Putting these expansions into (2.D.96) and (2.D.98) and keeping only terms in $c$ up to second order, the solutions of these equations reads

$$\hat{b}_N \simeq S_1^{-1} \left( 1 - \frac{S_2}{2S_1} \hat{c}_N + \frac{3S_2^2 - 2S_1 S_3}{12 S_1^2} \hat{c}_N^2 \right), \quad \text{and} \quad \hat{c}_N \simeq \frac{\frac{1}{2} S_2 - S_1^2}{\frac{1}{2} S_1 S_2 - \frac{1}{3} S_3}.$$

(2.D.108)

Inserting these solutions into (2.D.97) and (2.D.94) gives

$$\sup_{b,c} L_1 = -N \left[ 1 + \log u - (\hat{c}_N - 1) S_1 + \log S_1 + \frac{\hat{c}_N}{2} \cdot \frac{S_2}{S_1} \right.$$
$$\left. - \frac{3S_2^2 - 4S_1 S_3}{24 S_1^2} \hat{c}_N^2 \right],$$

(2.D.109)

up to the second order in $\hat{c}_N$, and

$$\sup_{\beta} L_0 = -N \left[ 1 + \log u + S_1 + \log S_1 \right],$$

(2.D.110)

which obtains the explicit formula

$$W_N = 2 \left[ \sup_{b,c} L_1 - \sup_{\beta} L_0 \right],$$

(2.D.111)

$$\simeq 2N \hat{c}_N^2 \cdot \left( \frac{S_3}{6S_1} - \frac{S_2}{2} + \frac{1}{8} \left( \frac{S_2}{S_1} \right)^2 \right).$$

(2.D.112)

Now by the law of large numbers, $S_1$ converges to $1/\beta$, $S_2$ converges to $2/\beta^2$ and $S_3$ converges to $6/\beta^3$ with probability 1 as $N$ goes to infinity. Thus, by the continuous mapping theorem

$$\frac{S_3}{6S_1} - \frac{S_2}{2} + \frac{1}{8} \left( \frac{S_2}{S_1} \right)^2 \xrightarrow{a.s} \frac{1}{2\beta^2},$$

(2.D.113)

so that defining the variables $\xi_1 = S_1 - \beta^{-1}$, $\xi_2 = S_2 - 2\beta^{-2}$, we can assert that

$$W_N = \beta^2 N \left( 2\xi_1 - \frac{\beta}{2} \xi_2 \right)^2 + o_p(1).$$

(2.D.114)

Now, accounting for the fact that

$$\sqrt{N}\beta \begin{pmatrix} \xi_1 \\ \xi_2 \end{pmatrix} \xrightarrow{law} \mathcal{N}\left(0, \begin{pmatrix} 1 & 4\beta^{-1} \\ 4\beta^{-1} & 20\beta^{-2} \end{pmatrix}\right) , \tag{2.D.115}$$

we can write

$$\sqrt{N}\beta\xi_2 = \frac{4}{\beta}(\sqrt{N}\beta\xi_1) + \frac{2}{\beta}\hat{\epsilon} , \tag{2.D.116}$$

where $\hat{\epsilon}$ is a Gaussian variable with zero mean and unit variance, independent from $\xi_1$. This implies that

$$W_N = \hat{\epsilon}^2 + o_p(1) , \tag{2.D.117}$$

which means that Wilks' statistic $W_N$ converges to a $\chi^2$-random variable with one degree of freedom.

# 3

# Notions of Copulas

In this chapter, we introduce the notion of copulas, which describes the dependence between several random variables. These variables can be the returns of different assets or the value of a given asset at different times, and more generally, any set of economic variables. We present some examples of classical families of copulas and provide several illustrations of the usefulness of copulas for actuarial,[1] economic, and financial applications.

Until relatively recently, the correlation coefficient was the measure of reference used to quantify the amplitude of dependence between two assets. From the old hypothesis or belief that the marginal distribution of returns is Gaussian, it was natural to extend this assumption of normality to the multivariate domain. Recall that only under the assumption of multivariate normality[2] is the correlation coefficient necessary and sufficient to capture the full dependence structure between asset returns. The growing attacks of the past three decades and the now overwhelming evidence against the Gaussian hypothesis also cast doubts on the relevance of the correlation coefficient as an adequate measure of dependence. See for instance [404] for a specific test of multivariate normality of asset returns. Actually, it is now clear that the correlation coefficient is grossly insufficient to provide an accurate description of the dependence between two assets [64, 148, 149] and that it is necessary to characterize the full joint multivariate distribution of asset returns. This is all the more important for rare large events whose deviations from normality are the most extreme both in amplitude and dependence.

Consider for simplicity the problem of characterizing the bivariate distribution of the returns of only two assets. It is essential to realize that the bivari-

---

[1] Actuarial science is a sister discipline of statistics. Actuaries play an important role in many of the financial plans that involve people, *e.g.*, life insurance, pension plans, retirement benefits, car insurance, unemployment insurance, and so on.

[2] To some extent, the correlation coefficient also adequately quantifies the dependence between elliptically distributed random variables, even if it may yield spurious conclusions – especially in the far tails – as we shall see in the next chapters.

ate distribution embodies two qualitatively different pieces of information on the two assets. On the one hand, it contains the two marginal distributions; on the other hand, it contains information on the dependence between the two assets irrespective of their individual (marginal) distributions. Only the introduction of the copula allows one to operate a clean dissection between these two pieces of information. The role of the copula of two random variables is precisely to offer a complete and unique description of the dependence structure existing between them, excluding all (parasiting) information on the marginal distribution of the random variables.

Such an approach in terms of copulas has witnessed a recent burst of interest and of activity spurred by its practical and theoretical implications. From an applied view point, determining the dependence between assets is at the core of risk management: the dependence governs (i) the optimization of diversification of risks by aggregation in portfolios, (ii) the hedging strategies based on derivatives, and (iii) the securitization[3] of different risky instruments to sell them to third parties. Specifically, the advantage of the copula formulation is to provide a better understanding and quantification of the interactions between assets by determining the diverse dependence structures between the various sources of risk. Applications to finance include the calculation of VaR (Value-at-Risk) and portfolio optimization [145], the calculation of option prices [99, 112], and credit risk [184, 186]. For various applications to insurance, see [110, 183, 478].

From a fundamental viewpoint, it is reasonable to think that the structure of dependence between assets reflects the underlying mechanisms at work in financial markets. In particular, the dependence between assets is in part the result of the interactions between the agents investing in the stock market[4]. Not only are investors responsible for the individual variations and fluctuations of assets but, by their asset allocation choices (buying or selling such or such security rather than another), they also create dependence between assets. It can thus be hoped that the study of the dependence between assets may complement the understanding of the important mechanisms at work in stock markets and therefore of the interactions between agents. It should also help in narrowing down the relevant macroscopic parameters influencing investors in their asset allocation.

Before presenting copulas and their fundamental properties, we should stress that this body of results applies also when the structure of dependence is time-varying. This remark is important since there is *a priori* no principle or reason for the dependence to be constant [380, 409, 412]. One should thus

---

[3] The process of aggregating similar instruments, such as loans or mortgages, into a negotiable security.

[4] Of course, the observed dependence between assets has also other inputs than just the action of economic agents on financial markets. The macroeconomic variables also play an important role, especially for assets belonging to the same economic sector, which are collectively sensitive to the same variations of the macroeconomic landscape. This is the stance taken by factor models described in Chap. 1.

study its dynamics in addition. However, such a study of the time-dynamics of the multivariate dependence structure between assets is extremely delicate both from an empirical and theoretical point of view. In addition, as we show in Chap. 6, some apparent time-varying dependence may appear as a spurious consequence of conditioning the measures of dependence on market phases with large volatility, for instance. This mechanism appears to explain a large part of the empirical observations on time-varying dependence, suggesting that it would be sufficient to model the time-dependent properties of volatility alone. We thus make the simplifying assumption that any possible time-dependence of the statistical properties of assets is entirely embedded in the evolution of the marginal distributions of their returns, while the dependence structure between assets remains invariant.

## 3.1 What is *Dependence?*

The notion of independence of random variables is very easy to define. From elementary probability theory, two random variables $X$ and $Y$ are *independent* if and only if, for any $x$ and $y$ in the supports of the distributions,

$$\Pr\left[X \le x; Y \le y\right] = \Pr\left[X \le x\right] \cdot \Pr\left[Y \le y\right] , \tag{3.1}$$

or equivalently

$$\Pr\left[X \le x \mid Y\right] = \Pr\left[X \le x\right] . \tag{3.2}$$

In other words, two random variables are independent if the knowledge of a piece of information about one of the random variables does not bring any new insight on the other one.

The notion of *dependence* is much more subtle to define, or at least to quantify. Let us start with the concept of *mutual complete dependence* [290]. It seems natural that two real random variables $X$ and $X'$ are mutually completely dependent if the knowledge of $X$ implies the knowledge of $X'$, and reciprocally. This statement simply means that there exists a one-to-one mapping $f$ such that:

$$X' = f(X), \quad \text{almost everywhere} , \tag{3.3}$$

which, as stressed in [270], implies the perfect predictability of one of the random variables from the other one. The mapping $f$ is either strictly increasing or strictly decreasing. In the first case, the random variables are said to be *comonotonic*.

In a second stage of our investigation of the concept of dependence, let us ask what could be the meaning of the following statement:

The random variables $X$ and $Y$ exhibit the same dependence as the random variables $X'$ and $Y'$.

A possible interpretation, explored in this chapter, is that the random variables $X$ and $X'$, on the one hand, and $Y$ and $Y'$, on the other hand, are comonotonic. In this case, all variables or functions describing the dependence between two (and more generally several) random variables should enjoy the property of invariance under an arbitrary increasing mapping. Let us assume that there exists a function $C$ describing the dependence of the random variables $X$ and $Y$ and a function $C'$ describing the dependence of the random variables $X'$ and $Y'$. Writing that $X$ and $X'$ (respectively $Y$ and $Y'$) are comonotonic,

$$X' = h_1(X) , \tag{3.4}$$
$$Y' = h_2(Y) , \tag{3.5}$$

where $h_1$ and $h_2$ are increasing functions on $\mathbb{R}$ (if we consider real-valued random variables), the property of invariance under strictly increasing mapping reads $C = C'$.

Let us now show how to build $C$. Does the usual correlation coefficient qualify? While the correlation coefficient measures some kind of dependence, it is only able to account for a linear dependence.[5] Therefore, it does not fulfill the requirement for a general concept of dependence which should involve any nonlinear monotonic structure. Thus, we must look for something else.

Let us consider the two random variables $X$ and $Y$ and their joint distribution function denoted by $H$:

$$H(x,y) = \Pr[X \le x; Y \le y] . \tag{3.6}$$

The marginal distributions of $X$ and $Y$ are respectively:

$$F(x) = \Pr[X \le x] = \lim_{t\to\infty} H(x,t) , \tag{3.7}$$
$$G(y) = \Pr[Y \le y] = \lim_{t\to\infty} H(t,y) . \tag{3.8}$$

For simplicity, let us assume that $F$ and $G$ are continuous and increasing, so that the usual inverses $F^{-1}$ and $G^{-1}$ exist. Then, let us define

$$C(u,v) = H\left(F^{-1}(u), G^{-1}(v)\right) , \quad \forall u, v \in [0,1] . \tag{3.9}$$

Let us now focus on the random variables $X'$ and $Y'$ given by (3.4–3.5) above. It is clear that their joint distribution function is

$$H'(x,y) = \Pr[X' \le x; Y' \le y] = \Pr\left[X \le h_1^{-1}(x); Y \le h_2^{-1}(y)\right]$$
$$= H\left(h_1^{-1}(x), h_2^{-1}(y)\right) , \tag{3.10}$$

---

[5] Indeed, consider the linear regression $Y = \beta X + \epsilon$ where $\beta$ is a constant and $X$ and $\epsilon$ are two independently distributed centered random variables with variances $\mathrm{Var}(X)$ and $\mathrm{Var}(\epsilon)$ respectively. Then, the knowledge of the covariance $\mathrm{Cov}(X,Y)$ and of the variance $\mathrm{Var}(X)$ of $X$ is equivalent to the knowledge of the linear dependence between $X$ and $Y$: $\mathrm{Cov}(X,Y) = \beta \, \mathrm{Var}(X)$. The correlation coefficient, $\mathrm{Corr}(X,Y) \equiv \mathrm{Cov}(X,Y)/\sqrt{\mathrm{Var}(X)\mathrm{Var}(Y)} = \left[1 + \mathrm{Var}(\epsilon)/(\beta^2\mathrm{Var}(X))\right]^{-1/2}$, involves in addition an information on $\mathrm{Var}(\epsilon)$.

while their marginal distributions are:

$$F'(x) = \Pr\left[X' \leq x\right] = F\left(h_1^{-1}(x)\right), \tag{3.11}$$

$$G'(y) = \Pr\left[Y' \leq y\right] = G\left(h_2^{-1}(y)\right) . \tag{3.12}$$

Now, considering

$$C'(u,v) = H'\left(F'^{-1}(u), G'^{-1}(v)\right), \quad \forall u, v \in [0,1] , \tag{3.13}$$

elementary algebraic manipulations show that

$$C(u,v) = C'(u,v), \quad \forall u, v \in [0,1] . \tag{3.14}$$

It turns out that the function $C$ defined by (3.9) is the only object obeying the property of invariance under strictly increasing mapping and which entirely captures the full dependence between $X$ and $Y$.

The following properties follow from simple calculations:

- $C(u,1) = u$ and $C(1,v) = v$, $\forall u, v \in [0,1]$,
- $C(u,0) = C(0,v) = 0$, $\forall u, v \in [0,1]$,
- $C$ is 2-increasing, namely, for all $u_1 \leq u_2$ and $v_1 \leq v_2$:

$$C(u_2, v_2) - C(u_2, v_1) - C(u_1, v_2) + C(u_1, v_1) \geq 0 . \tag{3.15}$$

This last property is a simple translation of the nonnegativity of probabilities, specifically of the following expression:

$$\Pr\left[F^{-1}(u_1) \leq X \leq F^{-1}(u_2); G^{-1}(v_1) \leq Y \leq G^{-1}(v_2)\right] \geq 0 . \tag{3.16}$$

As we shall see in the sequel, these three properties define the mathematical object called *copula*, which has been introduced by A. Sklar in the late 1950s [443] in order to describe the general dependence properties of random variables.

## 3.2 Definition and Main Properties of Copulas

This section provides a brief survey of the main properties of copulas, emphasizing the most important definitions and theorems useful in the following. For exhaustive and general presentations, we refer to [248, 370] and to [74, 183] for introductions oriented to financial and actuarial applications.

The definition of a copula of $n$ random variables generalizes the intuitive definition (3.9) presented above for the bivariate copula.

**Definition 3.2.1 (Copula).** *A function $C : [0,1]^n \longrightarrow [0,1]$ is a $n$-copula if it enjoys the following properties :*

- $\forall u \in [0,1]$, $C(1, \ldots, 1, u, 1 \ldots, 1) = u$ ,
- $\forall u_i \in [0,1]$, $C(u_1, \ldots, u_n) = 0$ *if at least one of the $u_i$'s equals zero* ,

- $C$ is grounded and $n$-increasing, i.e., the $C$-volume of every box whose vertices lie in $[0,1]^n$ is positive.

It is clear from this definition that a copula is nothing but a multivariate distribution with support in $[0,1]^n$ and with uniform marginals. It immediately follows that a convex sum of copulas remains a copula. The fact that such mathematical objects can be very useful for representing multivariate distributions with arbitrary marginals has been suggested in the previous introductory section and is stated more formally in the following result [443].

**Theorem 3.2.1 (Sklar's Theorem).** *Given a $n$-dimensional distribution function $F$ with continuous[6] (cumulative) marginal distributions $F_1, \ldots, F_n$, there exists a unique $n$-copula $C : [0,1]^n \longrightarrow [0,1]$ such that:*

$$F(x_1, \ldots, x_n) = C(F_1(x_1), \ldots, F_n(x_n)) . \tag{3.17}$$

Thus, the copula combines the marginals to form the multivariate distribution. This theorem provides both a parameterization of multivariate distributions and a construction scheme for copulas. Indeed, given a multivariate distribution $F$ with marginals $F_1, \ldots, F_n$, the function

$$C(u_1, \ldots, u_n) = F\left(F_1^{-1}(u_1), \ldots, F_n^{-1}(u_n)\right) \tag{3.18}$$

is automatically an $n$-copula.[7] This copula is the copula of the multivariate distribution $F$. We will use this method in the sequel to derive the expressions of standard copulas such as the Gaussian copula or the Student's copula.

In addition to the copula itself, it is often very useful to consider the two following quantities:

**Definition 3.2.2.** *Given $n$ random variables $X_1, \ldots, X_n$ with marginal survival distributions $\bar{F}_1, \ldots, \bar{F}_n$ and joint survival distribution $\bar{F}$, the survival copula $\bar{C}$ is such that:*

$$\bar{C}\left(\bar{F}_1(x_1), \ldots, \bar{F}_n(x_n)\right) = \bar{F}(x_1, \ldots, x_n) . \tag{3.19}$$

*The dual copula $C^*$ of the copula $C$ of $X_1, \ldots, X_n$ is defined by:*

$$C^*(u_1, \ldots, u_n) = 1 - \bar{C}(1 - u_1, \ldots, 1 - u_n), \quad \forall u_1, \ldots, u_n \in [0,1] . \tag{3.20}$$

---

[6] When this assumption fails, Sklar's theorem still holds, but in a weaker sense: a representation like (3.17) still exists but is not unique anymore.

[7] The quantile function, or generalized inverse, $F_i^{-1}$ of the distribution $F_i$ can be defined by:

$$F_i^{-1}(u) = \inf\{x \mid F_i(x) \geq u\}, \quad \forall u \in (0,1).$$

When the distribution function $F_i$ is strictly increasing, $F_i^{-1}$ denotes the usual inverse of $F_i$. In fact, any quantile function can be chosen. But, for noncontinuous margins, the copula (3.18) depends upon the precise quantile function which is selected.

While the survival copula is indeed a true copula, the dual copula is not. However, it can be simply related to the probability that (at least) one of the $X_i$'s is less than or equal to $x_i$. Indeed, one can easily check that:

$$\Pr\left[\bigcup_{i=1}^{n}\{X_i \le x_i\}\right] = C^*\left(F_1(x_1), \ldots, F_n(x_n)\right) . \tag{3.21}$$

A very powerful property shared by all copulas is their invariance under arbitrary increasing mapping of the random variables (this has been shown for the case of the bivariate copulas in the derivation ending with (3.14)):

**Theorem 3.2.2 (Invariance Theorem).** *Consider $n$ continuous random variables $X_1, \ldots, X_n$ with copula $C$. Then, if $h_1(X_1), \ldots, h_n(X_n)$ are increasing on the ranges of $X_1, \ldots, X_n$, the random variables $Y_1 = h_1(X_1), \ldots, Y_n = h_n(X_n)$ have exactly the same copula $C$.*

Let us stress again that this result demonstrates that the full dependence between the $n$ random variables is completely captured by the copula, independent of the shape of the marginal distributions. In other words, the Invariance Theorem shows that the copula is an intrinsic measure of dependence between random variables. Under a monotonic change of variable from an old variable to a new variable, these two variables are comonotonic by definition. Intuitively, as explained in the previous section, it is natural that a measure of dependence between two random variables should be insensitive to the substitution of one of the variables by a comonotonic variable: if $X$ and $X'$ are two comonotonic variables, one expects the same dependence structure for the pair $(X, Y)$ and for the pair $(X', Y)$. This is precisely the content of the Invariance Theorem on copulas. In contrast, a measure of dependence such as the correlation coefficient which is function of both the copula and the marginal distribution is not invariant under a monotonic change of variable. It does not constitute an intrinsic measure of dependence (we will come back in detail on this point in Chap. 4). The benefit of using copulas is the decoupling between the marginal distribution and the dependence structure, which justifies the separate study of marginal distributions on the one hand and of the dependence on the other hand.

Let us now state several useful properties enjoyed by copulas. First, any copula is uniformly continuous:

**Proposition 3.2.1.** *Given an $n$-copula $C$, for all $u_1, \ldots, u_n \in [0, 1]$ and all $v_1, \ldots, v_n \in [0, 1]$:*

$$|C(v_1, \ldots, v_n) - C(u_1, \ldots, u_n)| \le |v_1 - u_1| + \cdots + |v_n - u_n| . \tag{3.22}$$

This result is a direct consequence of the property that copulas are $n$-increasing. Indeed, restricting ourselves to the bivariate case for the simplicity of the exposition, the triangle inequality implies

$$|C\left(v_1, v_2\right) - C\left(u_1, u_2\right)| = |C\left(v_1, v_2\right) - C\left(u_1, v_2\right) + C\left(u_1, v_2\right) - C\left(u_1, u_2\right)|$$
$$\leq |C\left(v_1, v_2\right) - C\left(u_1, v_2\right)| + |C\left(u_1, v_2\right) - C\left(u_1, u_2\right)|,$$

and by (3.15), with some of the arguments put equal to 0 or 1, we have

$$|C\left(v_1, v_2\right) - C\left(u_1, v_2\right)| \leq |v_1 - u_1| , \tag{3.23}$$

and

$$|C\left(u_1, v_2\right) - C\left(u_1, u_2\right)| \leq |v_2 - u_2| , \tag{3.24}$$

which leads to the expected result.

Besides, it follows that a copula is differentiable almost everywhere:

**Proposition 3.2.2.** *Let $C$ be an $n$-copula. For almost all $(u_1, \ldots, u_n) \in [0, 1]^n$, the partial derivative of $C$ with respect to $u_i$ exists and:*

$$0 \leq \frac{\partial C}{\partial u_i}(u_1, \ldots, u_n) \leq 1 . \tag{3.25}$$

These two properties show that copulas enjoy nice regularity (or smoothness) conditions. In fact, the later one will turn out to be very useful for numerical simulations, as we shall see in Sect. 3.5.

Due to the property that copulas are $n$-increasing, we can find an upper and a lower bound for any copula. Choosing $u_2 = v_2 = 1$ in (3.15), we obtain that any bivariate copula satisfies

$$C(u, v) \geq u + v - 1 . \tag{3.26}$$

Since, in addition, a copula is non-negative, we obtain a lower bound for any bivariate copula:

$$C(u, v) \geq \max\left(u + v - 1, 0\right) . \tag{3.27}$$

Similarly, choosing alternatively $(u_1 = 0, v_2 = 1)$ and $(u_2 = 1, v_1 = 0)$, we get an upper bound for any bivariate copula

$$C(u, v) \leq \min(u, v) . \tag{3.28}$$

It is clear that these two bounds fulfill all the requirements of copulas, qualifying the functions $\max\left(u + v - 1, 0\right)$ and $\min(u, v)$ as genuine bivariate copulas. These two bounds are thus the tightest possible bounds. Generalization to higher dimension is straightforward, so that we can state

**Proposition 3.2.3 (Fréchet-Hoeffding Upper and Lower Bounds).** *Given an $n$-copula $C$, for all $u_1, \ldots, u_n \in [0, 1]$:*

$$\max\left(u_1 + \ldots + u_n - n + 1, 0\right) \leq C\left(u_1, \ldots, u_n\right) \leq \min\left(u_1, \ldots, u_n\right) . \tag{3.29}$$

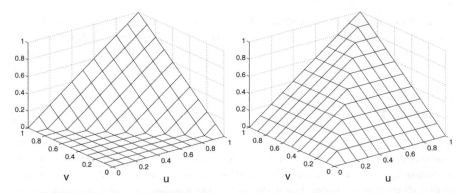

**Fig. 3.1.** The Fréchet-Hoeffding lower (*left panel*) and upper (*right panel*) bounds for bivariate copulas

These lower and upper bounds, which constitute the so-called *Fréchet-Hoeffding bounds*, are represented in Fig. 3.1 for the bivariate case. The upper bound is itself an $n$-copula, while the lower one is a copula only for $n = 2$. However, this lower bound remains the best possible insofar as, for any fixed point $(u_1, \ldots, u_n) \in [0, 1]^n$, there exists a copula $\tilde{C}$ such that, at this particular point:

$$\tilde{C}(u_1, \ldots, u_n) = \max(u_1 + \cdots + u_n - n + 1, 0) \ . \tag{3.30}$$

The Fréchet-Hoeffding upper bound represents the strongest form of dependence that several random variables can exhibit. In fact, it is nothing but the copula associated with comonotonicity. Similarly, when $n = 2$, the Fréchet-Hoeffding lower bound is nothing but the copula of countermonotonicity.

## 3.3 A Few Copula Families

As shown from Sklar's theorem 3.2.1, for each multivariate distribution, one can easily derive a copula. Notwithstanding their formidable number, a few copula families play a more important role.

### 3.3.1 Elliptical Copulas

Elliptical copulas derive from multivariate elliptical distributions [252]. Here, we give the two most important examples, the Gaussian and Student's copulas. By construction, these two copulas are close to each other in their central part, and become closer and closer in their tail only when the number of degrees of freedom of the Student's copula increases. As a consequence, it is sometimes difficult to distinguish between them, even with sensitive tests. However, as we shall see in Chap. 4, these two copulas may have drastically different behaviors with respect to the dependence between extremes.

Multiplicative factor models, which account for most of the stylized facts observed on financial time series, generate distributions with elliptical copulas. Multiplicative factor models contain in particular multivariate stochastic volatility models with a common stochastic volatility factor. They can be formulated as

$$\mathbf{X} = \sigma \cdot \mathbf{Y} \, , \tag{3.31}$$

where $\sigma$ is a positive random variable modeling the volatility, $\mathbf{Y}$ is a Gaussian random vector, independent of $\sigma$ and $\mathbf{X}$ is the vector of assets returns. In this framework, the multivariate distribution of asset returns $\mathbf{X}$ is an elliptical multivariate distribution. For instance, if the inverse $1/\sigma^2$ of the square of the volatility $\sigma$ is a constant times a $\chi^2$-distributed random variable with $\nu$ degrees of freedom, the distribution of asset returns will be the Student distribution with $\nu$ degrees of freedom. When the volatility follows ARCH or GARCH processes, then the asset returns are also elliptically distributed with fat-tailed marginal distributions. Such elliptical multivariate distribution ensures that each asset $X_i$ is asymptotically distributed according to a regularly varying distribution:[8] $\Pr\{|X_i| > x\} \sim L(x) \cdot x^{-\nu}$ – where $L(\cdot)$ denotes a slowly varying function – with the same exponent $\nu$ for all assets.

Elliptical copulas have the advantage of being easily synthesized numerically, which makes their use convenient for numerical simulations and for the study of scenarios. This results from the fact that it is easy to generate Gaussian or Student's distributed random variables which, upon appropriate monotonic changes of variables, give the correct marginal distributions while conserving the copula unchanged.

## The Gaussian Copula

The Gaussian copula is the copula derived from the multivariate Gaussian distribution. The Gaussian copula provides a natural setting for generalizing Gaussian multivariate distributions into so-called meta-Gaussian distributions. Meta-Gaussian distributions have been introduced in [283] (see [163] for a generalization to meta-elliptical distributions) and have been applied in many areas, from the analysis of experiments in high-energy particle physics [265] to finance [453]. These meta-Gaussian distributions have exactly the same dependence structure as the Gaussian distributions while differing in their marginal distributions which can be arbitrary.

Let $\Phi$ denote the standard Normal (cumulative) distribution and $\Phi_{\rho,n}$ the $n$-dimensional standard Gaussian distribution with correlation matrix $\rho$. Then, the Gaussian $n$-copula with correlation matrix $\rho$ is

$$C_{\rho,n}(u_1, \ldots, u_n) = \Phi_{\rho,n}\left(\Phi^{-1}(u_1), \ldots, \Phi^{-1}(u_n)\right) \, , \tag{3.32}$$

---

[8] See footnote 3 page 39.

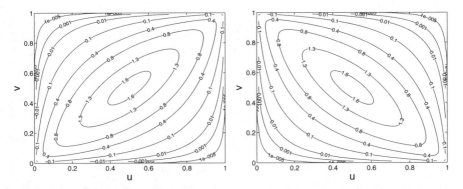

**Fig. 3.2.** Contour plot of the density (3.34) of the bivariate Gaussian copula with a correlation coefficient $\rho = 0.8$ (*left panel*) and $\rho = -0.8$ (*right panel*)

whose density (see Fig. 3.2)

$$c_{\rho,n}(u_1,\ldots,u_n) = \frac{\partial C_{\rho,n}(u_1,\ldots,u_n)}{\partial u_1 \ldots \partial u_n} \qquad (3.33)$$

reads

$$c_{\rho,n}(u_1,\ldots,u_n) = \frac{1}{\sqrt{\det \rho}} \exp\left(-\frac{1}{2}\boldsymbol{y}^t(\boldsymbol{u})(\rho^{-1}-\mathrm{Id})\boldsymbol{y}(\boldsymbol{u})\right) \qquad (3.34)$$

with $\boldsymbol{y}^t(\boldsymbol{u}) = (\Phi^{-1}(u_1),\ldots,\Phi^{-1}(u_n))$. Note that Theorem 3.2.1 and equation (3.18) ensure that $C_{\rho,n}(u_1,\ldots,u_n)$ in (3.32) is a copula.

The Gaussian copula is completely determined by the knowledge of the correlation matrix $\rho$. The parameters involved in the description of the Gaussian copula are simple to estimate, as we shall see in Chap. 5.

### Student's Copula

Student's copula is derived from Student's multivariate distribution. It provides a natural generalization of Student's multivariate distributions, in the form of meta-elliptical distributions [163]. These meta-elliptical distributions have exactly the same dependence structure as the Student's distributions while differing in their marginal distributions which can be arbitrary.

Given an $n$-dimensional Student distribution $T_{n,\rho,\nu}$ with $\nu$ degrees of freedom and a shape matrix $\rho$[9]

$$T_{n,\rho,\nu}(\mathbf{x}) = \frac{1}{\sqrt{\det \rho}} \frac{\Gamma\left(\frac{\nu+n}{2}\right)}{\Gamma\left(\frac{\nu}{2}\right)(\pi\nu)^{n/2}} \int_{-\infty}^{x_1}\cdots\int_{-\infty}^{x_n} \frac{d\mathbf{x}}{\left(1+\frac{\mathbf{x}^t\rho^{-1}\mathbf{x}}{\nu}\right)^{\frac{\nu+n}{2}}} , \qquad (3.35)$$

---

[9] Note that the shape matrix $\rho$ is nothing but the correlation matrix when the number of degrees of freedom $\nu$ is larger than 2, namely when the second moments of the variables $X_i$'s exist.

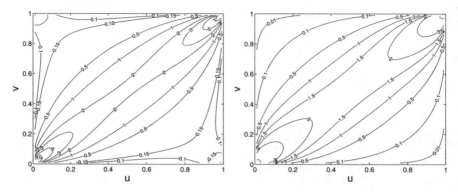

**Fig. 3.3.** Contour plot of the density (3.37) of a bivariate Student $t$ copula with a shape parameter $\rho = 0.8$ and $\nu = 2$ degrees of freedom (*left panel*) or $\nu = 10$ degrees of freedom (*right panel*). For small $\nu$'s, the difference between the Student copula and the Gaussian copula is striking on both diagonals. As $\nu$ increases, this difference decreases on the second diagonal but remains large (for $\nu = 10$) on the main diagonal, as can be observed by comparing the above right with the left panel of Fig. 3.2

the corresponding Student's copula reads:

$$C_{n,\boldsymbol{\rho},\nu}(u_1, \ldots, u_n) = T_{n,\boldsymbol{\rho},\nu}\left(T_\nu^{-1}(u_1), \ldots, T_\nu^{-1}(u_n)\right) , \qquad (3.36)$$

where $T_\nu$ is the univariate Student's distribution with $\nu$ degrees of freedom. The density of the Student's copula is thus

$$c_{n,\boldsymbol{\rho},\nu}(u_1, \ldots, u_n) = \frac{1}{\sqrt{\det \boldsymbol{\rho}}} \frac{\Gamma\left(\frac{\nu+n}{2}\right) \left[\Gamma\left(\frac{\nu}{2}\right)\right]^{n-1} \prod_{k=1}^{n} \left(1 + \frac{y_k^2}{\nu}\right)^{\frac{\nu+1}{2}}}{\left[\Gamma\left(\frac{\nu+1}{2}\right)\right]^n \left(1 + \frac{\boldsymbol{y}^t \boldsymbol{\rho}^{-1} \boldsymbol{y}}{\nu}\right)^{\frac{\nu+n}{2}}} , \qquad (3.37)$$

where $\boldsymbol{y}^t = (T_\nu^{-1}(u_1), \ldots, T_\nu^{-1}(u_n))$. See also Fig. 3.3.

Since Student's distribution tends to the normal distribution when $\nu$ goes to infinity, Student's copula tends to the Gaussian copula as $\nu \to +\infty$ [350]:

$$\sup_{\boldsymbol{u} \in [0,1]^n} |C_{n,\boldsymbol{\rho},\nu}(\boldsymbol{u}) - C_{\boldsymbol{\rho},n}(\boldsymbol{u})| \longrightarrow 0, \quad \text{as } \nu \to +\infty . \qquad (3.38)$$

The description of a Student copula relies on two parameters: the shape matrix $\boldsymbol{\rho}$, as in the Gaussian case, and in addition the number of degrees of freedom $\nu$. An accurate estimation of the parameter $\nu$ is rather difficult and this can have an important impact on the estimated value of the shape matrix.[10] As a consequence, the Student's copula may be more difficult to calibrate and to use than the Gaussian copula.

### 3.3.2 Archimedean Copulas

The importance of this class of copulas lies in that it encompasses a very large number of copulas while enjoying a certain number of interesting properties. In addition, as pointed out by Frees and Valdez [183], a large number of models developed to account for the dependence between various sources of risks in the theory of insurance lead to Archimedean copulas. The factor models constitute, however, a notable exception. While linear factor models play a fundamental role in the phenomenological description of interactions between financial assets, Archimedean copulas are not adequate to describe their corresponding dependence structure. In the same vein, the Gaussian and Student's copulas, as well as any elliptical copula, are not Archimedean.

An Archimedean copula is defined as follows:

**Definition 3.3.1 (Archimedean Copula).** *Let $\varphi$ be a continuous strictly decreasing, convex, function from $[0,1]$ onto $[0,\infty]$ and such that $\varphi(1) = 0$. Let $\varphi^{[-1]}$ be the pseudo-inverse of $\varphi$ :*

$$\varphi^{[-1]}(t) = \begin{cases} \varphi^{-1}(t), & \text{if } 0 \le t \le \varphi(0) , \\ 0, & \text{if } t \ge \varphi(0) , \end{cases} \tag{3.39}$$

*then the function*

$$C(u,v) = \varphi^{[-1]}(\varphi(u) + \varphi(v)) \tag{3.40}$$

*is an Archimedean copula with generator $\varphi$.*

The generalization to an $n$-copula seems straightforward:

$$C_n(u_1,\ldots,u_n) = \varphi^{[-1]}(\varphi(u_1) + \cdots + \varphi(u_n)) . \tag{3.41}$$

However, this formulation holds— *i.e.*, $C_n$ is actually an $n$-Archimedean copula – if and only if $\varphi^{[-1]}$ is $n$-monotonic:

$$(-1)^k \frac{d^k \varphi^{[-1]}(t)}{dt^k} \ge 0, \quad \forall k = 0, 1, \ldots, n . \tag{3.42}$$

When this later relation holds for all $n \in \mathbb{N}$, $\varphi^{[-1]}$ is said *completely monotonic*. In such a case, the bivariate Archimedean copula can be generalized to any dimension.

---

[10] Lindskog *et al.* [307] have recently introduced a robust estimation technique for the calibration of the shape matrix of any elliptical copula, which is described in Chap. 5.

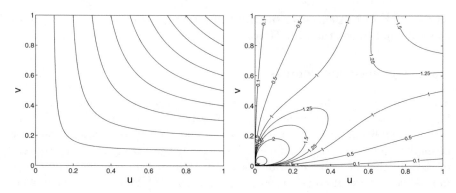

**Fig. 3.4.** Contour plot of Clayton's copula (*left panel*) and contour plot of its density (*right panel*) for parameter value $\theta = 1$

The complexity of the dependence structure between $n$ variables usually described by a function of $n$ variables is reduced and embedded, for Archimedean copulas, into the function of a single variable, the generator $\varphi$. This transforms a multidimensional formulation into a much simpler one-dimensional one.

Among the large number of copulas in the Archimedean family, the following copulas can be mentioned:

- Clayton's copula, which plays the role of a limit copula (see (3.61)):

$$C_\theta^{Cl}(u,v) = \max\left( \left[ u^{-\theta} + v^{-\theta} - 1 \right]^{-1/\theta}, 0 \right) , \quad \theta \in [-1, \infty) \qquad (3.43)$$

with generator $\varphi(t) = \dfrac{1}{\theta}\left( t^{-\theta} - 1 \right)$,

- Gumbel's copula, which plays a special role in the description of dependence using extreme value theory (see next Sect. 3.3.3):

$$C_\theta^{G}(u,v) = \exp\left( -\left[ (-\ln u)^\theta + (-\ln v)^\theta \right]^{1/\theta} \right) , \quad \theta \in [1, \infty) \qquad (3.44)$$

with generator $\varphi(t) = (-\ln t)^\theta$,

- Frank's copula:

$$C_\theta^{F}(u,v) = -\frac{1}{\theta} \ln\left( 1 + \frac{(e^{-\theta u} - 1)(e^{-\theta v} - 1)}{e^{-\theta} - 1} \right), \quad \theta \in \mathbb{R} \qquad (3.45)$$

with generator $\varphi(t) = -\ln \dfrac{e^{-\theta t} - 1}{e^{-\theta} - 1}$.

Note that the bivariate Fréchet-Hoeffding lower bound is an Archimedean copula, while the upper bound copula is not. For an overview of the members of the Archimedean family, we refer to Table 4.1 in [370].

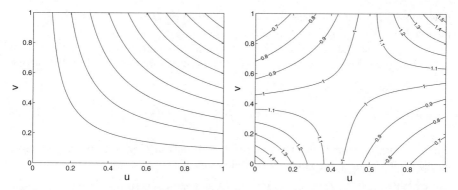

**Fig. 3.5.** Contour plot of Frank's copula (*left panel*) and contour plot of its density (*right panel*) for parameter value $\theta = 1$

A general procedure for constructing generators of the Archimedean copula has been proposed by Marshal and Olkin [348]. They have proved that, given a distribution function $F$ defined on $\mathbb{R}^+$ such that $F(0) = 0$, the inverse $\varphi(t) = \phi^{-1}(t)$ of the Laplace transform of $F$

$$\phi(t) = \int_0^\infty e^{-t \cdot x} \, dF(x) \tag{3.46}$$

is the generator of an Archimedean copula.

This suggests that frailty models [236, 480] can provide a natural mechanism for generating random variables with Archimedean copulas. Such models are common in actuarial science, because they offer a simple way to study the joint mortality of a group of individuals sharing common risk factors (see [103, 182, 237] among many others). In finance, they can also model the joint distribution of defaults of different obligators subjected to the same set of economic factors.

In each case, one focuses on the continuous random variables $T_i$ representing the survival time of the $i$th individual or company, *i.e.*, the time before death or default. Their individual survival distributions are defined by

$$S_i(t) = \Pr\left[T_i > t\right] \,, \tag{3.47}$$

with hazard rate:

$$h_i(t) = -\frac{d}{dt} \ln S_i(t) \,. \tag{3.48}$$

Conditional on a $p$-dimensional random vector $Z$ representing the risk factors, one can use a proportional hazard model [111], with the $i^{\text{th}}$ conditional hazard rate given by

$$h_i(t|Z) = e^{\boldsymbol{\beta} \cdot Z} b_i(t) \,, \tag{3.49}$$

where the $b_i(t)$'s are the base-line hazard rates and $\beta$ is the vector of regression parameters (the same for all individuals).

Defining the frailty variable $V = e^{\beta \cdot Z}$ and integrating the conditional hazard rates, one obtains the expression of the conditional survival distributions:

$$S_i(t|V = v) = e^{-v \cdot f_i(t)}, \quad \text{where} \quad f_i(t) = \int_0^t b_i(s)ds . \tag{3.50}$$

Then, assuming that $V$ has the distribution function $F$ with Laplace transform $\phi$ (cf. (3.46)), the joint survival distribution of the $T_i$'s is given by

$$\begin{aligned}
\Pr[T_1 > t_1, \ldots, T_n > t_n] &= \mathrm{E}^V \left[ S_1\left(t_1|V\right) \cdots S_n\left(t_n|V\right) \right], \\
&= \mathrm{E}^V \left[ e^{-V \cdot (f_1(u_1) + \cdots + f_n(u_n))} \right], \\
&= \int_0^\infty e^{-v \cdot (f_1(u_1) + \cdots + f_n(u_n))} \, dF(v), \\
&= \varphi^{-1}\left(f_1(u_1) + \cdots + f_n(u_n)\right) .
\end{aligned} \tag{3.51}$$

Since the unconditional marginal survival function of a given $T_i$ reads

$$S_i(t_i) = E^V\left[S_i\left(t_i|V\right)\right] = \varphi^{-1}\left(f_i(u_i)\right) , \tag{3.52}$$

Sklar's theorem shows that the (survival) copula of all the $T_i$'s is:

$$\bar{C}(u_1, \ldots, u_n) = \varphi^{-1}\left(\varphi(u_1) + \cdots + \varphi(u_n)\right) , \tag{3.53}$$

which is Archimedean, as expected.

As an example, let us consider Clayton's copula. Equation (3.43) shows that its generator is $\varphi(t) = t^{-\theta} - 1$, so that $\phi(t) = (1+t)^{-1/\theta}$, which is precisely the Laplace transform of a Gamma distribution $\Gamma(\theta^{-1}, 1)$ with parameter $1/\theta$, $\theta > 0$. As a consequence, considering a frailty variable $V$ following a Gamma distribution with parameters $(1/\theta, 1)$, $\theta > 0$ and $n$ conditionally independent random variables $U_i|V$, with conditional law:

$$\Pr[U_i \leq u_i|V = v] = e^{-v \cdot \left(u_i^{-\theta} - 1\right)}, \quad u_i \in [0, 1] , \tag{3.54}$$

one obtains $n$ uniformly distributed random variable $U_i$, whose dependence structure is the Clayton copula with parameter $\theta$.

Archimedean copulas enjoy the important property of *associativity*:

$$C_3(u, v, w) = C_2(u, C_2(v, w)) = C_2(C_2(u, v), w) , \tag{3.55}$$

where $C_2$ and $C_3$ respectively denote the bivariate and trivariate form of the copula under consideration. This property derives straightforwardly from (3.41). In other words, given three random variables $U, V$ and $W$, the dependence between the first two random variables taken together and the third one alone is the same as the dependence between the first random variable taken alone and the two last ones together. Therefore, if the dependence of the three random variables is described by an Archimedean copula, this implies a strong symmetry between the different variables in that they are exchangeable. As a consequence, when there is no reason to expect a breaking of symmetry between the random variables, an Archimedean copula may be a good choice to model their dependence. Such an assumption is often used in modeling large credit baskets. *A contrario*, when the random variables play very different roles, namely when they are not exchangeable, Archimedean copulas do not provide valid models of their dependence.

Another interesting property of Archimedean copulas is that their values $C(u, u)$ on the first bisectrix verify the following inequality:

$$C(u, u) < u, \quad \text{for all } u \in (0, 1) . \tag{3.56}$$

Reciprocally, one can demonstrate [370, Theorem 4.1.6] that any copula possessing these two properties (associativity and $C(u, u) < u$) are Archimedean. This provides an intuitive understanding of the nature of Archimedean copulas. It also allows one to understand why the Fréchet–Hoeffding upper bound copula is not Archimedean. Indeed, although it enjoys the associativity property, the Fréchet-Hoeffding upper bound is such that $C(u, u) = u$ for all $u \in [0, 1]$ (note that it is the only copula with this property).

Archimedean copulas obey an important limit theorem [260] of the type of the Gnedenko-Pikand-Balkema-de Haan (GPBH) theorem (see Chap. 2). Consider two random variables, $X$ and $Y$, distributed uniformly on $[0, 1]$, and whose dependence structure can be described by an Archimedean copula $C$. Then, the copula associated with the distribution of left-ordered quantiles tends, in most cases, to Clayton's copula (3.43) in the limit where the probability level of the quantiles goes to zero. To be more specific, let us denote by $\varphi$ the generator of the copula $C$, *assumed differentiable*. Let us define the conditional distribution

$$F_u(x) = \Pr[X \le x | X \le u, \, Y \le u] = \frac{C(x \wedge u, u)}{C(u, u)}, \quad \forall x \in [0, 1] , \tag{3.57}$$

where $x \wedge u$ means the minimum of $x$ and $u$, and the conditional copula

$$\begin{aligned}
C_u(x, y) &= \Pr[X \le F_u^{-1}(x), Y \le F_u^{-1}(y) | X \le u, \, Y \le u] \\
&= \frac{C\left(F_u^{-1}(x), F_u^{-1}(y)\right)}{C(u, u)} .
\end{aligned} \tag{3.58}$$

One can first show that, provided that $\varphi$ is a strick generator (that is, $\varphi(0)$ is infinite such that $\varphi^{[-1]} = \varphi^{-1}$), $C_u$ is a strict Archimedean copula with generator:

$$\varphi_u(t) = \varphi\left(F_u^{-1}(t)\right) - \varphi(u), \tag{3.59}$$

$$= \varphi\left(t \cdot \varphi^{-1}\left(2\varphi(u)\right)\right) - 2\varphi(u) , \tag{3.60}$$

from which it follows that the limiting behavior of $C_u$, as $u$ goes to zero, is:

$$\lim_{u \to 0} C_u(x, y) = C_\theta^{Cl}(x, y), \quad \forall (x, y) \in [0, 1] \times [0, 1] , \tag{3.61}$$

provided that $\varphi$ is regularly varying[11] at zero, with index $\theta \in \mathbb{R}_+$. When $\theta = 0$, $C_u$ tends to the independent copula while it tends to the Fréchet-Hoeffding upper bound copula when $\theta = \infty$.

Thus, Clayton's copula plays, in some sense, a role similar in $n$ dimensions to the generalized Pareto distribution in one dimension:

$$G_\xi(x) = 1 - (1 + \xi \cdot x)^{-1/\xi} . \tag{3.62}$$

This result is of particular relevance in the study of multivariate statistics of extremes.

### 3.3.3 Extreme Value Copulas

Another family of copulas which is of common use is that of extreme value copulas. These copulas are derived from the dependence structure of multivariate generalized extreme value (GEV) distributions, which provide the limit distributions of the component-wise maxima of $n$-dimensional random vectors, after a suitable normalization.

Consider $T$ iid $n$-dimensional random vectors $\boldsymbol{X_k} = (X_{k,1}, \ldots, X_{k,n})$, $k = 1, \ldots, T$ with distribution function $F$, and their component-wise maxima

$$M_{j,T} = \max_{1 \le k \le T} X_{k,j} . \tag{3.63}$$

For suitably chosen norming sequences $(a_{k,T}, b_{k,T})$, the limit distribution

$$\lim_{T \to \infty} \Pr \left( \frac{M_{1,T} - b_{1,T}}{a_{1,T}} \le z_1, \ldots, \frac{M_{n,T} - b_{n,T}}{a_{n,T}} \le z_n \right)$$

$$= \lim_{T \to \infty} F^T (a_{1,T} \cdot z_1 + b_{1,T}, \ldots, a_{n,T} \cdot z_n + b_{n,T}) , \tag{3.65}$$

if it exists, is given by

$$C (H_{\xi_1}(z_1), \ldots, H_{\xi_n}(z_n)) , \tag{3.66}$$

where $H_\xi$ is a GEV distribution (see Chap. 2), and $C$ is – by definition – an extreme value copula. Therefore, accounting for the general representation of multivariate extreme value (MEV) distributions (see [107]), we can state that:

---

[11] See footnote 3 page 39.

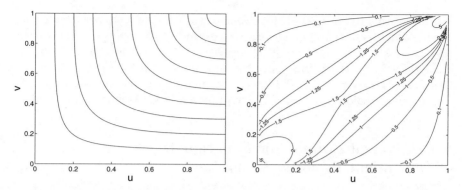

**Fig. 3.6.** Contour plot of Gumbel's copula (*left panel*) and of its density (*right panel*) for the parameter value $\theta = 2$

**Definition 3.3.2 (Extreme Value Copula).** *Any copula which admits the representation:*

$$C(u_1, \ldots, u_n) = \exp\left[-V\left(-\frac{1}{\ln u_1}, \ldots, -\frac{1}{\ln u_n}\right)\right] , \qquad (3.67)$$

*with*

$$V(x_1, \ldots, x_n) = \int_{\Pi_n} \max_i \left(\frac{w_i}{x_i}\right) \, dH(w) , \qquad (3.68)$$

*where $H$ is any positive finite measure such that $\int_{\Pi_n} w_i \, dH(w) = 1$ and $\Pi_n$ is the (n-1)-dimensional unit simplex:*

$$\Pi_n = \left\{w \in \mathbb{R}_+^n : \sum_{i=1}^{n} w_i = 1\right\} , \qquad (3.69)$$

*is an extreme value copula.*

One immediately observes that $V$ is a homogeneous function of degree $-1$. Thus, any extreme value copula satisfies [248]:

$$C(u_1{}^\alpha, \ldots, u_n{}^\alpha) = [C(u_1, \ldots, u_n)]^\alpha , \qquad (3.70)$$

for all $u \in [0, 1]^n$ and all $\alpha > 0$.

It is now easy to check that Gumbel's copula (3.44) belongs to the class of extreme value copula. It is depicted in Fig. 3.6: apart from a slight asymmetry with respect to the second bisectrix, it looks similar to a Student's copula.

The Fréchet-Hoeffding upper bound copula is also an extreme value copula since

$$\min(u_1{}^\alpha, \ldots, u_n{}^\alpha) = \min(u_1, \ldots, u_n)^\alpha . \qquad (3.71)$$

It is interesting to notice that this copula is the only associative extreme value copula which is not Archimedean. Indeed, due to the relation (3.70), either $C(u, \ldots, u) = u$ for all $u \in [0,1]$ or $C(u, \ldots, u) < u$ for all $u \in (0,1)$.[12] Since the Fréchet-Hoeffding upper bound copula is the only copula such that $C(u, \ldots, u) = u$, for all $u \in [0,1]$ [370], we can conclude that any extreme value copula, which enjoys the associativity property, is an Archimedean copula.

## 3.4 Universal Bounds for Functionals of Dependent Random Variables

In many situations, one has to consider various non-linear operations on sets of dependent random variables. For instance, one has to aggregate several risky positions in a portfolio, or to evaluate the pay-off of a derivative on a basket of several underlying assets. Very often, the actual dependence of the random variables under consideration is not known with sufficient accuracy. It is therefore interesting to ask whether it would be possible to obtain (sharp) bounds for the distribution of aggregated losses of a portfolio or of the pay-offs of a derivative constructed on a basket of assets. We will discuss in detail these two important examples in Sect. 3.6. For the time being, let us focus on the following mathematical result.

Consider $n$ random variables $X_1, \ldots, X_n$ with margins $F_1, \ldots, F_n$ and unknown copula $C$. Let $\psi : \mathbb{R}^n \longrightarrow \mathbb{R}$ and let $Y = \psi(X_1, \ldots, X_n)$. The most general result on bounds for $\Pr[Y \leq y]$ [13] has been recently derived by Embrechts, Höing and Juri [145]:

**Theorem 3.4.1.** *Let* $X_1, \ldots, X_n$ *be* $n$ *random variables with margins* $F_1, \ldots, F_n$ *and copula* $C$. *Let* $\psi : \mathbb{R}^n \longrightarrow \mathbb{R}$ *be an increasing function, left continuous in its last argument. Provided that there exists two functions* $C_{inf}$ *and* $C_{sup}$, *increasing in each of their arguments, such that* $C \geq C_{inf}$ *and* $C^* \leq C^*_{sup}$ *(where the expression of the dual copula* $C^*$ *is given in Definition 3.2.2), then*

$$F_{inf}(y) \leq \Pr[\psi(X_1, \ldots, X_n) \leq y] \leq F_{sup}(y),  \tag{3.72}$$

*where*

---

[12] Assuming that there exists a number $u^* \in (0,1)$ such that $C(u^*, \ldots, u^*) = u^*$, and raising this equation to the power $\alpha$, it follows that $C(u^{*\alpha}, \ldots, u^{*\alpha}) = u^{*\alpha}$ for any positive $\alpha$, by (3.70). Note that $u^{*\alpha}$ spans the entire interval $(0,1)$ when $\alpha$ ranges from zero to infinity. Thus, for all $u \in (0,1)$, $C(u, \ldots, u) = u$, and since this equality still holds when $u = (0, \ldots, 0)$ and $u = (1, \ldots, 1)$, we have:

$$\exists u^* \in (0,1), \ C(u^*, \ldots, u^*) = u^* \implies C(u, \ldots, u) = u, \ \forall u \in [0,1],$$

so that either $C(u, \ldots, u) = u$ for all $u \in [0,1]$ or $C(u, \ldots, u) < u$ for all $u \in (0,1)$.

[13] Former results concerning the case where $\psi$ is a sum of $n$ terms or where $\psi$ is an increasing continuous function can be found for instance in [327, 179, 486, 126].

$$F_{inf}(y) = \sup_{x_1,\ldots,x_{n-1}\in\mathbb{R}} C_{inf}\left(F_1(x_1),\ldots,F_{n-1}(x_{n-1}),F_n\left(\xi(x_1,\ldots,x_{n-1},y)\right)\right) ,$$

(3.73)

$$F_{sup}(y) = \inf_{x_1,\ldots,x_{n-1}\in\mathbb{R}} C_{sup}^*\left(F_1(x_1),\ldots,F_{n-1}(x_{n-1}),F_n\left(\xi(x_1,\ldots,x_{n-1},y)\right)\right) ,$$

(3.74)

*with*

$$\xi(x_1,\ldots,x_{n-1},y) = \sup\left\{t\in\mathbb{R};\ \psi\left(x_1,\ldots,x_{n-1},t\right)\le y\right\} .$$ (3.75)

A heuristic proof of this result can be found in Appendix 3.A.

In this theorem, $C_{inf}$ and $C_{sup}$ can be copulas, but this is not necessary. In particular, since any copula is larger than the Fréchet-Hoeffding lower bound, in the absence of any information on the dependence between the random variables, one can always resort to

$$C_{sup}(u_1,\ldots,u_n) = C_{inf}(u_1,\cdots,u_n) = \max(u_1+\ldots+u_n-n+1,0) . \quad (3.76)$$

This allows one to derive a universal bound for the probability that $\psi\left(X_1,\ldots,X_n\right)$ be less than $y$. Obviously, when additional information on the dependence is available, the bound can be improved. For instance, when the random variables are known to be positive orthant dependent – we will come back in Chap. 4 on this notion – we can choose the independence (or product) copula[14] for $C_{inf}$ and $C_{sup}$.

The bound provided by Theorem 3.4.1 is point-wise the best possible. Indeed, as shown in [145, Theorem 3.2], there always exists a copula $\tilde{C}$ for $X_1,\ldots,X_n$ such that the distribution of $\psi(X_1,\ldots,X_n)$ reaches the bound, at least at one point. Therefore, on the entire set of distribution functions, it is not possible to improve on this bound.

To conclude this section, let us state a straightforward bound implied by Theorem 3.4.1 for expectations. Denoting by $X_{inf}$ and $X_{sup}$ two random variables with distribution functions $F_{inf}$ and $F_{sup}$ respectively, and a non-decreasing function $G$, we obviously have:

$$\mathrm{E}\left[G\left(X_{sup}\right)\right] \le \mathrm{E}\left[G\left(\psi\left(X_1,\ldots,X_n\right)\right)\right] \le \mathrm{E}\left[G\left(X_{inf}\right)\right] .$$ (3.78)

Similar bound exists – *mutatis mutandis* – for any non-increasing function.

---

[14] Recall that the independence (or product) copula is:

$$C(u,v) = u\cdot v, \quad \forall u,v\in[0,1] ,$$

so that:

$$\begin{aligned}\Pr\left[X\le x, Y\le y\right] = F(x,y) &= C\left(F_X(x),F_Y(y)\right)\\ &= F_X(x)\cdot F_Y(y) = \Pr\left[X\le x\right]\cdot\Pr\left[Y\le y\right].\end{aligned}$$

# 3.5 Simulation of Dependent Data with a Prescribed Copula

An important practical application of copulas consists in the simulation of random variables with prescribed margins and various dependence structures in order to perform Monte-Carlo studies [171, 244], to generate scenarios for stress-testing investigations or to analyze the sensitivity of portfolio allocations to various parameters. We will come back to these various applications in the next section.

Here, we present several algorithms for the simulation of random variables with copulas characterizing a large class of dependences. The conceptually simplest approach is the acceptance-rejection method [218, 243]. However, this method is relatively slow in large dimensions, and therefore becomes unreliable due to the smallness of the size of obtainable statistical samples. In addition, it does not lend itself well to the study of the impact of the dependence structure on the optimal allocation of assets, for instance. As a consequence, another approach is desirable.

In fact, Sklar's theorem shows that the generation of $n$ random variables $X_1, \ldots, X_n$ with margins $F_1, \ldots, F_n$ and copula $C$ can be performed as follows:

1. Generate $n$ random variables $u_1, \ldots, u_n$ with uniform margins and copula $C$.
2. Apply the inversion method to each $u_i$, in order to generate each $x_i$:

$$x_i = F_i^{-1}(u_i) , \tag{3.79}$$

where $F_i^{-1}$ denotes the (generalized) inverse of $F_i$.

Therefore, the main difficulty in generating $n$ random variables following the joint distribution $H(x_1, \ldots, x_n) = C\left(F_1^{-1}(x_1), \ldots, F_n^{-1}(x_n)\right)$ lies in the generation of $n$ auxiliary random variables with uniform margins and dependence structure given by the copula $C$. We will now present two methods to simulate $n$-dimensional random vectors: the first one is specific to elliptical copulas while the second one applies to a wide range of copulas.

### 3.5.1 Simulation of Random Variables Characterized by Elliptical Copulas

The simulation of random variables whose dependence structure is given by an elliptical copula is particularly simple. This is one of the many appeals of this family of copulas.

By virtue of the invariance Theorem 3.2.2, the simulation of random variables with an elliptical copula is equivalent to the problem of the simulation of elliptically distributed random variables. Therefore, simulating an $n$-dimensional vector $\mathbf{X} = (X_1, \ldots, X_n)$ following an $n$-Gaussian copula with correlation matrix $\boldsymbol{\rho}$ is particularly easy. Indeed, one has just to use the following algorithm.

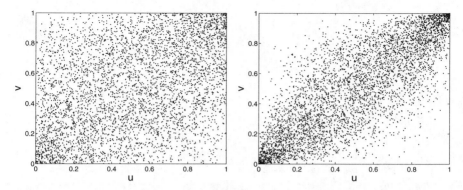

**Fig. 3.7.** Five thousand realizations of two random variables whose distribution function is given by the Gaussian copula with correlation coefficient $\rho = 0.4$ (*left panel*) and $\rho = 0.8$ (*right panel*)

### Algorithm 1

1. Generate $n$ independent standard Gaussian random variables: $\mathbf{u} = (u_1, \ldots, u_n)$ using the Box-Müller algorithm [77], for instance,
2. find the Cholevsky composition of $\boldsymbol{\rho}$: $\boldsymbol{\rho} = \mathbf{A} \cdot \mathbf{A}^t$, where $\mathbf{A}$ is a lower-triangular matrix,
3. set $\mathbf{y} = \mathbf{A} \cdot \mathbf{u}$,
4. and finally evaluate $x_i = \Phi(y_i)$, $i = 1, \ldots, n$, where $\Phi$ denotes the univariate standard Gaussian distribution function.

To generate an $n$-dimensional random vector drawn from a more complicated elliptical copula, it is useful to recall that any centered and elliptically distributed random vector $\mathbf{X}$ admits the following stochastic representation [252]:

$$\mathbf{X} = R \cdot \mathbf{N}, \tag{3.80}$$

where $\mathbf{N}$ is a centered Gaussian vector with covariance matrix $\boldsymbol{\Sigma}^2$ and $R$ is a positive random variable independent of $\mathbf{N}$. As an example, to generate an $n$-dimensional random vector drawn from a Student copula with $\nu$ degrees of freedom and shape matrix $\boldsymbol{\rho}$, one has to follow

### Algorithm 2

1. Generate $n$ independent standard Gaussian random variables: $\mathbf{u} = (u_1, \ldots, u_n)$,
2. find the Cholevsky composition of $\boldsymbol{\rho}$: $\boldsymbol{\rho} = \mathbf{A} \cdot \mathbf{A}^t$,
3. set $\mathbf{z} = \mathbf{A} \cdot \mathbf{u}$,
4. generate a random variable $r$, independent of $\mathbf{z} = (z_1, \ldots, z_n)$ and following a $\chi^2$-distribution with $\nu$ degrees of freedom,
5. set $\mathbf{y} = \sqrt{\nu \cdot r^{-1}} \cdot \mathbf{z}$,

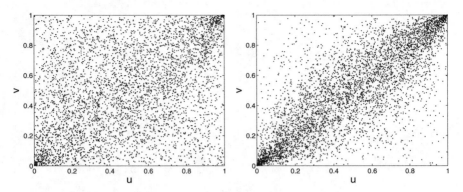

**Fig. 3.8.** Five thousand realizations of two random variables whose distribution function is given by Student's copula with shape coefficient $\rho = 0.4$ (*left panel*) and $\rho = 0.8$ (*right panel*) and $\nu = 3$ degrees of freedom

6. and finally, evaluate $x_i = T_\nu(y_i)$, $i = 1, \ldots, n$, where $T_\nu$ denotes the univariate standard Student's distribution function with $\nu$ degrees of freedom.

When the representation (3.80) is known explicitly, as in the example involving the Gaussian or Student copulas, the generation of the $n$-dimensional random vector by Algorithm 2 is straightforward. However, the law of the random variable $R$ is difficult to derive for most elliptical distributions. In that case, the general algorithm described in the next paragraph is much more useful.

### 3.5.2 Simulation of Random Variables Characterized by Smooth Copulas

The second general method is based upon the simple fact that:

$$\Pr\left[U_1 \leq u_1, \ldots, U_n \leq u_n\right] = \Pr\left[U_n \leq u_n | U_1 = u_1, \ldots, U_{n-1} = u_{n-1}\right]$$
$$\times \Pr\left[U_1 \leq u_1, \ldots, U_{n-1} \leq u_{n-1}\right],$$

which gives

$$\Pr\left[U_1 \leq u_1, \ldots, U_n \leq u_n\right] = \Pr\left[U_n \leq u_n | U_1 = u_1, \ldots, U_{n-1} = u_{n-1}\right]$$
$$\times \Pr\left[U_{n-1} \leq u_{n-1} | U_1 = u_1, \ldots, U_{n-2} = u_{n-2}\right]$$
$$\vdots$$
$$\times \Pr\left[U_2 \leq u_2 | U_1 = u_1\right] \cdot \Pr\left[U_1 \leq u_1\right] \quad (3.81)$$

by a straightforward recursion.

Therefore, applying this reasoning to the $n$-copula $C$, and denoting by $C_k$ the copula of the $k$ first variables, this yields:

$$C(u_1, \ldots, u_n) = C_n(u_n | u_1, \ldots, u_{n-1}) \ldots C_2(u_2 | u_1) \cdot \underbrace{C_1(u_1)}_{=u_1}, \quad (3.82)$$

where we define:

$$C_k\left(u_k|u_1,\ldots,u_{k-1}\right)=\frac{\partial_{u_1}\ldots\partial_{u_{k-1}}C_k\left(u_1,\ldots,u_k\right)}{\partial_{u_1}\ldots\partial_{u_{k-1}}C_{k-1}\left(u_1,\ldots,u_{k-1}\right)}.\tag{3.83}$$

As a consequence, in order to simulate $n$ random variables with copula $C$, one just has to

1. generate $n$ uniform and independent random variables: $v_1,\ldots,v_n$,
2. set $u_1=v_1$,
3. set $u_2=C_2^{-1}\left(v_2|u_1\right)$,
   $\vdots$

n+1. set $u_n=C_n^{-1}\left(v_n|u_1,\ldots,u_{n-1}\right)$.

This algorithm is particularly efficient when one considers the Archimedean copula. Genest and MacKay [198] have shown that, in such a case, it is very simple to generate pairs of random variables whose distribution function is given by the copula $C$ with generator $\varphi$. Indeed, the previous algorithm simply leads to

1. generate two uniform and independent random variables: $v_1,v_2$,
2. set $u_1=v_1$,
3. set $u_2=\varphi^{[-1]}\left[\varphi\left(\varphi'^{-1}\left(\frac{\varphi'(u_1)}{v_2}\right)\right)-\varphi(u_1)\right]$.

Applying this simplified algorithm to simulate Frank's copula leads to the following algorithm:

1. generate two uniform and independent random variables: $v_1,v_2$,
2. set $u_1=v_1$,
3. set $u_2=-\frac{1}{\theta}\ln\left(1+\frac{v_2\left(e^{-\theta}-1\right)}{v_2+(1-v_2)\cdot e^{-\theta\cdot v_1}}\right)$.

The same scheme can also be used to simulate Clayton's copula. However, Devroye [129] has proposed a somewhat simpler method for Clayton's copula with positive parameter $\theta$:

1. generate two standard exponential random variables: $v_1,v_2$,
2. generate a random variable $x$ following the distribution $\Gamma\left(\theta^{-1},1\right)$,
3. set $u_1=\left(1+\frac{v_1}{x}\right)^{-1/\theta}$ and $u_2=\left(1+\frac{v_2}{x}\right)^{-1/\theta}$.

This approach is in fact related to Marshall and Olkin's work [348]. Indeed, it is straightforward to check that, with the specification above, one has:

$$\Pr\left[U_i\le u_i|X=x\right]=e^{-x\cdot\left(u_i^{-\theta}-1\right)},\tag{3.84}$$

as in (3.54). Figure 3.9 provides an example of the realizations obtained by the use of these two algorithms.

A similar algorithm works for Gumbel's copula with parameter $\theta>1$, but it requires the generation of a random variable following a positive stable

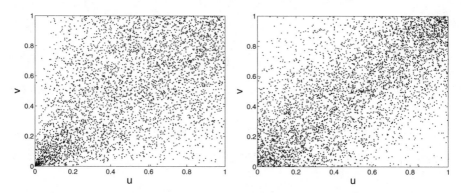

**Fig. 3.9.** Five thousand realizations of two random variables whose distribution function is given by Clayton's copula with parameter $\theta = 1$ (*left panel*) and by Frank's copula with parameter $\theta = 5$ (*right panel*)

law with tail index $1/\theta$, since the inverse of the generator of such a copula is $\phi(t) = e^{-t^{1/\theta}}$, $t \geq 0$. For an overview and softwares to generate random Lévy variables, see the Web pages of Professors J. Huston McCulloch (`http://economics.sbs.ohio-state.edu/jhm/jhm.html`) and John P. Nolan (`http://academic2.american.edu/~jpnolan/stable/stable.html`).

To conclude on the question concerning the simulation of dependent random variables, the second approach is sometimes more appropriate for $n$-copulas, with $n > 2$, because the algorithm based upon the inversion of the conditional copulas can rapidly become intractable for large $n$.

## 3.6 Application of Copulas

This section reviews several applications of copulas to risk assessment, in particular to tail risks in the presence of dependence [145, 179, 336], to option pricing [99, 417] and also to default risks [184, 302, 303]. In view of the growing importance of copulas in financial applications [100], an exhaustive presentation is not realistic. We thus restrict our discussion to examples that we have found particularly illustrative. For many other examples, see the following references concerning portfolio theory [228, 338], performance measurements [240], insurance applications [183, 272] or decision theory [258], among many others.

### 3.6.1 Assessing Tail Risk

One of the most important activities in the financial as well as in the actuarial worlds consists in assessing the risk of uncertain aggregated positions. This risk is often measured by the Value-at-Risk $\mathrm{VaR}_\alpha$ at probability level $\alpha$. $\mathrm{VaR}_\alpha$ is the lower $\alpha$-quantile of the net risk position $Y$, as illustrated in Fig. 3.10:

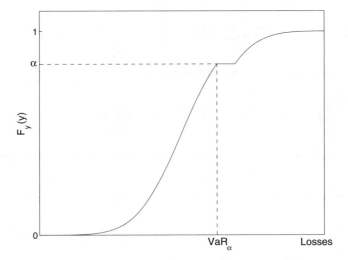

**Fig. 3.10.** Value-at-Risk at probability level $\alpha$ for the loss $Y$ with distribution function $F_Y$. We show the case where the distribution has a gap to exemplify how the Value-at-Risk is defined in such a degenerate case

$$\text{Var}_\alpha = \inf \{t \in \mathbb{R}; \; \Pr[Y \leq t] \geq \alpha\} \; . \tag{3.85}$$

In this definition of the Value-at-Risk, we take the convention of counting *losses* as positive. In this section, we show how to bound the Value-at-Risk of a portfolio using copulas.

Considering $n$ risky investments or insurance losses $X_1, \ldots, X_n$, the net risk of the position is:

$$Y = \sum_{1}^{n} w_i \cdot X_i \; , \tag{3.86}$$

where $w_i$ denotes the weight of position $i$ in the portfolio. It is convenient to define $\tilde{X}_i = w_i \cdot X_i$, so that $Y$ simply becomes the sum of the $\tilde{X}_i$'s. If $F_i$ is the distribution function of each $X_i$, the distribution function of $\tilde{X}_i$ is $\tilde{F}_i(\cdot) = F_i\left(\frac{\cdot}{w_i}\right)$. Now, applying Theorem 3.4.1 with $\psi\left(\tilde{X}_1, \ldots, \tilde{X}_n\right) = \tilde{X}_1 + \ldots + \tilde{X}_n$, we obtain – using slightly different notations:

$$F_{min}(y) \leq \Pr[Y \leq y] \leq F_{max}(y) \; , \tag{3.87}$$

with

$$F_{min}(y) = \sup_{x \in A(y)} \max \left\{ \sum_{i=1}^{n} \tilde{F}_i^-(x_i) - (n-1), 0 \right\} \tag{3.88}$$

and

$$F_{max}(y) = \inf_{x \in A(y)} \min \left\{ \sum_{i=1}^{n} \tilde{F}_i^-(x_i), 1 \right\} , \qquad (3.89)$$

where $\tilde{F}_i^-$ denotes the left limit of $\tilde{F}_i$ and

$$A(y) = \left\{ x = (x_1, \ldots, x_n) \in \mathbb{R}^n; \ \sum_{i=1}^{n} x_i = y \right\} . \qquad (3.90)$$

Therefore, a tight bound for the Value-at-Risk of the aggregated position $Y$ is:

$$\text{VaR}_\alpha^{min} \leq \text{VaR}_\alpha(Y) \leq \text{VaR}_\alpha^{max} , \qquad (3.91)$$

with:

$$\text{VaR}_\alpha^{min} = \inf \{ t \in \mathbb{R}; \ F_{max}(t) \geq \alpha \} , \qquad (3.92)$$

and

$$\text{VaR}_\alpha^{max} = \inf \{ t \in \mathbb{R}; \ F_{min}(t) \geq \alpha \} . \qquad (3.93)$$

These two relations have a clear economic meaning: they represent respectively the most optimistic and pessimistic outcomes one can expect in the absence of any information on the actual dependence structure between the different sources of risk.

A closed-form expression for $F_{min}$ and $F_{max}$ is almost impossible to obtain in the general case where the marginal distributions of each of the assets are different. However, when all the risks can be described by distributions belonging to the same class, some general results have been obtained [126]. As an example, let us consider the case of a portfolio made of $n$ risks (with the set of weights $\{w_i, i = 1, \ldots, n\}$) following shifted-Pareto distributions with the same tail index $\beta > 0$:

$$\Pr[X_i \leq x] = 1 - \left[ \frac{\lambda_i}{\lambda_i + (x - \theta_i)} \right]^\beta , \qquad x \geq \theta_i . \qquad (3.94)$$

This model provides a reasonable description of the tails of the distribution of returns of financial assets, such as stocks returns or FX (foreign exchange) rates as discussed in Chap. 2, with $\beta \simeq 3 - 4$. Shifted-Pareto distributions are also relevant for modeling insurance claims associated with industrial [496, 497] as well as natural disasters like, earthquakes [274, 411], floods [386] or fires [328]. Using (3.94), the upper and lower bounds for the Value-at-Risk are given by:

$$\text{VaR}_\alpha^{min} = \sum_{i=1}^{n} w_i \cdot \theta_i + \max_i \{ w_i \cdot \lambda_i \} \cdot \left[ \frac{1}{(1 - \alpha)^{1/\beta}} - 1 \right] , \qquad (3.95)$$

and

$$\text{VaR}_\alpha^{max} = \sum_{i=1}^{n} w_i \cdot (\theta_i - \lambda_i) + \frac{\tilde{\lambda}}{(1-\alpha)^{1/\beta}} , \tag{3.96}$$

where

$$\tilde{\lambda} = \left[ \sum_{i=1}^{n} (w_i \cdot \lambda_i)^{\frac{\beta}{1+\beta}} \right]^{\frac{1+\beta}{\beta}} . \tag{3.97}$$

These relations have been obtained by recursion. Indeed, as emphasized by Frank *et al.* [179], (3.88–3.89) involve searching an extremum over the hyperplane $A(y)$. Such an extremum can be found recursively according to

$$F_{max}^{(n)}(y) = \inf_{x \in A(y)} \min \left\{ \sum_{i=1}^{n} \tilde{F}_i(x_i), 1 \right\} , \tag{3.98}$$

$$= \inf_{x \in \mathbb{R}} \min \left\{ F_{max}^{(n-1)}(x) + \tilde{F}_n(y-x), 1 \right\} , \tag{3.99}$$

and equivalently

$$F_{min}^{(n)}(y) = \sup_{x \in A(y)} \max \left\{ \sum_{i=1}^{n} \tilde{F}_i^-(x_i) - (n-1), 0 \right\} , \tag{3.100}$$

$$= \sup_{x \in \mathbb{R}} \max \left\{ F_{min}^{(n-1)}(x) + \tilde{F}_n(y-x) - 1, 0 \right\} . \tag{3.101}$$

Unfortunately, this approach is efficient only as long as $F_{max}$ and $F_{min}$ remain of the same class as the distributions $F_i$'s, as occurs in the shifted-Pareto example (3.94). In general, the $F_i$'s are different and one has to rely on numerical procedures to derive the bounds of real portfolio risks.

An efficient numerical algorithm has been proposed by Williamson and Downs [486]. Starting with $T - 1$ observations of the risks $X_1, \ldots, X_n$, one first evaluates the upper and lower bounds for the $\text{VaR}_\alpha$ of a portfolio made of $X_1$ and $X_2$. Let $q_i(k/T)$ denote the empirical quantiles of order $k/T$ of $X_i$. Let us set $-\infty < q_i(0) < q_i(1/T)$ and $q_i(1 - 1/T) < q_i(1) < \infty$. It can be shown that convergent estimators of $\text{VaR}_\alpha^{min}$ and $\text{VaR}_\alpha^{max}$ are given by:

$$\hat{\text{VaR}}_{k/T}^{min} = \max_{0 \le j \le k} \left\{ q_1(j/T) + q_2((k-j)/T) \right\} , \tag{3.102}$$

$$\hat{\text{VaR}}_{k/T}^{max} = \min_{k \le j \le T} \left\{ q_1(j/T) + q_2(1 - (j-k)/T) \right\} . \tag{3.103}$$

In practice, the convergence of $\hat{\text{VaR}}_{k/T}$ is very fast. Using the same kind of arguments as in (3.98–3.100), it appears that this method can be used iteratively, making possible the calculation of the bounds for (reasonably) large portfolios. An illustration of this method for three portfolios made of large capitalization US stocks is depicted in Fig. 3.11. For a portfolio of ten stocks and $T = 1500$, only a few seconds are required to obtain the Value-at-Risk bounds.

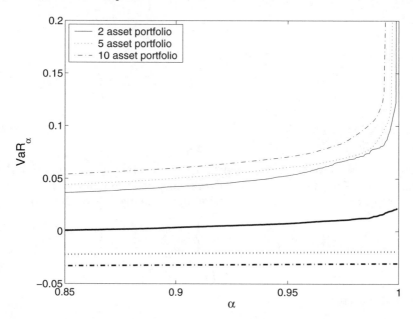

**Fig. 3.11.** *Upper* and *lower* bounds for the VaR of a portfolio over the period from 25 January, 1995 to 29 December, 2000 made of two assets (Applied Materials Inc. and Coca Cola Co: plain lines), five assets (the two above plus E.M.C Corp MA, General Electric Co, General Motors Corp: *dotted lines*) and ten assets (the five above plus Hewlett Packard Co, I.B.M Corp, Intel Corp, Medtronic Inc. and Merck & Co Inc.: *dash-dotted lines*). We find practically identical results when exchanging these assets with others from the largest capitalization stocks. The lower negative bounds for portfolios of 5 and 10 assets correspond to the favourable situation where diversification has removed the risks of losses

### 3.6.2 Asymptotic Expression of the Value-at-Risk

In several special cases, the tail risk of a portfolio, made of assets exhibiting nontrivial dependence, can be approximately calculated by a linear or quadratic approximation [472] or by using an asymptotic expansion. Here, we follow this later approach and provide an example borrowed from [336].

Consider a portfolio of $N$ assets whose dependence structure is given by the Gaussian copula. We will discuss the relevance and the limits of this assumption in Chap. 5. In addition, we assume that the returns of each asset are distributed according to a so-called *modified*-Weibull distribution characterized by its density

$$ p(x) = \frac{1}{2\sqrt{\pi}} \frac{c}{\chi^{\frac{c}{2}}} |x|^{\frac{c}{2}-1} e^{-\left(\frac{|x|}{\chi}\right)^c} , \tag{3.104} $$

or more generally

$$p(x) = \frac{1}{2\sqrt{\pi}} \frac{c_+}{\chi_+^{\frac{c_+}{2}}} |x|^{\frac{c_+}{2}-1} e^{-\left(\frac{|x|}{\chi_+}\right)^{c_+}} \qquad \text{if } x \geq 0 \qquad\qquad (3.105)$$

$$p(x) = \frac{1}{2\sqrt{\pi}} \frac{c_-}{\chi_-^{\frac{c_-}{2}}} |x|^{\frac{c_-}{2}-1} e^{-\left(\frac{|x|}{\chi_-}\right)^{c_-}} \qquad \text{if } x < 0 \;, \qquad\qquad (3.106)$$

when it is desirable to take into account a possible asymmetry between negative and positive values (thus leading to possible nonzero mean and skewness of the returns). This parameterization has the remarkable property that if the random variable $X$ follows a modified-Weibull law with exponent $c$, then the variable

$$Y = \text{sgn}(X) \; \sqrt{2} \left(\frac{|X|}{\chi}\right)^{\frac{c}{2}} \qquad\qquad (3.107)$$

follows a standard Gaussian law. This offers a simple visual test of the hypothesis that the returns are distributed according to the *modified*-Weibull distribution: starting from the empirical returns, one transforms them by converting the empirical distribution into a Gaussian one. Then, plotting the transformed variables as a function of the raw returns should give the power law (3.107) if the *modified*-Weibull distribution is a good model. Figure 3.12 shows the (negative) transformed returns of the S&P's 500 index as a function of the raw returns over the time interval from 03 January, 1995 to 29 December,

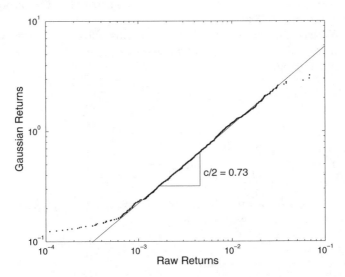

**Fig. 3.12.** Graph of the Normalized returns $Y$ of the Standard & Poor's 500 index (as explained in the text) versus its raw returns $X$, from 03 January, 1995 to 29 December, 2000 for the negative tail of the distribution. The double logarithmic scales clearly show a straight line over an extended range of data, qualifying the power law relationship (3.107)

2000. The double logarithmic scales of Fig. 3.12 qualifies a power law with exponent $c/2 = 0.73$ over an extended range of data.

For such a portfolio constituted of assets with returns distributed according to *modified*-Weibull distributions with the same exponent $c > 1$, it can be shown that the distribution of its returns is still given by a modified-Weibull law, in the asymptotic regime of large losses (counted as negative). Specifically, the distribution function $F_\pi$ of the portfolio losses is asymptotically equivalent to a *modified*-Weibull distribution function $F_Z$,

$$F_\pi(x) \sim \lambda \cdot F_Z(x), \quad \text{as} \quad x \longrightarrow -\infty , \tag{3.108}$$

where $\lambda$ is a constant, with the same exponent $c$ and with a scale factor $\hat{\chi}$ given by:

$$\hat{\chi} = \left( \sum_i w_i \chi_i \sigma_i \right)^{\frac{c-1}{c}} , \tag{3.109}$$

where $\chi_i$ is the scale factor of asset $i$, $w_i \geq 0$ is its relative weight in the portfolio, $\sigma_i$ is the solution of

$$\sum_{k=1}^N V_{ik} \ w_k \chi_k \ \sigma_k^{1-c/2} = \sigma_i^{c/2} , \qquad \forall i = 1, \ldots, N , \tag{3.110}$$

and $\mathbf{V}$ is the correlation matrix of the Gaussian copula. The proof of this result can be found in [336].

For two particular cases, the above equations allow us to retrieve simple closed-form formulas. For independent assets, one has $\mathbf{V} = \text{Id}$, so that the solution of (3.110) is

$$\sigma_i = (w_i \chi_i)^{\frac{1}{c-1}}, \quad \forall i = 1, \ldots, N \tag{3.111}$$

and thus

$$\hat{\chi} = \left( \sum_{i=1}^N (w_i \chi_i)^{\frac{c}{c-1}}, \right)^{\frac{c-1}{c}} , \quad \text{and} \quad \lambda = \left[ \frac{c}{2(c-1)} \right]^{\frac{N-1}{2}} , \tag{3.112}$$

(see Appendix 3.B for a direct proof of this result). For comonotonic assets, $V_{ij} = 1$ for all $i, j = 1, \ldots N$, which leads to

$$\sigma_i = \left( \sum_{k=1}^N w_k \chi_k \right)^{\frac{1}{c-1}} , \quad \forall i = 1, \ldots, N \tag{3.113}$$

and thus

$$\hat{\chi} = \sum_{i=1}^N w_i \chi_i, \quad \text{and} \quad \lambda = 1 . \tag{3.114}$$

This result is obvious and can be directly retrieved from the comonotonicity between the assets. In fact, in such a case the distribution of the portfolio is a modified-Weibull law, not only asymptotically but *exactly* over the whole range.

Denoting by $W(0)$ the initial amount of money invested in the risky portfolio, the asymptotic Value-at-Risk, at probability level $\alpha$, can easily be computed with the formula

$$\text{VaR}_\alpha \simeq W(0) \, \frac{\hat{\chi}}{2^{1/c}} \left[ \Phi^{-1} \left( 1 - \frac{\alpha}{\lambda} \right) \right]^{2/c} , \tag{3.115}$$

$$\simeq \xi(\alpha)^{2/c} \, W(0) \cdot \hat{\chi} , \tag{3.116}$$

where the function $\Phi(\cdot)$ denotes the cumulative Normal distribution function and

$$\xi(\alpha) \equiv \frac{1}{\sqrt{2}} \Phi^{-1} \left( 1 - \frac{\alpha}{\lambda} \right) . \tag{3.117}$$

The example provided here for a portfolio made of assets whose dependence is described by a Gaussian copula can be easily extended to more complex cases. For instance, the same kind of asymptotic expansion can be performed for the Student's copula. This illustrates the simplification brought by the use of copula for some parametric calculations of tail risks.

### 3.6.3 Options on a Basket of Assets

As suggested in [99, 417], copulas offer a useful framework for pricing multivariate contingent claims. Indeed, they provide natural pricing kernels that allow one to determine the price of options defined on a basket of assets by simply gathering the prices of options written on each individual asset.

Following [99], let us consider a market with two risky assets $S_1$ and $S_2$ and a risk-free asset $B$. For simplicity – but without loss of generality – the risk-free interest rate is set to zero. Let us assume the existence of two digital options $O_1$ on $S_1$ and $O_2$ on $S_2$ respectively, with maturity $T$. They pay one monetary unit at time $T$ if the value $S_i(T)$ of the underlying asset at time $T$ is more than $K_i$. Their price $P_i$ is:

$$P_i = \text{E}^{\mathbb{Q}} \left[ 1_{\{S_i(T) > K_i\}} \right] = \text{Pr}^{\mathbb{Q}} \left[ S_i(T) > K_i \right] , \tag{3.118}$$

where $\mathbb{Q}$ denotes a risk-neutral probability measure, equivalent to the historical probability measure $\mathbb{P}$. $\mathbb{Q}$ is unique when the market is complete.

Now, consider the *bivariate* digital option $O$ which pays one monetary unit at time $T$ if the value $S_1(T)$ is larger than $K_1$ *and* the value $S_2(T)$ is larger than $K_2$. The price of such an option on a basket of two assets is

$$P = \text{E}^{\mathbb{Q}} \left[ 1_{\{S_1(T) > K_1, S_2(T) > K_2\}} \right] = \text{Pr}^{\mathbb{Q}} \left[ S_1(T) > K_1, S_2(T) > K_2 \right] . \tag{3.119}$$

By Sklar's Theorem (3.2.1), we can write the price of the bivariate digital option as a function of the price of each individual digital option:

$$P = C^Q (P_1, P_2) , \qquad (3.120)$$

where $C^Q$ is a risk-neutral (survival) copula. Just as the individual risk-neutral density embodies traders' expectations on future asset prices and therefore represents a forward-looking indicator of market risk [44], the risk-neutral copula contains the expectations on future co-movements of the basket of assets [56].

Accounting for Fréchet-Hoeffding bounds, we can assert that the price of any bivariate digital option must satisfy

$$\max\{P_1 + P_2 - 1, 0\} \le P \le \min\{P_1, P_2\} . \qquad (3.121)$$

This relation can be interpreted as a direct consequence of the no-arbitrage principle, as we now show. The considered market exhibits four states, denoted by HH, HL, LH and LL. In the first state, both $S_1(T)$ and $S_2(T)$ are larger than $K_1$ and $K_2$ respectively. In the second state, only $S_1(T)$ is larger than $K_1$, while in the third state, only $S_2(T)$ is larger than $K_2$. In the fourth state, both $S_1(T)$ and $S_2(T)$ are smaller than $K_1$ and $K_2$ respectively. It is convenient to introduce the vector $p$ whose components are the price of the bivariate digital option, of the risk-free asset, and of the two digital options,

$$p = \begin{pmatrix} P \\ 1 \\ P_1 \\ P_2 \end{pmatrix} . \qquad (3.122)$$

Let us introduce the matrix indicator defined by

$$\Pi = \begin{pmatrix} 1_{HH}(O) & 1_{HL}(O) & 1_{LH}(O) & 1_{LL}(O) \\ 1_{HH}(1) & 1_{HL}(1) & 1_{LH}(1) & 1_{LL}(1) \\ 1_{HH}(O_1) & 1_{HL}(O_1) & 1_{LH}(O_1) & 1_{LL}(O_1) \\ 1_{HH}(O_2) & 1_{HL}(O_2) & 1_{LH}(O_2) & 1_{LL}(O_2) \end{pmatrix} , \qquad (3.123)$$

where

$$1_{HH}(O) = 1_{\{S_1(T)>K_1, S_2(T)>K_2\}|HH} = 1 ,$$
$$1_{HL}(O) = 1_{\{S_1(T)>K_1, S_2(T)>K_2\}|HL} = 0 ,$$
$$1_{LH}(O) = 1_{\{S_1(T)>K_1, S_2(T)>K_2\}|LH} = 0 ,$$
$$1_{LL}(O) = 1_{\{S_1(T)>K_1, S_2(T)>K_2\}|LL} = 0 ,$$

$1_{HH}(1) = 1_{HL}(1) = 1_{LH}(1) = 1_{LL}(1) = 1$ for the risk-free asset, and

$$1_{HH}(O_1) = 1_{\{S_1(T)>K_i\}|HH} = 1 \, ,$$
$$1_{HL}(O_1) = 1_{\{S_1(T)>K_i\}|HL} = 1 \, ,$$
$$1_{LH}(O_1) = 1_{\{S_1(T)>K_i\}|LH} = 0 \, ,$$
$$1_{LL}(O_1) = 1_{\{S_1(T)>K_i\}|LL} = 0 \, ,$$
$$1_{HH}(O_2) = 1_{\{S_2(T)>K_i\}|HH} = 1 \, ,$$
$$1_{HL}(O_2) = 1_{\{S_2(T)>K_i\}|HL} = 0 \, ,$$
$$1_{LH}(O_2) = 1_{\{S_2(T)>K_i\}|LH} = 1 \, ,$$
$$1_{LL}(O_2) = 1_{\{S_2(T)>K_i\}|LL} = 0 \, .$$

The first row of $\Pi$ in (3.123) corresponds to the bivariate digital option, the second row to the risk-free asset, the third row to the option on $S_1$ and the fourth row to the option on $S_2$. The first column corresponds to state HH, the second column to state HL, the third column to state LH and the fourth column to state LL. This yields

$$\Pi = \begin{pmatrix} 1\,0\,0\,0 \\ 1\,1\,1\,1 \\ 1\,1\,0\,0 \\ 1\,0\,1\,0 \end{pmatrix} . \tag{3.124}$$

In short, the matrix $\Pi$ allows one to obtain the value of the four assets in each of the four states of the world.

The absence of arbitrage opportunity amounts to the existence of a vector $\tilde{p}$ with positive components such that the vector $p$ of prices can be written as follows [103, 214]

$$p = \Pi \cdot \tilde{p} \, . \tag{3.125}$$

Since, in the present case, the market is complete by construction, the matrix $\Pi$ can be inverted and we have

$$\tilde{p} = \Pi^{-1} \cdot p = \begin{pmatrix} P \\ P_1 - P \\ P_2 - P \\ P - P_1 - P_2 + 1 \end{pmatrix} . \tag{3.126}$$

Writing that all the components of $\tilde{p}$ are positive is equivalent to:

$$\max\{P_1 + P_2 - 1, 0\} < P < \min\{P_1, P_2\} \, . \tag{3.127}$$

This retrieves (3.121) except for the fact that the Fréchet-Hoeffding bounds are now excluded. In fact, as recalled earlier, the Fréchet-Hoeffding upper and lower bounds are associated with the comonotonicity and the counter-monotonicity. These two situations are obviously excluded from the formulation in terms of the pricing kernel since the market cannot be considered as complete in those cases. Therefore, the prices associated with the Fréchet-Hoeffding bounds are nothing but the static super-replication[15] prices of the bivariate digital option. Indeed, selling for instance the bivariate digital option for the price $P = \min\{P_1, P_2\}$, the trader can buy the least expensive of the two digital options, say $O_1$ if $P_1 \leq P_2$. Then, at maturity, she can pay one monetary unit to the buyer of the binary digital option with certainty since the binary option generates a cash-flow of one monetary unit if and only if the world is in the state $HH$ for which $O_1$ also generates a cash-flow of one monetary unit.

It is straightforward to extend the previous calculations to the case of multivariate digital options written on a larger basket of underlying assets. The restriction to *bivariate* digital options presented here is only for notational convenience.

More generally, let us consider an option written on a basket of $N$ underlying assets $S_1, \ldots, S_N$. Let the pay-off of such an option be

$$G\left[\psi\left(S_1(T), \ldots, S_N(T)\right)\right] , \tag{3.128}$$

where $T$ still denotes the maturity. $G$ is typically the univariate pay-off characterizing the contract. For instance, for a European call with strike $K$, we have:

$$G(x) = [x - K]^+ . \tag{3.129}$$

The function $\psi$ describes how the $N$ underlying assets $S_i$ determine the terminal cash-flow. For instance, one can consider an option on the minimum of the $N$ assets

$$\psi\left(S_1(T), \ldots, S_N(T)\right) = \min\{S_1(T), \ldots, S_N(T)\} , \tag{3.130}$$

or on a weighed sum (a portfolio) of these assets

$$\psi\left(S_1(T), \ldots, S_N(T)\right) = \sum_{i=1}^{N} w_i \cdot S_i(T) . \tag{3.131}$$

The fair price of such a contract is, as usual, given by

$$P = \mathrm{E}^{\mathbb{Q}}\left[G\left(\psi\left(S_1(T), \ldots, S_N(T)\right)\right)\right] . \tag{3.132}$$

Using Theorem 3.4.1 and (3.78), we can assert that

$$\mathrm{E}^{\mathbb{Q}}\left[G\left(S_{sup}\right)\right] \leq P \leq \mathrm{E}^{\mathbb{Q}}\left[G\left(S_{inf}\right)\right] , \tag{3.133}$$

where $S_{inf}$ and $S_{sup}$ are two random variables with distribution functions $F_{inf}$ and $F_{sup}$ respectively (see Theorem. 3.4.1).

---

[15] To super-replicate means to hedge with certainty.

As an example, let us consider a rainbow call[16] on the minimum of the $N$ assets $S_1, \ldots, S_N$, with strike $K$ and maturity $T$ [466]. For simplicity, we assume a zero interest rate. The value of such a contract is:

$$P = \mathrm{E}^{\mathbb{Q}} \left[ \min\{S_1(T), \ldots, S_N(T)\} - K \right]^+ . \tag{3.134}$$

Denoting $\psi = \min\{S_1(T), \ldots, S_N(T)\}$, we have

$$\mathrm{Pr}^{\mathbb{Q}} \left[ \psi \leq x \right] = \mathrm{Pr}^{\mathbb{Q}} \left[ \min\{S_1(T), \ldots, S_N(T)\} \leq x \right], \tag{3.135}$$

$$= 1 - \mathrm{Pr}^{\mathbb{Q}} \left[ \min\{S_1(T), \ldots, S_N(T)\} > x \right], \tag{3.136}$$

$$= 1 - \mathrm{Pr}^{\mathbb{Q}} \left[ S_1(T) > x, \ldots, S_N(T) > x \right], \tag{3.137}$$

$$= 1 - \overline{C^{\mathbb{Q}}} \left( P_1(x), \ldots, P_N(x) \right) , \tag{3.138}$$

where $P_i(x) = \mathrm{Pr}^{\mathbb{Q}} \left[ S_i(T) > x \right]$ is the price of a digital option written on the underlying asset $S_i$, which pays one monetary unit if $S_i(T)$ is larger than $x$. This immediately yields

$$1 - \min\{P_1(x), \ldots, P_N(x)\} \leq \mathrm{Pr}^{\mathbb{Q}} \left[ \psi \leq x \right] \tag{3.139}$$

and

$$\mathrm{Pr}^{\mathbb{Q}} \left[ \psi \leq x \right] \leq 1 - \max\{P_1(x) + \cdots + P_N(x) - (N-1), 0\} . \tag{3.140}$$

Thus, defining $S_{inf}$ and $S_{sup}$ as two random variables such that:

$$\mathrm{Pr}^{\mathbb{Q}} \left[ S_{inf} \leq x \right] = 1 - \min\{P_1(x), \ldots, P_N(x)\}, \tag{3.141}$$

$$\mathrm{Pr}^{\mathbb{Q}} \left[ S_{sup} \leq x \right] = 1 - \max\{P_1(x) + \cdots + P_N(x) - (N-1), 0\} \tag{3.142}$$

it follows from (3.133) that

$$\mathrm{E}^{\mathbb{Q}} \left[ S_{sup} - K \right]^+ \leq P \leq \mathrm{E}^{\mathbb{Q}} \left[ S_{inf} - K \right]^+ . \tag{3.143}$$

The quantitative values of these two bounds are obtained after calibration and numerical integration.

To obtain more accurate information on the price of options defined on a basket of assets, it is necessary to specify the nature of the risk-neutral copula. The problem comes from the fact that there exists no general relation between the historical copula $C^{\mathbb{P}}$ and the risk-neutral $C^{\mathbb{Q}}$. However, in some special cases, one can obtain this relation. For instance, in the multivariate Black-Scholes model, both the historical and the risk-neutral copulas are Gaussian copulas, with the same correlation matrix. This result generalizes to the case where asset prices follow diffusion processes with deterministic drifts and volatilities [112].

In the more realistic case where one considers a stochastic volatility model (under $\mathbb{P}$) like

---

[16] Rainbow options get their name from the fact that their underlying is two or more assets rather than one.

$$\frac{dS_i(t)}{S_i(t)} = \mu_i\left(t, \sigma_i(t)\right) dt + \sigma_i(t) dB_i(t), \quad i = 1, \dots, N \tag{3.144}$$

$$d\sigma_i(t) = a_i\left(t, \sigma_i(t)\right) dt + b_i\left(t, \sigma_i(t)\right) dW_i(t), \tag{3.145}$$

for instance, where $B_i(t)$ and $W_i(t)$ denote standard Wiener processes and where $a_i(\cdot, \cdot)$ and $b_i(\cdot, \cdot)$ are chosen such that the $\sigma_i(t)$'s remain positive almost surely, one cannot express $C^{\mathbb{P}}$ and $C^{\mathbb{Q}}$ explicitly. In addition, since individual volatilities are a non-traded assets, the market is incomplete, and the choice of a risk-neutral measure $\mathbb{Q}$ – which amounts to choosing the market prices of volatility risks $\lambda_i$ – is not unique. One has to set additional constraints in order to select an appropriate $\mathbb{Q}$. Many methods have been developed for univariate stochastic volatility models, which can be extended to the multivariate case. Let us mention the minimal martingale measure [176, 434, 435], the minimal entropy measure [192, 403] or the variance-optimal measure [54, 177, 226, 292, 384], for instance. All these examples are, in fact, particular cases of $q$-optimal measures for $q = 0$, 1 and 2, respectively) [125, 234], i.e. measures which are the closest to the objective (or historical) measure $\mathbb{P}$ in the sense of the $q$th moment of their relative density. Such measures minimize the functional

$$H_q\left(\mathbb{P}, \mathbb{Q}\right) = \begin{cases} \mathrm{E}\left[\frac{q}{q-1}\left(\frac{d\mathbb{Q}}{d\mathbb{P}}\right)^q\right], & \text{if } \mathbb{Q} \ll \mathbb{P} \\ +\infty, & \text{otherwise}, \end{cases} \tag{3.146}$$

for $q \in \mathbb{R} \setminus \{0, 1\}$ and

$$H_q\left(\mathbb{P}, \mathbb{Q}\right) = \begin{cases} \mathrm{E}\left[(-1)^{q+1}\left(\frac{d\mathbb{Q}}{d\mathbb{P}}\right)^q \cdot \ln \frac{d\mathbb{Q}}{d\mathbb{P}}\right], & \text{if } \mathbb{Q} \ll \mathbb{P} \\ +\infty, & \text{otherwise}, \end{cases} \tag{3.147}$$

for $q \in \{0, 1\}$. The symbol "$\ll$" means *absolutely continuous*, i.e., the sets of zero measure for $\mathbb{P}$ are also sets of zero measure for $\mathbb{Q}$.

Such measures have the additional advantage of allowing an interpretation in terms of utility maximizing agents. Indeed, asset prices obtained under $q$-optimal measures represent the marginal utility indifferent prices for investors with HARA[17] utility functions [230].

Using the risk-neutral probability measure $\mathbb{Q}$ which amounts to taking a vanishing market price of the volatility risk, and if in addition the rates of return $\mu_i\left(t, \sigma_i(t)\right)$ do not depend on $\sigma_i$, then it can be shown that $C^{\mathbb{P}} = C^{\mathbb{Q}}$ (see Appendix 3.C). In such a case, the calibration of the copula under historical data provides the risk-neutral copula.

Unfortunately, when these conditions are not met, or when one considers more general diffusion models of the form

$$dS_i(t) = \mu_i\left(t, S_i(t)\right) dt + \sigma_i\left(t, S_i(t)\right) dW_i(t), \quad i = 1, \dots, N, \tag{3.148}$$

it is in general impossible to obtain a relation between $C^{\mathbb{P}}$ and $C^{\mathbb{Q}}$. In this case, the risk-neutral copula can only be and has to be determined directly

---

[17] Hyperbolic absolute risk aversion.

from options prices. In practice, when one deals with contracts which are not actively traded, or contracts negotiated OTC,[18] data may be rare, leading to serious restrictions for the calibration of the risk-neutral copula and showing the limit of the approach.

### 3.6.4 Basic Modeling of Dependent Default Risks

Default risk models are basically of two kinds. The first class contains models which are close to many actuarial models. They rely on the assumption that, conditional on a set of economic factors, the individual default probabilities of each obligator are independent. Such models are known as *mixture models* [248]. They include frailty models, presented page 113, as well as professional models like CreditRisk$^+$ [114]. It is in general difficult to obtain an analytical expression of their dependence structure.

The second class of default risk models are based on Merton's seminal work on firm value [358]. In particular, industry standards like Moody's KMV [273] and RiskMetrics [406] are extensions of this original model. They consider that the default of an obligator occurs when a latent variable, which usually represents the firm's asset value, goes below some level usually representing the value of the firm's liabilities. In the more recent model by Li [303], the latent variables account for the time-to-default of an obligator and the crossing level represents the time horizon of interest. These approaches are equivalent since, once a dynamics is specified for the assets, one can derive, in principle, the law of the time-to-default.

These models assume the same dependence structure for the latent variables, characterized by a Gaussian copula. Hence, the joint probability of default is closely related to the Gaussian copula. Indeed, let us consider $N$ obligators and let $D_i$ be the default indicator of obligator $i$. $D_i$ equals one if obligator $i$ has defaulted and zero otherwise. Let $(X_1, \ldots, X_N)$ denote the vector of latent variables and $(T_1, \ldots, T_N)$ the vector of thresholds below which default occurs:

$$D_i = 1 \iff X_i \leq T_i \,. \tag{3.149}$$

The joint probability that obligators $i_1, \ldots, i_k$ $(k \leq N)$ default is

$$\begin{aligned}
\Pr\left[D_{i_1} = 1, D_{i_2} = 1, \ldots, D_{i_k} = 1\right] &= \Pr\left[X_{i_1} \leq T_{i_1}, \ldots X_{i_k} \leq T_{i_k}\right], \\
&= C\left(\Pr\left[X_{i_1} \leq T_{i_1}\right], \ldots, \Pr\left[X_{i_k} \leq T_{i_k}\right]\right), \\
&= C\left(\pi_{i_1}, \ldots, \pi_{i_k}\right) \,, \tag{3.150}
\end{aligned}$$

where $C$ denotes the (Gaussian) copula of the latent variables $X_{i_1}, \ldots, X_{i_k}$ and $\pi_{i_1}, \ldots, \pi_{i_k}$ are the individual default probabilities of obligators $i_1, \ldots, i_k$.

---

[18] Over-the-counter: a market for securities made up of dealers who may or may not be members of a formal securities exchange. The over-the-counter market is conducted over the telephone and is a negotiated market rather than an auction market such as the NYSE.

In the KMV methodology, the variables $\{X_i\}$ model the return processes of the assets. They are assumed multivariate Gaussian, and their correlations are set by a factor model representing the various underlying macroeconomic variables impacting the dynamics of the asset returns. Each threshold $T_i$ is determined by an option technique applied to the historical data of the $i$th firm.

CreditMetrics' approach is also based upon the assumption that the $X_i$'s are multivariate Gaussian random variables. However, they do not represent the evolution of the asset value itself but the evolution of the rating of the firm. The range of each $X_i$ is divided into classes which represent the possible rating classes of the firm. The classes are determined so that they agree with historical data. This procedure allows one to fix simultaneously all the values of the thresholds $\{T_i\}$. Again, the correlations are calibrated by assuming a factor model.

In Li's model, the latent variable $X_i$ is interpreted as the time-to-default of obligator $i$ and the thresholds $T_i$'s are all equal to $T$, the time horizon over which the credit portfolio is monitored. Here, the multivariate distribution of the $X_i$'s is not Gaussian anymore (since, now, the $X_i$'s are positive random variables). The marginal distribution of each $X_i$ is exponential with parameter $\lambda_i$:

$$\Pr\left[X_i \leq x_i\right] = 1 - e^{-\lambda_i \cdot x_i} \,, \tag{3.151}$$

while the copula remains Gaussian. Again, the correlations between the $X_i$'s can be determined from a factor model.

This recurrent use of a Gaussian factor model which is equivalent to describing the dependence between the latent variables in terms of a Gaussian copula has been ratified by the recommendations of the BIS [42] concerning credit risk modeling. However, there are many indications suggesting that this Gaussian copula approach may be grossly inadequate to account for large credit risk (see [186] for instance), since the Gaussian copula might – by construction – underestimate the largest concomitant risks. We will come back in more detail on this crucial point in the next chapter (Chap. 4) where we will present and contrast the different available measures of dependence and address more precisely how to assess the dependence in the tails of the distribution.

# Appendix

## 3.A Simple Proof of a Theorem on Universal Bounds for Functionals of Dependent Random Variables

Here, we provide a simple heuristic proof of Theorem 3.4.1. For simplicity, we restrict ourselves to the bivariate case: we consider a random vector $(X, Y)$

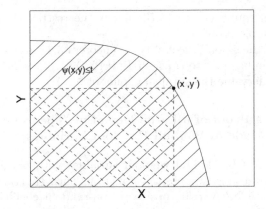

**Fig. 3.13.** The area hatched with plain lines represents the set of points $(x,y)$ such that $\psi(x,y) \le t$. The area hatched with *dashed lines* represents the set of points $(x',y')$ such that $x' \le x^*$ and $y' \le y^*$ for some $(x^*,y^*)$ satisfying $\psi(x^*,y^*) = t$. By definition, the $F$-measure of this area is $\Pr[X \le x^*, Y \le y^*] = F(x^*,y^*)$

with joint distribution function $F$, and continuous margins $F_X$ and $F_Y$, respectively. In addition, we assume that the function $\psi$ is continuous and increasing in each argument. In such a case, provided that $t$ belongs to the range of $\psi$, the set of points $(x,y)$ such that $\psi(x,y)$ is less than $t$ has a typical shape represented by the area hatched with plain lines in Fig. 3.13.

By definition, $\Pr[\psi(X,Y) \le t]$ is the $F$-measure of this hatched area:

$$\Pr[\psi(X,Y) \le t] = \int_{\psi(x,y)\le t} dF(x,y) . \tag{3.A.1}$$

For any couple $(x^*,y^*)$ such that $\psi(x^*,y^*) = t$,

$$\Pr[\psi(X,Y) \le t] \ge \Pr[X \le x^*, Y \le y^*] = F(x^*,y^*) , \tag{3.A.2}$$

since $F(x^*,y^*) = \int_{\{x \le x^*, y \le y^*\}} dF(x,y)$ is the $F$-measure of the area hatched with dashed lines in Fig. 3.13, which is included within the area representing the set of points $\{(x,y): \psi(x,y) \le t\}$. Given any copula $C_{inf}$ such that

$$C_{inf}(u,v) \le C(u,v), \quad \forall(u,v) \in [0,1]^2 , \tag{3.A.3}$$

where $C$ denotes the copula of the random vector $(X,Y)$, we can write:

$$\Pr[\psi(X,Y) \le t] \ge C_{inf}(F_X(x^*), F_Y(y^*)) , \tag{3.A.4}$$

for all $(x^*,y^*)$ such that $\psi(x^*,y^*) = t$, which finally allows us to assert that:

$$\Pr[\psi(X,Y) \le t] \ge \sup_{\psi(x^*,y^*)=t} C_{inf}(F_X(x^*), F_Y(y^*)) . \tag{3.A.5}$$

This equation is equivalent to (3.73) under the restrictive assumptions retained in this simple proof.

The proof of the second inequality of Theorem 3.4.1 follows the same line of reasoning. One has just to consider $\Pr[\psi(X,Y) \geq t]$, which leads, *mutatis mutandis*, to undervalue the survival copula of $(X,Y)$.

## 3.B Sketch of a Proof of a Large Deviation Theorem for Portfolios Made of Weibull Random Variables

Let $X_1, X_2, \ldots, X_N$ be $N$ i.i.d random variables with density $p(\cdot)$. Let us denote by $f(\cdot)$ and $g(\cdot)$ two positive functions such that $p(\cdot) = g(\cdot)e^{-f(\cdot)}$. Let $w_1, w_2, \ldots, w_N$ be $N$ real (positive) non-random coefficients, and $S = \sum_{i=1}^{N} w_i x_i$.

Let $\chi = \{\boldsymbol{x} \in \mathbb{R}^N, \sum_{i=1}^{N} w_i x_i = S\}$. The density of the variable $S$ is given by

$$P_S(S) = \int_\chi d\boldsymbol{x} \; e^{-\sum_{i=1}^{N}[f(x_i) - \ln g(x_i)]} . \tag{3.B.6}$$

We will assume the following conditions on the function $f$:

1. $f(\cdot)$ is three times continuously differentiable and four times differentiable,
2. $f^{(2)}(x) > 0$, for $|x|$ large enough,
3. $\lim_{x \to \pm\infty} \frac{f^{(3)}(x)}{(f^{(2)}(x))^2} = 0$,
4. $f^{(3)}$ is asymptotically monotonic,
5. there is a constant $\beta > 1$ such that $\frac{f^{(3)}(\beta \cdot x)}{f^{(3)}(x)}$ remains bounded as $x$ goes to infinity,
6. there exists $C_1, C_2 > 0$ and some $\nu > 0$ such that $C_1 \cdot x^\nu \leq g(\cdot) \leq C_2 \cdot x^\nu$, as $x$ goes to infinity.

Under the assumptions stated above, the leading order expansion of $P_S(S)$ for large $S$ and finite $N > 1$ is obtained by a generalization of Laplace's method which assumes that the set of $x_i^*$'s that maximize the integrand in (3.B.6) are a solution of

$$f_i'(x_i^*) = \sigma(S)w_i , \tag{3.B.7}$$

where $\sigma(S)$ is nothing but a Lagrange multiplier introduced to minimize the expression $\sum_{i=1}^{N} f_i(x_i)$ under the constraint $\sum_{i=1}^{N} w_i x_i = S$. This constraint shows that at least one $x_i$, for instance $x_1$, goes to infinity as $S \to \infty$. Since $f'(x_1)$ is an increasing function by Assumption 2, which goes to infinity as $x_1 \to +\infty$ (Assumption 3), expression (3.B.7) shows that $\sigma(S)$ goes to infinity with $S$, as long as the weight of the asset 1 is not zero. Putting the divergence of $\sigma(S)$ with $S$ in expression (3.B.7) for $i = 2, \ldots, N$ ensures that each $x_i^*$ increases when $S$ increases and goes to infinity when $S$ goes to infinity.

Expanding $f_i(x_i)$ around $x_i^*$ yields

$$f(x_i) = f(x_i^*) + f'(x_i^*) \cdot h_i + \int_{x_i^*}^{x_i^*+h_i} dt \int_{x_i^*}^t du \, f''(u) \,, \tag{3.B.8}$$

where the set of $h_i = x_i - x_i^*$ obey the condition

$$\sum_{i=1}^N w_i h_i = 0 \,. \tag{3.B.9}$$

Summing (3.B.8) over $i$ in the presence of relation (3.B.9), we obtain

$$\sum_{i=1}^N f(x_i) = \sum_{i=1}^N f(x_i^*) + \sum_{i=1}^N \int_{x_i^*}^{x_i^*+h_i} dt_i \int_{x_i^*}^{t_i} du_i \, f''(u_i) \,. \tag{3.B.10}$$

Thus $\exp(-\sum f(x_i))$ can be rewritten as follows:

$$\exp\left[-\sum_{i=1}^N f(x_i)\right] = \exp\left[-\sum_{i=1}^N f(x_i^*) - \sum_{i=1}^N \int_{x_i^*}^{x_i^*+h_i} dt_i \int_{x_i^*}^{t_i} du_i \, f''(u_i)\right] \,. \tag{3.B.11}$$

Let us now define the compact set $\mathcal{A}_C = \{\mathbf{h} \in \mathbb{R}^N, \sum_{i=1}^N f''(x_i^*)^2 \cdot h_i^2 \le C^2\}$ for any given positive constant $C$ and the set $\mathcal{H} = \{\mathbf{h} \in \mathbb{R}^N, \sum_{i=1}^N w_i h_i = 0\}$. We can thus write

$$P_S(S) = \int_{\mathcal{H}} d\mathbf{h} \, e^{-\sum_{i=1}^N [f(x_i) - \ln g(x_i)]} \,, \tag{3.B.12}$$

$$= \int_{\mathcal{A}_C \cap \mathcal{H}} d\mathbf{h} \, e^{-\sum_{i=1}^N [f(x_i) - \ln g(x_i)]}$$

$$+ \int_{\overline{\mathcal{A}_C} \cap \mathcal{H}} d\mathbf{h} \, e^{-\sum_{i=1}^N [f(x_i) - \ln g(x_i)]} \,. \tag{3.B.13}$$

Let us analyze in turn the two integrals of the right-hand side of (3.B.13). Concerning the first integral, it can be shown that

$$\lim_{S \to \infty} \frac{\int_{\mathcal{A}_C \cap \mathcal{H}} d\mathbf{h} \, e^{-\sum_{i=1}^N \int_{x_i^*}^{x_i^*+h_i} dt \int_{x_i^*}^t du \, f''(u) - \ln g(x_i^*+h_i)}}{(2\pi)^{\frac{N-1}{2}} \prod_i g(x_i^*) \sqrt{\sum_{i=1}^N \frac{w_i^2 \prod_{j=1}^N f_j''(x_j^*)}{f_i''(x_i^*)}}} = 1, \text{ for some positive } C. \tag{3.B.14}$$

The cumbersome proof of this assertion is found in [336]. It is based upon the fact that

1. by Assumptions 1, 3, 4 and 5 for all $\mathbf{h} \in \mathcal{A}_C$ and all $\epsilon_i > 0$

$$\left| \frac{\sup_{\xi \in \mathcal{G}_i} |f^{(3)}(\xi)|}{f''(x_i^*)} \right| \leq \epsilon_i, \quad \text{for } x_i^* \text{ large enough,} \tag{3.B.15}$$

where $\mathcal{G}_i = \left[ x_i^* - \frac{C}{f''(x_i^*)}, x_i^* + \frac{C}{f''(x_i^*)} \right]$,

2. for all $\epsilon_i > 0$ and $x_i^*$ large enough:

$$\forall \mathbf{h} \in \mathcal{A}_C, \quad (1 - \epsilon_i)^\nu \leq \frac{g(x_i^* + h_i)}{g(x_i^*)} \leq (1 + \epsilon_i)^\nu, \tag{3.B.16}$$

by Assumptions 1 and 6.

Now, for the second integral on the right-hand side of (3.B.13), we have to show that

$$\int_{\overline{\mathcal{A}_C \cap \mathcal{H}}} d\mathbf{h} \, e^{-\sum f(x_i^* + h_i) - g(x_i^* + h_i)} \tag{3.B.17}$$

can be neglected. This is obvious since, by Assumption 2 and 6, the function $f(x) - \ln g(x)$ remains convex for $x$ large enough, which ensures that $f(x) - \ln g(x) \geq C_1 |x|$ for some positive constant $C_1$ and $x$ large enough. Thus, choosing the constant $C$ in $\mathcal{A}_C$ large enough, we have

$$\int_{\overline{\mathcal{A}_C \cap \mathcal{H}}} d\mathbf{h} \, e^{-\sum_{i=1}^N f(x_i) - \ln g(x_i)} \leq \int_{\overline{\mathcal{A}_C \cap \mathcal{H}}} d\mathbf{h} \, e^{-C_1 \sum_{i=1}^N |x_i* + h_i|} \sim \mathcal{O}\left( e^{-\frac{\alpha}{f''(x^*)}} \right) \tag{3.B.18}$$

for some positive $\alpha$. Thus, for $S$ large enough, the density $P_S(S)$ is asymptotically equal to

$$P_S(S) = \prod_i g(x_i^*) \frac{(2\pi)^{\frac{N-1}{2}}}{\sqrt{\sum_{i=1}^N \frac{w_i^2 \prod_{j=1}^N f_j''(x_j^*)}{f_i''(x_i^*)}}} \exp\left[ -\sum_{i=1}^N f(x_i^*) \right]. \tag{3.B.19}$$

In the case of the *modified* Weibull variables, we have

$$f(x) = \left( \frac{|x|}{\chi} \right)^c, \tag{3.B.20}$$

and

$$g(x) = \frac{c}{2\sqrt{\pi}\chi^{c/2}} \cdot |x|^{\frac{c}{2}-1}, \tag{3.B.21}$$

which satisfies our assumptions if and only if $c > 1$. In such a case, we obtain

$$x_i^* = \frac{w_i^{\frac{1}{c-1}}}{\sum_{j=i}^{N} \omega_j^{\frac{c}{c-1}}} \cdot S \,, \tag{3.B.22}$$

which, after some simple algebraic manipulations, yields

$$P(S) \sim \left[\frac{c}{2(c-1)}\right]^{\frac{N-1}{2}} \frac{c}{2\sqrt{\pi}} \frac{1}{\hat{\chi}^{c/2}} |S|^{\frac{c}{2}-1} e^{-\left(\frac{|S|}{\hat{\chi}}\right)^c} \tag{3.B.23}$$

with

$$\hat{\chi} = \left(\sum_{i=1}^{N} w_i^{\frac{c}{c-1}}\right)^{\frac{c-1}{c}} \cdot \chi \,. \tag{3.B.24}$$

Let us now consider $N$ independent random variables $X_1, X_2, \ldots, X_N$ with *modified*-Weibull pdfs with the same exponent $c > 1$ but different scale factors $\chi_i$. Let $w_1, w_2, \ldots, w_N$ be $N$ non-random real coefficients. Then, the variable

$$S_N = w_1 X_1 + w_2 X_2 + \cdots + w_N X_N \tag{3.B.25}$$

follows asymptotically a modified-Weibull with scale factor

$$\hat{\chi} = \left(\sum_{i=1}^{N} |w_i \chi_i|^{\frac{c}{c-1}}\right)^{\frac{c-1}{c}} \,, \quad c > 1 \,. \tag{3.B.26}$$

Indeed, let $Y_1, Y_2, \ldots, Y_N$ be $N$ independent and identically distributed random variables with modified-Weibull pdfs with same exponent $c > 1$ and scale factor $\chi = 1$. Then,

$$(X_1, X_2, \ldots, X_N) \overset{law}{=} (\chi_1 Y_1, \chi_2 Y_2, \ldots, \chi_N Y_N) \,, \tag{3.B.27}$$

which yields

$$S_N \overset{d}{=} w_1 \chi_1 \cdot Y_1 + w_2 \chi_2 \cdot Y_2 + \cdots + w_N \chi_N \cdot Y_N \,. \tag{3.B.28}$$

Thus, (3.B.26) immediately follows from (3.B.24).

## 3.C Relation Between the Objective and the Risk-Neutral Copula

Assuming that we have a filtered probability space $(\Omega, \mathcal{F}, (\mathcal{F}_t)_{0 \leq t \leq T}, \mathbb{P})$ – $\mathbb{P}$ denotes the objective or historical probability measure – generated by a $2N$-dimensional Brownian motion $(B_1, W_1, \ldots, B_N, W_N)$ with (constant) correlation matrix $\boldsymbol{\rho}$, let us consider the $N$-dimensional stochastic volatility model:

$$\frac{dS_i(t)}{S_i(t)} = \mu_i(t, \sigma_i(t))\, dt + \sigma_i(t) dB_i(t), \quad i = 1, \ldots, N \tag{3.C.29}$$

$$d\sigma_i(t) = a_i(t, \sigma_i(t))\, dt + b_i(t, \sigma_i(t))\, dW_i(t), \tag{3.C.30}$$

where $S_i$ is the price of asset $i$, while $a_i(\cdot, \cdot)$ and $b_i(\cdot, \cdot)$ are chosen so that the volatility $\sigma_i(t)$ of each asset remains positive almost surely. As an example, one can choose

$$a_i(t, \sigma_i) = \kappa_i \left( \frac{m_i}{\sigma_i} - \sigma_i \right), \quad \text{and} \quad b_i(t, \sigma_i) = \beta_i \ . \tag{3.C.31}$$

This stochastic volatility model is equivalent to the Heston model [232] written for the squared volatility instead of the volatility itself. In the present case, the condition $\kappa_i \cdot m_i \geq \beta_i{}^2$, together with $\kappa_i, m_i > 0$, ensures the positivity of $\sigma_i(t)$, provided that $\sigma_i(0) > 0$.

The solution of (3.C.29) with $S_i(0) = S_i^0$ is:

$$S_i(t) = S_i^0 \exp \left[ \int_0^t \left( \mu_i(s, \sigma_i(s)) - \frac{1}{2}\sigma_i(s)^2 \right) ds + \int_0^t \sigma_i(s) dB_i(s) \right] ,$$
$$\tag{3.C.32}$$

where $\sigma_i(t)$ is solution of (3.C.30). Denoting by $Z_i(t)$ the random variable

$$\int_0^t \left( \mu_i(s, \sigma_i(s)) - \frac{1}{2}\sigma_i(s)^2 \right) ds + \int_0^t \sigma_i(s) dB_i(s) , \tag{3.C.33}$$

we can assert that the copula $C^{\mathbb{P}}$ of $(S_1(t), \ldots, S_N(t))$ is the same as the copula of $(Z_1(t), \ldots, Z_N(t))$, since each $S_i(t) = S_i^0 \cdot \exp[Z_i(t)]$ is an increasing transform of the corresponding $Z_i(t)$.

Assuming that the usual conditions are satisfied, Girsanov Theorem[19] allows us to assert that there exists a probability measure $\mathbb{Q}$, equivalent to $\mathbb{P}$ on $\mathcal{F}_T$, such that

$$\frac{d\mathbb{Q}}{d\mathbb{P}} = \exp \left[ -\sum_{i=1}^N \left( \int_0^t \frac{\mu_i(s, \sigma_i(s))}{\sigma_i(s)} dB_i(s) + \frac{1}{2} \int_0^t \left[ \frac{\mu_i(s, \sigma_i(s))}{\sigma_i(s)} \right]^2 ds \right) \right.$$
$$\left. -\sum_{i=1}^N \left( \int_0^t \lambda_i(s, \sigma_i(s)) dW_i(s) + \frac{1}{2} \int_0^t \lambda_i(s, \sigma_i(s))^2 ds \right) \right] , \tag{3.C.34}$$

for any suitable processes $(\lambda_1, \ldots, \lambda_N)$, and that

$$\tilde{B}_i(t) = B_i(t) + \int_0^t \frac{\mu_i(s, \sigma_i(s))}{\sigma_i(s)} ds, \quad i = 1, \ldots, N \tag{3.C.35}$$

$$\tilde{W}_i(t) = W_i(t) + \int_0^t \lambda_i(s, \sigma_i(s)) ds, \quad i = 1, \ldots, N \tag{3.C.36}$$

---

[19] In the theory of probability, the Girsanov Theorem specifies how stochastic processes change under changes in measure. The theorem is especially important in the theory of asset pricing as it allows one to convert the physical measure which describes the probability that an underlying (such as a share price or interest rate) will take a particular value into the risk-neutral measure used for evaluating the derivatives on the underlying.

are Brownian motions under $\mathbb{Q}$, with correlation matrix $\boldsymbol{\rho}$. Since the volatility is a non-traded asset, the problem of market incompleteness arises, so that there is not a unique risk-neutral measure such that discounted assets prices are martingale.

For simplicity, let us assume that the risk-free interest rate is vanishing so that asset prices are directly discounted prices. Under any $\mathbb{Q}$, using (3.C.35) and (3.C.36), (3.C.29–3.C.30) can be written

$$\frac{dS_i(t)}{S_i(t)} = \sigma_i(t)d\tilde{B}_i(t), \quad i = 1, \ldots, N \tag{3.C.37}$$

$$d\sigma_i(t) = [a_i(t, \sigma_i(t)) - \lambda_i(t, \sigma_i(t)) \cdot b_i(t, \sigma_i(t))] \, dt + b_i(t, \sigma_i(t)) \, d\tilde{W}_i(t), \tag{3.C.38}$$

which shows that $S_i(t)$ is a $\mathbb{Q}$-martingale. The solution of (3.C.37) with $S_i(0) = S_i^0$, under $\mathbb{Q}$, is:

$$S_i(t) = S_i^0 \exp\left[-\frac{1}{2}\int_0^t \sigma_i(s)^2 \, ds + \int_0^t \sigma_i(s)d\tilde{B}_i(s)\right], \tag{3.C.39}$$

where $\sigma_i(t)$ is now the solution of (3.C.38). Denoting by $\tilde{Z}_i(t)$ the random variable

$$-\frac{1}{2}\int_0^t \sigma_i(s)^2 ds + \int_0^t \sigma_i(s)d\tilde{B}_i(s), \tag{3.C.40}$$

we can assert that the copula $C^{\mathbb{Q}}$ of $(S_1(t), \ldots, S_N(t))$ is the same as the copula of $\left(\tilde{Z}_1(t), \ldots, \tilde{Z}_N(t)\right)$.

Therefore, $C^{\mathbb{P}} = C^{\mathbb{Q}}$ if and only if the copula of $\left(\tilde{Z}_1(t), \ldots, \tilde{Z}_N(t)\right)$ is the same as the copula of $(Z_1(t), \ldots, Z_N(t))$. In the general case, the $Z_i(t)$'s and $\tilde{Z}_i(t)$'s are not simple increasing transforms of each other. Therefore, their copulas are not identical and $C^{\mathbb{P}} \neq C^{\mathbb{Q}}$. But in the particular case where the rates $\mu_i$ are deterministic functions – *i.e.*, independent of $\sigma_i(t)$ – the copula $C^{\mathbb{P}}$ is nothing but the copula of the random variables:

$$Z_i^*(t) = -\frac{1}{2}\int_0^t \sigma_i(s)^2 ds + \int_0^t \sigma_i(s)dB_i(s), \quad i = 1, \ldots, N, \tag{3.C.41}$$

where $\sigma_i(t)$ is the solution of (3.C.30), since the maps

$$x \longmapsto S_i^0 e^x \tag{3.C.42}$$

are monotonous increasing functions of their argument. If the market prices $\lambda_i$'s of volatility risks are vanishing, the vectors $(Z_1^*(t), \ldots, Z_N^*(t))$ and $\left(\tilde{Z}_1(t), \ldots, \tilde{Z}_N(t)\right)$ are equal in law, since (3.C.30) and (3.C.38) are then the same. Thus, in this case, $(Z_1^*(t), \ldots, Z_N^*(t))$ and $\left(\tilde{Z}_1(t), \ldots, \tilde{Z}_N(t)\right)$ have the same copula, and therefore $C^{\mathbb{P}} = C^{\mathbb{Q}}$.

# 4

# Measures of Dependences

In the previous chapter, we have shown how to describe with copulas the general dependence structure of several random variables, with the goal of modeling baskets of asset returns, or more generally, any multivariate financial risk. However, the general framework provided by copulas does not exclude more specific measures of dependences that can be useful to target particular ranges of variations of the random variables.

This chapter presents and describes in detail the most important *dependence measures*. Starting with the description of the basic concept of linear dependence, through linear correlation and canonical $N$-correlation coefficients, we then focus on concordance measures and on more interesting families of dependence measures. We then turn to measures of extreme dependence. In each case, we underline their relationship with copulas.

## 4.1 Linear Correlations

### 4.1.1 Correlation Between Two Random Variables

The linear correlation is probably still the most widespread measure of dependence, both in finance and insurance. Given two random variables $X$ and $Y$, the *linear correlation coefficient* is defined as:

$$\rho(X, Y) = \frac{\text{Cov}[X, Y]}{\sqrt{\text{Var}[X] \cdot \text{Var}[Y]}}, \tag{4.1}$$

provided that the variances $\text{Var}[X]$ and $\text{Var}[Y]$ exist. $\text{Cov}[X, Y]$ is the covariance of $X$ and $Y$. The coefficient $\rho(X, Y)$ is called a *linear* correlation coefficient because its knowledge is equivalent to that of the coefficient $\beta$ of the linear regression $Y = \beta X + \epsilon$, where $\epsilon$ is the residual which is linearly uncorrelated with $X$. We have indeed $\rho = \beta \sqrt{\frac{\text{Var}[X]}{\text{Var}[Y]}}$.

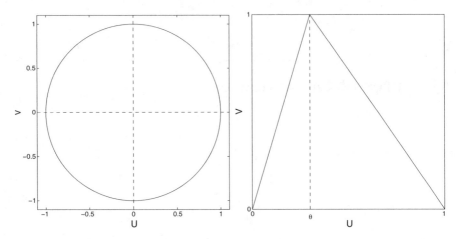

**Fig. 4.1.** Graph of the variable $V = \sin \omega$ versus $U = \cos \omega$ for $\omega \in [0, 2\pi]$ (*left panel*) and graph of the variable $V = \frac{U}{\theta} \cdot 1_{U \in [0,\theta]} + \frac{1-U}{1-\theta} \cdot 1_{U \in [\theta,1]}$ (*right panel*)

Regularly varying random variables (power-like random variables) with a tail index less than two do not have finite variances; they thus do not admit a correlation coefficient. In addition, when the tail index belongs to the interval $(2, 4]$, the correlation coefficient exists but its Pearson estimator, based on a sample of size $T$ $\{(X_i, Y_i)\}_{i=1}^T$:

$$\hat{\rho}_T = \frac{\dfrac{1}{T} \sum_{i=1}^T \left(X_i - \bar{X}\right) \cdot \left(Y_i - \bar{Y}\right)}{\sqrt{\dfrac{1}{T} \sum_{i=1}^T \left(X_i - \bar{X}\right)^2 \cdot \dfrac{1}{T} \sum_{i=1}^T \left(Y_i - \bar{Y}\right)^2}}, \tag{4.2}$$

where $\bar{X}$ and $\bar{Y}$ denote the sample means of $X$ and $Y$ respectively, performs rather poorly, insofar as its asymptotic distribution is not Gaussian but Lévy stable [356]. Therefore, a sample correlation coefficient may exhibit large deviations from its true value, providing very inaccurate estimates. This is particularly problematic for financial purposes since, as recalled in Chap. 2, the existence of the fourth moment for the distribution of stock returns is still a topic of active debate.

Considering two independent random variables, it is well known that their correlation coefficient equals zero. However, the converse does not hold. Indeed, given a random variable $\omega$ uniformly distributed in $[0, 2\pi]$, let us define the couple of random variables:

$$(U, V) = (\cos \omega, \sin \omega) \ . \tag{4.3}$$

It is easy to check that $\rho(U, V) = 0$, even though the two random variables are not independent, as shown in the left panel of Fig. 4.1 which plots the variable $V$ as a function of $U$.

More striking is the case where the knowledge of one of the variables completely determines the other one. As an example, consider a random variable U, uniformly distributed on $[0, 1]$ and the random variable $V$ defined by:

$$\begin{cases} V = \frac{U}{\theta} & U \in [0, \theta], \\ V = \frac{1-U}{1-\theta} & U \in [\theta, 1] , \end{cases} \tag{4.4}$$

for some $\theta \in [0, 1]$ (see right panel of Fig. 4.1). One can easily show that $V$ is also uniformly distributed on $[0, 1]$ and that

$$\rho(U, V) = 2\theta - 1 , \tag{4.5}$$

so that $U$ and $V$ are uncorrelated for $\theta = 1/2$ while $V$ remains perfectly predictable from $U$.

When two random variables, $X$ and $Y$, are linearly dependent:

$$Y = \alpha + \beta \cdot X , \tag{4.6}$$

the correlation coefficient $\rho(X, Y)$ equals $\pm 1$, depending on whether $\beta$ is positive or negative (in the previous example, this corresponds to $\theta = 1$ or $0$, respectively). Here, the converse holds. This derives from the representation:

$$\rho(X, Y)^2 = 1 - \min_{\alpha, \beta} \frac{\mathrm{E}\left[(Y - (\alpha + \beta \cdot X))^2\right]}{\mathrm{Var}\,[Y]} , \tag{4.7}$$

where $\mathrm{E}[\,]$ denotes the expectation with respect to the joint distribution of $X$ and $Y$. $\rho(X, Y)^2$ is called the coefficient of determination and gives the proportion of the variance of one variable $(Y)$ that is predictable from the other variable $(X)$.

By Cauchy-Schwartz inequality, (4.1) allows one to show that $\rho \in [-1, 1]$. But, given two random variables $X$ and $Y$ with fixed marginal distribution functions $F_X$ and $F_Y$, it is not always possible for the correlation coefficient to reach the bounds $\pm 1$. Indeed, Chap. 3 has shown that any bivariate distribution function $F$ is bracketed by the Fréchet-Hoeffding bounds:

$$\max\{F_X(x) + F_Y(y) - 1, 0\} \le F(x, y) \le \min\{F_X(x), F_Y(y)\} . \tag{4.8}$$

Therefore, applying Hoeffding identity [130]

$$\rho(X, Y) = \int \int [F(x, y) - F_X(x) \cdot F_Y(y)] \; dx \; dy , \tag{4.9}$$

one can now conclude that, given $F_X$ and $F_Y$, the correlation coefficient $\rho$ lies between $\rho_{\min}$ and $\rho_{\max}$, where $\rho_{\min}$ is attained when $X$ and $Y$ are countermonotonic random variables while $\rho_{\max}$ is attained when $X$ and $Y$ are comonotonic random variables.

As an illustration, let us consider the following example from Embrechts et al. [149]. Given two random variables with log-normal marginal distributions: $X \sim \log \mathcal{N}(0,1)$ and $Y \sim \log \mathcal{N}(0,\sigma)$, the upper and lower bounds for $\rho(X,Y)$ are given by

$$\rho_{\min} = \rho\left(e^Z, e^{-\sigma Z}\right) \quad \text{and} \quad \rho_{\max} = \rho\left(e^Z, e^{\sigma Z}\right) , \qquad (4.10)$$

where $Z$ is a standard Gaussian random variable. A straightforward calculation, based upon the fact that

$$\mathrm{E}\left[e^{\alpha \cdot Z}\right] = e^{\frac{\alpha^2}{2}} , \qquad (4.11)$$

gives

$$\rho_{\min} = \frac{e^{-\sigma} - 1}{\sqrt{(e-1)\left(e^{\sigma^2} - 1\right)}} \quad \text{and} \quad \rho_{\max} = \frac{e^{\sigma} - 1}{\sqrt{(e-1)\left(e^{\sigma^2} - 1\right)}} . \qquad (4.12)$$

Figure 4.2 represents these two bounds as a function of $\sigma$. As $\sigma$ becomes of the order of or larger than 3, $\rho_{\min}$ becomes extremely close to zero, so that in this case an (almost) vanishing correlation coefficient corresponds to a countermonotonic relation between the two random variables. For $\sigma$ larger than 4, both the lower and the upper bounds can hardly be distinguished from zero. Thus, a very small value of the correlation coefficient cannot (must not) be always considered as the signature of a weak dependence between two random variables.

The correlation coefficient is invariant under an increasing affine change of variable of the form

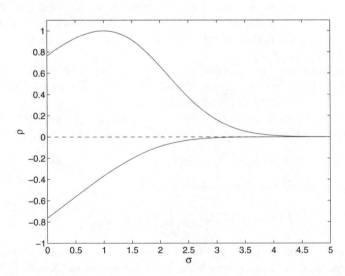

**Fig. 4.2.** Graph of $\rho_{\min}$ and $\rho_{\max}$ given by (4.12) versus $\sigma$ for two random variables with log-normal marginal distributions: $\log \mathcal{N}(0,1)$ and $\log \mathcal{N}(0,\sigma)$

$$X' = a \cdot X + b, \quad a > 0 , \tag{4.13}$$

$$Y' = c \cdot Y + d, \quad c > 0 , \tag{4.14}$$

since $\rho(X', Y') = \rho(X, Y)$. However, this property does not generalize to any (nonlinear) increasing transformation. As a consequence, the correlation coefficient does not give access to the dependence between two random variables in the sense of Chap. 3. This lack of invariance with respect to nonlinear changes of variables is due to the fact that the correlation coefficient aggregates information on both the marginal behavior of each random variable and on their true dependence structure given by the copula.

### 4.1.2 Local Correlation

Instead of focusing on the overall correlation, one can look at the *local* linear dependence between two random variables. This idea, introduced by Doksum *et al.* [58, 134], enables one to probe the changes of the correlation strength as a function of the value of the realizations of the random variables. It allows, for instance, to address the question of whether the correlation remains constant or vary when the realizations of the random variables are typical or not. This is particularly useful when dealing with contagions of crises (see Chap. 6) or when investigating whether flight-to-quality actually occurs between stock and bond markets, for instance.

The definition of the local correlation coefficient is quite natural. It starts from the remark that, in a linear framework, if the two random variables $X$ and $Y$ are related by

$$Y = \alpha + \beta X + \epsilon , \tag{4.15}$$

where $\epsilon$ is independent from (or at least uncorrelated with) $X$, the correlation coefficient reads

$$\rho = \frac{\beta \cdot \sigma_X}{\sqrt{\beta^2 \cdot \sigma_X^2 + \sigma_\epsilon^2}} , \tag{4.16}$$

where $\sigma_X^2$ and $\sigma_\epsilon^2$ denote respectively the variance of $X$ and of the error term $\epsilon$.

Let us now assume that the more general relation

$$Y = f(X) + \sigma(X) \cdot \epsilon \tag{4.17}$$

holds between $X$ and $Y$, with $\sigma_\epsilon = 1$ and $f$ differentiable. In the neighborhood of $X = x_0$, one can linearize the relation above as follows:

$$Y = [f(x_0) - x_0 \cdot f'(x_0)] + f'(x_0)X + \sigma(x_0) \epsilon \tag{4.18}$$

and, by analogy with (4.16), define the local linear correlation coefficient by

$$\rho(x_0) = \frac{f'(x_0) \cdot \sigma(x_0)}{\sqrt{f'(x_0)^2 \cdot \sigma(x_0)^2 + \sigma(x_0)^2}} \, . \tag{4.19}$$

It is straightforward to check that the local correlation coefficient reduces to the usual linear correlation coefficient when $f$ is an affine mapping and $\sigma(x)$ remains constant. In addition, the local correlation coefficient $\rho(x)$ fulfills the same main properties as the linear correlation coefficient $\rho$:

1. $\rho(x) \in [-1, 1]$,
2. $\rho(x)$ is invariant under (increasing) linear mappings in both $X$ and $Y$,
3. $\rho(x) = 0$ for all $x$ if $X$ and $Y$ are independent.

Beside and in constrast with the linear correlation coefficient, the local correlation coefficient equals $\pm 1$ only if $\sigma(x)$ is zero (the sign depends on that of the derivative of $f$), so that $Y = f(X)$. Thus, the local correlation coefficient avoids the drawback of the linear correlation coefficient that a vanishing value can be found even when $X$ and $Y$ are deterministically related to each other.

### 4.1.3 Generalized Correlations Between $N > 2$ Random Variables

The (overall) correlation coefficient $\rho$ is a linear measure of dependence between *two* random variables. We now present a natural generalization to $N$ random variables, whose exposition borrows from [324].

Let us denote by $\boldsymbol{X}(t)$ a random vector of $N$ components, for instance the vector of returns of $N$ assets in a portfolio. The mean values of the components of $\boldsymbol{X}(t)$ are first estimated and then subtracted to each vector $\boldsymbol{X}(t)$ for $t = 1, \ldots, L$, where $L$ denotes the sample size, equal for instance to the chosen length of the time interval used for the estimations. For ease of notation, we keep $\boldsymbol{X}(t)$ to represent the now centered vectors. The sample estimate of the covariance matrix of these $N$ random variables over some interval of length $L$ is

$$\boldsymbol{S}_{X(t)} = \frac{1}{L} \sum_{t=1}^{L} \boldsymbol{X}(t) \cdot \boldsymbol{X}(t)^T \, , \tag{4.20}$$

where $^T$ denotes the transpose.

Let us now divide the $N$ components of the vectors $\boldsymbol{X}(t)$ into two parts: a scalar $X_i(t)$ constituted of one of the components and an $(N-1)$-dimensional column vector $\boldsymbol{\xi}_i(t) = [X_1(t), \ldots, X_{i-1}(t), X_{i+1}(t), \ldots, X_N(t)]^T$ made of the other components. By multiplying (scalar product) each vector $\boldsymbol{\xi}_i$ by some still unknown vector $\boldsymbol{\phi}$, we obtain a set of scalar values $\zeta_i = \boldsymbol{\phi}^T \cdot \boldsymbol{\xi}_i$. Let us now search for the vector $\boldsymbol{\phi}$ which makes the square of the correlation coefficient between the two random variables $X_i$ and $\zeta_i$ maximum. This procedure constitutes an example of the implementation of the classical solution developed by Hotelling [235, 398] on canonical correlations: the vector $\boldsymbol{\phi}$ is defined as the eigenvector corresponding to the maximal eigenvalue (which is equal to the maximal correlation coefficient between the two random variables $X_i$ and $\zeta_i$) of the following matrix of size $(N-1) \times (N-1)$:

$$S_{\xi_i \xi_i}^{-1} \, S_{\xi_i X_i} \, S_{X_i X_i}^{-1} \, S_{X_i \xi_i} \qquad (4.21)$$

where

$$S_{X_i X_i} = \mathrm{Cov}(X_i, X_i)\,, \quad S_{X_i \xi_i} = S_{\xi_i X_i}^T = \mathrm{Cov}(X_i, \xi_i^T)\,, \quad S_{\xi_i \xi_i} = \mathrm{Cov}(\xi_i, \xi_i)\,. \qquad (4.22)$$

The matrices in formulas (4.21) and (4.22) are submatrices of the general $N \times N$ covariance matrix $S_{XX} = \mathrm{Cov}(X, X^T)$ (whose estimation is given in (4.20)). Thus, replacing the matrix $S_{XX}$ (and its submatrices) in (4.21) and (4.22) by its sample estimate (4.20) allows one to compute the vector $\phi$ and the set of scalar values $\zeta_i$ for $i = 1, \ldots, N$. One can call the maximum eigenvalue of the matrix (4.21) the "canonical coefficient of $N$-correlation" between the random variable $X_i$ and the other $N-1$ variables, which captures the common factors between $X_i$ and all the other $N-1$ variables. Performing similar operations with all other components of the vector $X$, one thus obtains a $N$-dimensional vector of canonical coefficients of $N$-correlation equal to the largest eigenvalues of the matrices (4.21) for $i = 1, \ldots, N$. For $N = 2$, the $(N-1)$-dimensional matrix (4.21) reduces to the square of the standard correlation coefficient between the $N = 2$ variables.

A slightly different but equivalent formulation is as follows. Consider the regression of a random variable $X_i$ on the $(N-1)$-dimensional random vector $\xi_i(t) = [X_1(t), \ldots, X_{i-1}(t), X_{i+1}(t), \ldots, X_N(t)]^T$, *i.e.*, the evaluation of a vector $\phi$ of regression coefficients in the linear formula:

$$X_i = \sum_{j \neq i} \phi_j X_j \; + \; \epsilon_i = \phi^T \cdot \xi_i + \epsilon_i \,, \qquad (4.23)$$

where $\epsilon_i$ is a regression residual. If the vector $\phi$ is defined by the least-squares method of minimizing

$$\sum_{t=1}^{L} \left( \phi^T \cdot \xi_i - X_i \right)^2 \qquad (4.24)$$

with respect to $\phi$, then its estimate is easily obtained as

$$\hat{\phi} = S_{\xi_i \xi_i}^{-1} \cdot S_{\xi_i X_i} \,. \qquad (4.25)$$

Let $\hat{\xi}_i = \hat{\phi}^T \cdot \xi_i$ denote the contribution to the regression (4.23) for this estimate (4.25). Since

$$\mathrm{Cov}(X_i, \hat{\xi}_i) = \mathrm{Cov}(X_i, S_{\xi_i \xi_i}^{-1} \cdot S_{\xi_i X_i} \cdot \xi_i) = S_{X_i \xi_i} \cdot S_{\xi_i \xi_i}^{-1} \cdot S_{\xi_i X_i} \,, \qquad (4.26)$$

it follows that the correlation coefficient between $\hat{\phi}$ given by (4.25) and $X_i$ is equal to the scalar $\boldsymbol{S}_{X_i\xi_i} \cdot \boldsymbol{S}_{\xi_i\xi_i}^{-1} \cdot \boldsymbol{S}_{\xi_i X_i} \cdot S_{X_i X_i}^{-1}$, which is nothing but the maximum eigenvalue of the matrix (4.21) [398]. This shows that the canonical coefficient of $N$-correlation:

$$\rho_N^i = \boldsymbol{S}_{X_i\xi_i} \cdot \boldsymbol{S}_{\xi_i\xi_i}^{-1} \cdot \boldsymbol{S}_{\xi_i X_i} \cdot S_{X_i X_i}^{-1} \tag{4.27}$$

can be determined from the solution of the regression problem (4.23, 4.24). This correspondence between the two formulations is rooted in the equivalence between linear correlation and the coefficient of linear regression, as pointed out above.

Again, this canonical coefficient of $N$-correlation is, by construction, invariant under linear transformations of each $X_i$ individually. However, it is not left unchanged under nonlinear monotonic transformations. It is therefore necessary to look for other measures of dependence which are only functions of the copula. The concordance measures described below enjoy this property.

## 4.2 Concordance Measures

### 4.2.1 Kendall's Tau

A fundamental question for financial risk management is the following:

"Do the prices of two (or more) assets tend to rise or fall together?"

If the answer is affirmative, the diversification of risks will probably be difficult, since diversification is based upon the fact that the fall of an asset is statistically balanced by the rise of another one. A natural way to quantify the propensity of assets to move together is to compare the probability that they rise (or fall) together with the probability that one of the two assets rises (respectively falls) while the other one falls (respectively rises). This can be translated mathematically as follows. Starting with two independent realizations $(X_1, Y_1)$ and $(X_2, Y_2)$ of the same pair of random variables $(X, Y)$, let us consider the quantity

$$\tau = \Pr\left[(X_1 - X_2) \cdot (Y_1 - Y_2) > 0\right] - \Pr\left[(X_1 - X_2) \cdot (Y_1 - Y_2) < 0\right] . \tag{4.28}$$

The left-most term in the r.h.s. (right-hand side) gives the probability of *concordance*, *i.e.*, the probability that $X$ and $Y$ move together upward or downward. In contrast, the right-most term in the r.h.s. represents the probability of discordance, *i.e.*, the probability that the two random variables move in opposite directions.

The expression (4.28) defines the population version of the so-called Kendall's $\tau$. This quantity is invariant under increasing transformation of the marginal distributions. Indeed, given any increasing mapping $G_X$ and $G_Y$, we have

$$X_1 \geq X_2 \quad \Longleftrightarrow \quad G_X(X_1) \geq G_X(X_2) \ , \tag{4.29}$$
$$Y_1 \geq Y_2 \quad \Longleftrightarrow \quad G_Y(Y_1) \geq G_Y(Y_2) \ . \tag{4.30}$$

As a consequence, Kendall's $\tau$ depends only on the copula of $(X, Y)$. For continuous random variables, expression (4.28) can be transformed into

$$\tau = 2\Pr\left[(X_1 - X_2) \cdot (Y_1 - Y_2) > 0\right] - 1 \ , \tag{4.31}$$

which yields the following expression in terms of a functional of the copula $C$ of the two random variables:

$$\tau(C) = 4 \int\int C(u,v) \, dC(u,v) - 1 \ . \tag{4.32}$$

From this equation, one easily checks that Kendall's $\tau$ varies between $-1$ and $+1$. The lower bound is reached if and only if the variables $(X, Y)$ are countermonotonic, while the upper bound is attained if and only if $(X, Y)$ are comonotonic. In addition, $\tau$ equals zero for independent random variables. However, as for the (linear) correlation coefficient, $\tau$ may vanish even for non-independent random variables.

In spite of its attractive structure, (4.32) is not always very useful for calculations and one often has to resort to numerical integration (by use of quadrature, for instance). However, some more tractable expressions have been found for particular families of copulas.

## Archimedean Copulas

Genest and McKay [198] have shown that, for generators $\varphi$ which are strictly decreasing functions from $[0, 1]$ onto $[0, \infty]$ with $\varphi(1) = 0$, Kendall's $\tau$ of the Archimedean copula

$$C(u,v) = \varphi^{-1}\left(\varphi(u) + \varphi(v)\right) \tag{4.33}$$

is given by

$$\tau = 1 + 4 \int_0^1 \frac{\varphi(t)}{\varphi'(t)} dt \ . \tag{4.34}$$

This expression relies on the general fact that (4.32) can be rewritten as

$$\tau = 4 \cdot \mathrm{E}\left[C(U,V)\right] - 1 \ , \tag{4.35}$$

where $U$ and $V$ are uniform random variables with joint distribution function $C$. Now, in the particular case of an Archimedean copula, one can show that [370]

$$\Pr\left[C(U,V) \leq t\right] = t - \frac{\varphi(t)}{\varphi'(t^+)} , \tag{4.36}$$

which immediately yields the results given by (4.34). Table 4.1 provides closed form expressions for Kendall's $\tau$'s of Clayton's copula, Gumbel's copula and Frank's copula, which are shown in Fig. 4.3 as a function of their corresponding form parameters $\theta$.

**Table 4.1.** Expression of Kendall's $\tau$ for three Archimedean copulas. $D_1$ denotes the Debye function $D_1(x) = \frac{1}{x} \int_0^x dt \, \frac{t}{e^t - 1}$

| Copula | $\varphi(t)$ | Kendall's $\tau$ | Range |
|--------|-----------|---------------|-------|
| Clayton | $\frac{1}{\theta}\left(t^{-\theta} - 1\right)$ | $\dfrac{\theta}{\theta + 2}$ | $\theta \in [-1, \infty]$ |
| Gumbel | $(-\ln t)^\theta$ | $\dfrac{\theta - 1}{\theta}$ | $\theta \in [1, \infty]$ |
| Frank | $-\ln \dfrac{e^{-\theta t} - 1}{e^{-\theta} - 1}$ | $1 - \dfrac{4}{\theta}\left[1 - D_1(\theta)\right]$ | $\theta \in [-\infty, \infty]$ |

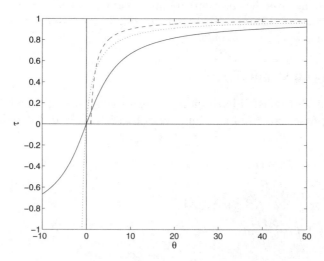

**Fig. 4.3.** Graph of Kendall's $\tau$'s as a function of the form parameter $\theta$ defined in Table 4.1, for Clayton's copula (*dotted line*), Gumbel's copula (*dashed line*) and Frank's copula (*plain line*). Kendall's $\tau$ for Frank's copula is symmetric with respect to the origin

## Elliptical Copulas

This particularly useful family of copulas also allows for tractable calculation of Kendall's $\tau$. Generalizing the result originally obtained by Stieltjes [115] for the Gaussian distribution, Lindskog *et al.* [307] have shown that the relation

$$\tau = \frac{2}{\pi} \arcsin \rho \tag{4.37}$$

holds for any pair of random variables whose dependence structure is given by an elliptical copula. The parameter $\rho$ denotes the shape coefficient (or correlation coefficient, when it exists) of the elliptical distribution naturally associated with the considered elliptical copula.

This result is particularly interesting because it provides a robust estimation method for the shape parameter $\rho$. Of course, when the elliptical distribution associated with the elliptical copula admits a second moment, the correlation coefficient exists and $\rho$ can be estimated from Pearson's coefficient (4.2). However, when the elliptical distribution does not admit a second moment, this approach fails. In this case, Kendall's $\tau$ has the advantage of always existing and of being easily estimated. In fact, its superiority is even greater, as demonstrated by Fig. 4.4 which shows that estimates of $\tau$ yield more robust estimates of $\rho$ via (4.37). This is especially true when the tails of the marginals associated with the elliptical distributions are heavy. In the example depicted in Fig. 4.4, we have considered two Student's distributions with three and ten degrees of freedom respectively. While the estimates of $\rho$ provided by Kendall's $\tau$ (dashed curve) remain approximately equally efficient in both cases, the efficiency of the estimates of $\rho$ provided by Pearson's

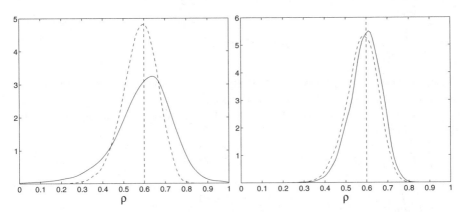

**Fig. 4.4.** Probability density function of the correlation coefficient $\rho$ estimated from synthetic realizations generated with a student distribution with three degrees of freedom (*left panel*) and ten degrees of freedom (*right panel*) both with a true value of $\rho = 0.6$ for a sample size equal to 100. The continuous curve represents the pdf obtained from Pearson's estimator while the *dashed curve* gives the pdf obtained when estimating Kendall's $\tau$ and applying (4.37)

coefficient (continuous curve) drops dramatically as the number of degrees of freedom of the Student's distributions decreases. This phenomenon can be ascribed to the fact that, with three degrees of freedom, the correlation coefficient still exists but the asymptotic distribution of Pearson's coefficient is not Gaussian but has a heavy tail, as recalled in Sect. 4.1.

### 4.2.2 Measures of Similarity Between Two Copulas

Consider two copulas $C_1$, $C_2$ and the copula $C = w \cdot C_1 + (1 - w) \cdot C_2$, with $w \in [0, 1]$. Chapter 3 has recalled that the convex sum of several copulas remains a copula. Kendall's $\tau$ of copula $C$ can be written as

$$\tau_C = w^2 \cdot \tau_{C_1} + 2w(1 - w) \cdot Q(C_1, C_2) + (1 - w)^2 \cdot \tau_{C_2} , \tag{4.38}$$

with

$$Q(C_1, C_2) = 4 \int\int_{[0,1]^2} C_1(u, v) \, dC_2(u, v) - 1 , \tag{4.39}$$

$$= 4 \int\int_{[0,1]^2} C_2(u, v) \, dC_1(u, v) - 1 . \tag{4.40}$$

To provide an intuitive interpretation of $Q(C_1, C_2)$, let us consider two copulas $C_1$ and $C_2$ with identical Kendall's $\tau$: $\tau_{C_1} = \tau_{C_2} = \tau$. This means that, through the prism of Kendall's $\tau$, these two copulas $C_1$ and $C_2$ have the same dependence. Kendall's $\tau$ of the copula $C$ formed by their convex sum will also be equal to $\tau$, for all values of $w$, if and only if $Q(C_1, C_2) = \tau$, that is, if expression (4.40) is equal to expression (4.32) obtained for either $C_1$ or $C_2$. The difference

$$4 \int\int_{[0,1]^2} [C_1(u, v) - C_2(u, v)] \, dC_2(u, v) \tag{4.41}$$

between these two expressions (4.40) and (4.32) therefore allows one to define the notion of proximity between two copulas.

Since any copula is bounded by the Fréchet-Hoeffding upper and lower bounds, we have

$$\int\int_{[0,1]^2} [\max(u + v - 1, 0) - C_2(u, v)] \, dC_2(u, v)$$

$$\leq \int\int_{[0,1]^2} [C_1(u, v) - C_2(u, v)] \, dC_2(u, v) \tag{4.42}$$

$$\leq \int\int_{[0,1]^2} [\min(u, v) - C_2(u, v)] \, dC_2(u, v) ,$$

which can be rewritten as

$$\int_0^1 C_2(u, 1-u)\ du \ - \ \frac{\tau+1}{4}$$

$$\leq \int\int_{[0,1]^2} [C_1(u,v) - C_2(u,v)]\ dC_2(u,v)$$

$$\leq \int_0^1 C_2(u,u)\ du \ - \ \frac{\tau+1}{4}\ . \tag{4.43}$$

The left-most term is always negative while the right-most one is positive. In the case where the left-most term is the opposite of the right-most one,

$$\int_0^1 C_2(u,u)du - \frac{\tau+1}{4} = -\left[\int_0^1 C_2(u,1-u)du - \frac{\tau+1}{4}\right] > 0\ , \tag{4.44}$$

one can renormalize expression (4.43) to obtain

$$-1 \leq \frac{\int\int_{[0,1]^2} [C_1(u,v) - C_2(u,v)]\ dC_2(u,v)}{\int_0^1 C_2(u,u)du - \frac{\tau+1}{4}} \leq 1\ . \tag{4.45}$$

Choosing a fixed copula $C_2$ as a reference, this provides a new dependence measure, allowing to assess the similarity between any copula $C_1$ and the reference copula $C_2$. Two particular choices of $C_2$ have been studied in the literature:

## Spearman's Rho

Let us choose $C_2$ as the product copula $\Pi(u,v) = u \cdot v$, describing independence. One easily checks that

$$\int_0^1 u^2\ du - \frac{\tau+1}{4} = -\left[\int_0^1 u \cdot (1-u)\ du - \frac{\tau+1}{4}\right] = \frac{1}{12}\ , \tag{4.46}$$

while

$$\int_0^1 \int_0^1 u \cdot v\ dudv = \frac{1}{4}\ , \tag{4.47}$$

so that the central fraction in (4.45) leads to define the so-called *Spearman's rho*:

$$\rho_s(C) = 12 \int\int_{[0,1]^2} C(u,v)\ dudv - 3\ . \tag{4.48}$$

This equation can be interpreted as the difference between the probability of concordance and the probability of discordance for the two pairs of random variables $(X_1, Y_1)$ and $(X_2, Y_3)$, where the pairs $(X_1, Y_1)$, $(X_2, Y_2)$ and $(X_3, Y_3)$ are three independent realizations drawn from the same distribution:

$$\rho_s = 3\left(\Pr[(X_1 - X_2)(Y_1 - Y_3) > 0] - \Pr[(X_1 - X_2)(Y_1 - Y_3) < 0]\right) . \quad (4.49)$$

By definition, Spearman's rho equals zero for independent random variables while the lower (resp. upper) bound is reached if and only if the random variables are countermonotonic (resp. comonotonic). An alternative expression is

$$\rho_s(C) = 12 \int\int_{[0,1]^2} u \cdot v \, dC(u, v) - 3 . \quad (4.50)$$

It enlightens the fact that Spearman's rho is related to the linear correlation of the rank. Indeed, considering two random variables $X$ and $Y$, with marginal distributions $F_X$ and $F_Y$, it is straightforward to check that

$$\rho_s = \frac{\mathrm{Cov}\left(F_X(X), F_Y(Y)\right)}{\sqrt{\mathrm{Var}F_X(X) \cdot \mathrm{Var}F_Y(Y)}} . \quad (4.51)$$

Our introduction of Spearman's rho, motivated from Kendall's $\tau$, shows that they are closely related. In fact, given any copula $C$, Kruskal [281] has shown that

$$\frac{3\tau - 1}{2} \leq \rho_S \leq -\frac{\tau^2 - 2\tau - 1}{2}, \quad \tau \geq 0 , \quad (4.52)$$

$$\frac{\tau^2 + 2\tau - 1}{2} \leq \rho_S \leq \frac{3\tau + 1}{2}, \quad \tau \leq 0 . \quad (4.53)$$

Figure 4.5 shows that the area of accessible values for the couple $(\tau, \rho_S)$ represents a relatively narrow strip, reflecting the strong relation between Kendall's $\tau$ and Spearman's rho.

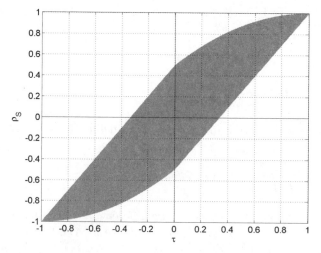

**Fig. 4.5.** The *shaded area* represents the allowed values for the couple $(\tau, \rho_S)$

## Gini's Gamma

Instead of choosing the reference copula $C_2(u, v) = \Pi(u, v) = u \cdot v$, one can consider an equally weighted mixture of the two copulas expressing monotonic dependence:

$$C_2(u, v) = \frac{1}{2}\min(u, v) + \frac{1}{2}\max(u + v - 1, 0) \ . \tag{4.54}$$

The central fraction in (4.45) then measures how far is a given copula $C_1$ from the monotonous dependence. Simple algebraic manipulations show that

$$\int_0^1 C_2(u, u) \, du - \frac{\tau + 1}{4} = - \left[ \int_0^1 C_2(u, 1 - u) \, du - \frac{\tau + 1}{4} \right] = \frac{1}{8} \ , \tag{4.55}$$

and

$$\int_0^1 \int_0^1 C_2(u, v) \, dC_2(u, v) = \frac{1}{4} \ . \tag{4.56}$$

The central fraction in (4.45) then yields the so-called *Gini's gamma*:

$$\gamma(C) = 4 \left[ \int_0^1 C(u, u) \, du + \int_0^1 C(u, 1 - u) \, du - \frac{1}{2} \right] \ . \tag{4.57}$$

Note that this measure of dependence only relies on the values taken by $C$ on its main diagonals. The alternative expression

$$\gamma(C) = 4 \left[ \int_0^1 C(u, u) \, du - \int_0^1 [u - C(u, 1 - u)] \, du \right] \tag{4.58}$$

shows that Gini's gamma represents the difference of the area between the values of $C(u, v)$ and $\max(u + v - 1, 0)$ on the first diagonal and between the value of $C(u, v)$ and $\min(u, v)$ on the second diagonal (see the shaded areas in Fig. 4.6).

### 4.2.3 Common Properties of Kendall's Tau, Spearman's Rho and Gini's Gamma

The three measures of dependence – Kendall's tau, Spearman's rho and Gini's gamma – presented in the previous paragraphs enjoy the same set of properties:

1. they are defined for any pair of continuous random variables $X$ and $Y$,
2. they are symmetric: for any pair $X$ and $Y$, $\tau(X, Y) = \tau(Y, X)$, for instance,
3. they range from $-1$ to $+1$, and reach these bounds when $X$ and $Y$ are countermonotonic and comonotonic respectively,

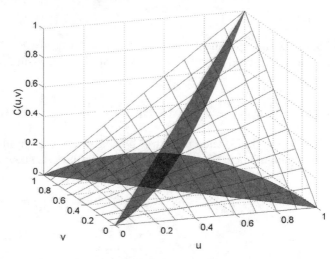

**Fig. 4.6.** The *shaded surface* represents the area between the values of $C(u, v)$ – here the product copula $\Pi(u, v) = u \cdot v$ – and $\max(u + v - 1, 0)$ on the first diagonal and between the value of $C(u, v)$ and $\min(u, v)$ on the second diagonal

4. they equal zero for independent random variables,
5. if the pair of random variables $(X_1, X_2)$ is more dependent than the pair $(Y_1, Y_2)$ in the following sense:

$$C_X(u, v) \geq C_Y(u, v), \quad \forall u, v \in [0, 1] , \tag{4.59}$$

then the same ranking holds for any of these three measures; for instance, $\tau(X_1, X_1) \geq \tau(Y_1, Y_2)$.

Any measure of dependence fulfilling these five properties is named a *concordance measure*. The central fraction in (4.45), with any exchangeable copula $C_2$ such that condition (4.44) is fulfilled together with $\rho_s(C_2) = 3\,\tau(C_2)$, ensuring that the numerator of the central term of (4.45) vanishes for $C_1(u, v) = u.v$, provides a measure of dependence which satisfies the five conditions above, and is thus a concordance measure.

## 4.3 Dependence Metric

Concordance measures fulfill most of the requirements expected from a measure of dependence. Following Granger *et al.* [214], one can impose slightly more demanding properties for a functional measure $\mathcal{F}[X, Y]$ of dependence between two random variables $X$ and $Y$, which strengthen properties 1–4 of concordance measures as follows:

1. $\mathcal{F}$ is well defined for both continuous and discrete random variables,

2. $\mathcal{F}$ is invariant under continuous and strictly increasing transformations of the random variables, $i.e.$, $\mathcal{F}$ depends only on the copula of $X$ and $Y$,
3. $\mathcal{F}$ is a distance,
4. $\mathcal{F}$ equals 0 if $X$ and $Y$ are independent, and varies between 0 and 1,
5. $\mathcal{F}$ equals 1 (or, at least, reaches a maximum) if there exist a measurable mapping between the random variables $X$ and $Y$: $X = f(Y)$,
6. $\mathcal{F}$ has a simple relationship with the (linear) correlation coefficient in the case of a bivariate normal distribution.

Dependence measures satisfying all these requirements are named *Dependence metrics*.

As an example, one can consider the measure introduced by Bhattacharya, Matusita and Hellinger:

$$S = \frac{1}{2} \int_{\mathbb{R}^2} \left[ 1 - \left( \frac{f(x)g(y)}{h(x,y)} \right)^{\frac{1}{2}} \right]^2 dH(x,y) , \tag{4.60}$$

where $f$ and $g$ denote the marginal densities of $X$ and $Y$ respectively, while $h$ and $H$ are the bivariate density and distribution functions of $(X,Y)$. Simple algebraic manipulations give

$$S = \int_{[0,1]^2} \left[ 1 - [c(u,v)]^{1/2} \right] dudv , \tag{4.61}$$

where $c$ is the density of the copula of $X$ and $Y$, showing that $S$ agrees with the second requirement. Properties 3–5 are easy to check while the last requirement has been established in [442]. Indeed, for two random variables with Gaussian copula and shape coefficient $\rho$, one has

$$S = 1 - \frac{\left( 1 - \rho^2 \right)^{5/4}}{\left( 1 - \frac{\rho^2}{2} \right)^{3/2}} . \tag{4.62}$$

This dependence metric is in fact related to a generalized relative entropy between the joint density $h$ and the product density $f \cdot g$. Consider the generalized Kullback-Leibler distance (obtained by symmetrization of the Kullback-Leibler divergence) for the $k$-class entropy family [225] defined by

$$H_k(f) = \frac{1}{k-1} \left( 1 - \mathrm{E} \left[ f^{k-1} \right] \right), \quad k \neq 1 , \tag{4.63}$$

$$= -\mathrm{E} \left[ \ln f \right], \quad k = 1 , \tag{4.64}$$

where $f$ is the density of the random variable (or vector) under consideration.[1] In the particular case $k = 1$, one retrieves the usual Shannon entropy. One

---

[1] This $k$-class entropy is also known as Tsallis entropy of order $k$ in the physical literature [476] and has many applications to characterize complex systems with nonseparable long-range space/time dependences.

can then show that $S$ is equal to one-fourth of the *symmetric* relative entropy of $h$ and $f \cdot g$ for the 1/2-class entropy.

Dependence metrics such as $S$ provide very useful tools to test the presence of complicated serial dependences. This is particularly important not only to analyze and forecast financial time series [214] but also to test the goodness-of-fit in copula modeling, as we shall see in Chap. 5.

## 4.4 Quadrant and Orthant Dependence

In practice, it is often useful to characterize the dependence of more than two variables. For instance, risk management deals with portfolios made of dozens up to tens of thousands of assets. The analysis of the risks associated with such portfolios requires the assessment of the dependence between many ($N$) variables. It is in general not true that the genuine multivariate dependence between the $N$ variables can be adequately quantified by $N(N-1)/2$ dependence measures between all possible pairs. A first approach to define generalized correlations between $N > 2$ random variables has been already described in Sect. 4.1.3. We now discuss other measures which can be shown to be pure copula properties.

First, note that the previous concept of *concordance* cannot be easily extended to more than two random variables. The intuition behind this statement can be obtained by taking the example of Kendall's $\tau$. One could think of generalizing the integral expression (4.32) of Kendall's $\tau$ to higher dimensions. However, this straightforward generalization loses several nice properties of the concordance measures. In particular, the concept of countermonotonicity cannot be used for more than two random variables. Indeed, consider three random variables $X, Y$ and $Z$, such that $(X, Y)$ and $(Y, Z)$ are countermonotonic; then $(X, Z)$ are necessarily comonotonic.[2] As a consequence, even if the extension to higher dimensions of Kendall's $\tau$ remains bounded by $-1$ (by Fréchet-Hoeffding inequality), it is not ascertained that this bound can still be reached. Therefore, the interpretation of a negative value for such a generalized Kendall's $\tau$ would not be obvious.

In order to provide measures of dependences which do not suffer from this problem, let us first introduce the notion of positive quadrant dependence [300]. Two random variables $X$ and $Y$ are *positive quadrant dependent* (PQD) if

$$\Pr[X \leq x, Y \leq y] \geq \Pr[X \leq x] \cdot \Pr[Y \leq y] , \quad \forall x, y . \tag{4.65}$$

This inequality means that the probability that the two random variables $X$ and $Y$ are simultaneously small is at least as large as it would be if these two

---

[2] This effect is related to the concept of "frustration" introduced in statistical physics to describe situations in which constraints tending to create opposite states in two interacting variables cannot be all obeyed in systems of three or more elements [475, 481]. Frustration leads in general to multiple equilibria [360].

random variables were independent. If $X$ and $Y$ represent the returns of two PQD assets, the probability that they undergo simultaneous large losses is not less than it would be if they were independent. As a consequence, one expects (and it can be shown [130]) that risk-averse investors prefer a portfolio $\tilde{X} + \tilde{Y}$ made of independent replications of assets $X$ and $Y$ to a portfolio $X + Y$ made of the actual PDQ assets. This means that, for any increasing concave utility function $U$,

$$\mathrm{E}\left[U(X+Y)\right] \leq \mathrm{E}\left[U\left(\tilde{X}+\tilde{Y}\right)\right] . \tag{4.66}$$

Inequality (4.65) can be rewritten as

$$\Pr[X > x, Y > y] \geq \Pr[X > x] \cdot \Pr[Y > y] , \quad \forall x, y . \tag{4.67}$$

This defines two random variables as PQD if the probability that they are simultaneously *large or small* is at least as large as it would be if these two random variables were independent. This definition is relevant for risk management purpose, since it amounts to ask whether large losses of individual assets tend to occur more frequently together than they would if the assets were independent.

Definition (4.65) implies that $X$ and $Y$ are PQD if and only if their copula $C$ satisfies

$$C(u, v) \geq \Pi(u, v) = u \cdot v , \quad \forall u, v \in [0, 1] . \tag{4.68}$$

This ensures that the PQD property depends only on the dependence structure of the random variables (and not on their marginals).

The PQD property and the concordance measures are intimately related. Indeed, as recalled in Sect. 4.2.3, if the pair of random variables $(X_1, X_2)$ is more dependent than the pair $(Y_1, Y_2)$, it is also more concordant. So, any PQD pair of random variables is more concordant than independent pairs of random variables. But, since any concordance measure equals zero for independent random variables, we can assert that, given any concordance measure, any pair of PQD random variables has a *positive* concordance measure. In particular, Kendall's tau, Spearman's rho or Gini's gamma are necessarily positive for PQD random variables. Besides, (4.48) shows that the Spearman's rho is a kind of averaged positive quadrant dependence.

To conclude this brief survey of the properties of PQD random variables, let us stress that the same result holds for the usual linear correlation coefficient. Indeed, by Hoeffding identity (4.9), any PQD random variables exhibit a nonnegative correlation coefficient. Unfortunately, the converse does not hold. However, given two random variables $X$ and $Y$ such that the linear correlation coefficient $\rho(f(X), g(Y))$ exists and is non-negative for any nondecreasing functions $f$ and $g$, then these two random variables are PQD [300].

Let us now generalize the bivariate concept of positive quadrant dependence to the multivariate concept of *positive orthant dependence*. We will say that $N$ random variables $X_1, X_2, \ldots, X_N$ are *Positive Lower Orthant Dependent* (PLOD) if

$$\Pr[X_1 \leq x_1, \ldots, X_N \leq x_N] \geq \Pr[X_1 \leq x_1] \cdots \Pr[X_N \leq x_N] , \qquad (4.69)$$

for all $x_i$'s. As in the bivariate case, this equation simply means that the probability that the $N$ random variables $X_1, \ldots, X_N$ are simultaneously small is at least as large as it would be if these $N$ random variables were independent.

Similarly, $N$ random variables $X_1, X_2, \ldots, X_N$ are *Positive Upper Orthant Dependent* (PUOD) if

$$\Pr[X_1 > x_1, \ldots, X_N > x_N] \geq \Pr[X_1 > x_1] \cdots \Pr[X_N > x_N] , \qquad (4.70)$$

for all $x_i$'s. Again, this equation has a simple interpretation: the probability that the $N$ random variables $X_1, \ldots, X_N$ are simultaneously large is at least as large as it would be if these $N$ random variables were independent. Note that the two definitions (4.69) and (4.70) are not equivalent anymore for $N > 2$.

Finally, $N$ random variables $X_1, X_2, \ldots, X_N$ are *Positive Orthant Dependent* (POD) if they are both PUOD and PLOD: the probability that the $N$ random variables $X_1, \ldots, X_N$ are simultaneously small or large is at least as large as it would be if these $N$ random variables were independent.

In terms of copulas, these definitions can be expressed as follows. Given a $N$-random vector $\mathbf{X} = (X_1, \ldots, X_N)$ with copula $C$,

$$\mathbf{X} \text{ is PLOD} \iff C(u_1, \ldots, u_N) \geq \prod_{i=1}^{N} u_i, \quad \forall u_i \in [0, 1] , \qquad (4.71)$$

and

$$\mathbf{X} \text{ is PUOD} \iff \bar{C}(u_1, \ldots, u_N) \geq \prod_{i=1}^{N} (1 - u_i) , \qquad (4.72)$$

where $\bar{C}$ denotes the survival copula of $C$.

For Archimedean copulas, the PLOD property can easily be related to the shape of its generator $\varphi$. In fact, in order for an Archimedean copula $C_\varphi$ to be PLOD, it is sufficient that the mapping

$$x \in \mathbb{R}_+ \longmapsto \varphi\left(e^{-x}\right) \qquad (4.73)$$

be concave, or at least sub-additive (the former implying the later). Indeed, for any $u_i \in [0, 1]$, the assumption that $\varphi\left(e^{-x}\right)$ is sub-additive allows us to write that

$$\varphi\left(\exp\left[-(-\ln u_1) - \cdots - (-\ln u_N)\right]\right) \leq \varphi\left(e^{\ln u_1}\right) + \cdots + \varphi\left(e^{\ln u_N}\right),$$
$$\leq \varphi(u_1) + \cdots + \varphi(u_N) , \qquad (4.74)$$

so that

$$\exp\left[-(-\ln u_1) - \cdots - (-\ln u_N)\right] \leq \varphi^{[-1]}\left(\varphi\left(u_1\right) + \cdots + \varphi\left(u_N\right)\right) , \qquad (4.75)$$

which is equivalent to

$$C_\varphi\left(u_1, \ldots, u_N\right) \geq \prod_{i=1}^{N} u_i . \qquad (4.76)$$

The proof that the subadditivity of $\varphi\left(e^{-x}\right)$ is in fact a *necessary and sufficient* condition for an Archimedean copula to be PLOD can be found in [147].

Let us remark that any completely monotonic generator fulfills the requirement that (4.73) be concave. Therefore, any Archimedean copula which admits a generalization to arbitrary dimension is PLOD. Archimedean copulas which exist in any dimension necessarily exhibit positive associations and their bivariate marginals cannot have negative concordance measures. In this respect, the bivariate Clayton or Frank copulas admit an $n$-dimensional generalization for positive parameter value $\theta$ only.

The property of POD is a reasonable assumption for most asset returns. This allows us to sharpen the (universal) bound for the VaR of the portfolios considered in Fig. 3.11. Instead of considering the Fréchet-Hoeffding lower bound in (3.76), one can choose $C_{inf} = C_{sup} = \Pi$, where $\Pi(u,v) = u \cdot v$ is the product copula.

The concept of POD is also appealing for testing whether some trading strategies are actually market neutral. Such strategies are very common in the alternative investment industry. They aim at decoupling portfolio moves from market moves, in order to ensure a better stability of the performance of portfolios. Portfolio managers often focus solely on their fund's *beta*, trying to keep it as small as possible while raising their *alpha* (the market-independent part of the expected return). However, if this approach allows them in principle to remove any linear dependence between the portfolio and the market, it totally neglects nonlinear and extreme dependences. Therefore, testing for POD seems necessary in order to check whether a fund is actually market neutral. Denuit and Scaillet [127] have proposed a nonparametric test for POD and, considering the HRF and CSFB/Tremont market neutral hedge fund indices, they have shown that both of them exhibit weak linear dependence with the S&P 500 index – as expected – but that POD cannot be rejected between the CSFB/Tremont market neutral index and the Standard & Poor's 500. Therefore, some funds contributing in the composition of the CSFB/Tremont index may exhibit nonlinear or extreme dependence with the Standard & Poor's 500. This teaches us that focusing on *beta* is clearly not sufficient to ensure market neutrality. We will come back to this problem at the end of this chapter when constructing portfolios which minimize the impact of extreme market moves.

## 4.5 Tail Dependence

### 4.5.1 Definition

Positive quadrant (and more generally orthant) dependence is a very strong property. It requires that the relation (4.68) holds for every point on the unit square for two variables (in the hypercube for more than two variables). It could be interesting to weaken this definition to focus on properties of *local* positive quadrant dependences. For instance, one could wish to focus on the lower left corner only, in order to assess whether joint losses occurring with (marginal) probability level less than $u$, say, appear more likely together than one could expect from statistically independent losses. Recall that the smaller the value of $u$, the more extreme are the losses. In this vein, the notion of tail dependence, aiming at quantifying the propensity of two random variables to exhibit concomitant extreme movements, has been introduced as a particularly interesting measure of extreme risks.

The concept of tail dependence is appealing in its simplicity. By definition, the (upper) tail dependence coefficient is

$$\lambda_U = \lim_{u \to 1} \Pr[X > F_X^{-1}(u) | Y > F_Y^{-1}(u)] , \qquad (4.77)$$

and quantifies the probability to observe a large $X$, assuming that $Y$ is large itself. In other words, given that $Y$ is very large (at some level of probability $u$), the probability that $X$ is very large at the same probability level $u$ defines asymptotically the tail dependence coefficient $\lambda$. As an example, if $X$ and $Y$ represent the volatilities of two different national markets, their coefficient of tail dependence $\lambda$ gives the probability that both markets exhibit together very high volatilities.

One can also interpret this expression (4.77) in terms of a Value-at-Risk. Indeed, the quantiles $F_X^{-1}(u)$ and $F_Y^{-1}(u)$ are nothing but the Values-at-Risk of assets (or portfolios) $X$ and $Y$ at the confidence level $u$, if we count losses as positive. Thus, the coefficient $\lambda_U$ simply provides the probability that $X$ exceeds the VaR at level $u$, assuming that $Y$ has exceeded the VaR at the same probability level $u$, when this level goes to one. As a consequence, the probability that both $X$ and $Y$ exceed their VaR at the level $u$ is asymptotically given by $\lambda_U \cdot (1 - u)$ as $u \to 1$. As an example, consider a daily VaR calculated at the 99% confidence level. Then, the probability that both $X$ and $Y$ undergo a loss larger than their VaR at the 99% level is approximately given by $\lambda_U/100$. Thus, when $\lambda_U$ is about 0.1, the typical recurrence time between such concomitant large losses is about 4 years, while for $\lambda_U \approx 0.5$ it is less than 10 months.

### 4.5.2 Meaning and Refinement of Asymptotic Independence

One of the appeals of this definition (4.77) of tail dependence is that it is a pure copula property, *i.e.*, it is independent of the margins of $X$ and $Y$.

Indeed, let $C$ be the copula of the variables $X$ and $Y$. If their bivariate copula $C$ is such that

$$\lim_{u \to 1} \frac{1 - 2u + C(u,u)}{1 - u} = \lim_{u \to 1} 2 - \frac{\log C(u,u)}{\log u} = \lambda_U \tag{4.78}$$

exists, then $C$ has an upper tail dependence coefficient $\lambda_U$ (see [106, 149, 147]).

In a similar way, one can define the coefficient of lower tail dependence:

$$\lambda_L = \lim_{u \to 0^+} \Pr\{X < F_X^{-1}(u) \mid Y < F_Y^{-1}(u)\} = \lim_{u \to 0^+} \frac{C(u,u)}{u} . \tag{4.79}$$

If $\lambda > 0$,[3] the copula presents tail dependence and large events tend to occur simultaneously, with (conditional) probability $\lambda$. On the contrary, when $\lambda = 0$, the copula has no tail dependence and the variables $X$ and $Y$ are said to be *asymptotically independent*. There is however a subtlety in this definition (4.77) of tail dependence. To make it clear, first consider the case where, for large $X$ and $Y$, the cumulative distribution function $F(x,y)$ factorizes such that

$$\lim_{x,y \to \infty} \frac{F(x,y)}{F_X(x)F_Y(y)} = 1 , \tag{4.80}$$

where $F_X(x)$ and $F_Y(y)$ are the margins of $X$ and $Y$ respectively. This means that, for $X$ and $Y$ sufficiently large, these two variables can be considered as independent. It is then easy to show that

$$\lim_{u \to 1} \Pr\{X > F_X^{-1}(u) | Y > F_Y^{-1}(u)\} = \lim_{u \to 1} 1 - F_X(F_X^{-1}(u)) \tag{4.81}$$

$$= \lim_{u \to 1} 1 - u = 0 , \tag{4.82}$$

so that independent variables really have no tail dependence $\lambda = 0$, as one can expect.

However, the result $\lambda = 0$ does not imply that the multivariate distribution can be automatically factorized asymptotically, as shown by the Gaussian example. Indeed, the Gaussian bivariate distribution cannot be factorized, even asymptotically for extreme values, since the non-diagonal term of the quadratic form in the exponential function does not become negligible in general as $X$ and $Y$ go to infinity together. Therefore, in a weaker sense, there may still be a dependence in the tail even when $\lambda = 0$.

To make this statement more precise, following [106], let us introduce the coefficient

$$\bar{\lambda}_U = \lim_{u \to 1} \frac{2 \log \Pr\{X > F_X^{-1}(u)\}}{\log \Pr\{X > F_X^{-1}(u), Y > F_Y^{-1}(u)\}} - 1 \tag{4.83}$$

$$= \lim_{u \to 1} \frac{2 \log(1 - u)}{\log[1 - 2u + C(u,u)]} - 1 . \tag{4.84}$$

---

[3] In the sequel, $\lambda$ without subscript will represent either $\lambda_U$ or $\lambda_L$.

It can be shown that the coefficient $\bar{\lambda}_U = 1$ if and only if the coefficient of tail dependence $\lambda_U > 0$, while $\bar{\lambda}_U$ takes values in $[-1, 1)$ when $\lambda_U = 0$, allowing us to refine the nature of the dependence in the tail in the case when the tail dependence coefficient is not sufficiently informative. It has been established that, when $\bar{\lambda} > 0$, the variables $X$ and $Y$ are simultaneously large more frequently than independent variables, while simultaneous large deviations of $X$ and $Y$ occur less frequently than under independence when $\bar{\lambda} < 0$. In the first case, the variables $X$ and $Y$ can be said to be locally PQD (positive quadrant dependent) in the neighborhood of the point $(0, 0)$ and/or $(1, 1)$ in probability space.

To summarize, independence (factorization of the bivariate distribution) implies no tail dependence ($\lambda = 0$). But $\lambda = 0$ is not sufficient to imply factorization and thus true independence. It also requires as a necessary condition that $\bar{\lambda} = 0$.

### 4.5.3 Tail Dependence for Several Usual Models

We present several general results allowing for the calculation of the tail dependence of Archimedean copulas, elliptical copulas and copulas derived from factor models.

**Archimedean Copulas**

The generator of an Archimedean copula fully embodies the properties of dependence (and therefore of extreme dependence). As a consequence, the coefficient of tail dependence of an Archimedean copula can be expressed solely in terms of its generator. A simple application of L'Hospital's rule shows that any Archimedean copula, with a *strict generator* $\varphi$ (that is, such that $\varphi(0)$ is infinite so that $\varphi^{[-1]} = \varphi^{-1}$), has a coefficient of upper tail dependence given by

$$\lambda_U = 2 - 2 \lim_{t \to 0} \frac{\varphi^{-1'}(2t)}{\varphi^{-1'}(t)} \, . \tag{4.85}$$

As a consequence, if $\varphi^{-1'}(0) > -\infty$, the coefficient of upper tail dependence is identically zero. For an Archimedean copula to present tail dependence, it is necessary that $\lim_{t \to 0} \varphi^{-1'}(t) = -\infty$.

Similarly, the coefficient of lower tail dependence is

$$\lambda_L = 2 \lim_{t \to \infty} \frac{\varphi^{-1'}(2t)}{\varphi^{-1'}(t)} \, , \tag{4.86}$$

so that $\varphi^{-1'}(\infty)$ must be equal to 0 in order for the Archimedean copula to have a nonzero lower tail dependence.

**Table 4.2.** Expressions of the coefficient of upper and lower tail dependence for three Archimedean copulas. Note that the usual range for the parameter $\theta$ of Clayton's copula has to be restricted to $[0, \infty]$ in order for the generator to be "strict"

| Copula | $\varphi(t)^{-1'}$ | $\lambda_L$ | $\lambda_U$ | Range |
|--------|-----------------|-------------|-------------|-------|
| Clayton | $(1 + \theta t)^{-1/\theta}$ | $2^{-1/\theta}$ | $0$ | $\theta \in [0, \infty]$ |
| Gumbel | $-\dfrac{t^{1/\theta-1}}{\theta} \exp\left(-t^{1/\theta}\right)$ | $0$ | $2 - 2^{1/\theta}$ | $\theta \in [1, \infty]$ |
| Frank | $-\dfrac{1}{\theta} \cdot \dfrac{\left(1 - e^{-\theta}\right) e^{-t}}{1 - (1 - e^{-\theta}e^{-t})}$ | $0$ | $0$ | $\theta \in [-\infty, \infty]$ |

Table 4.2 gives the coefficients of tail dependence of several Archimedean copulas. It illustrates the fact that some copulas have an upper tail dependence but no lower tail dependence (the Gumbel copula) or, on the contrary, some copulas have no upper tail dependence but have a lower tail dependence (Clayton copula). More precisely, the coefficient of lower tail dependence of Clayton's copula equals $2^{-1/\theta}$ while the coefficient of upper tail dependence of Gumbel's copula is $2 - 2^{1/\theta}$. In addition, the generator of Clayton's copula is regularly varying at $t = 0$ (see Table 4.1), with a tail index $-\theta$ while the generator of Gumbel's copula is regularly varying at $t = 1$, with tail index $\theta$.

In fact, one can show that any Archimedean copula, with a generator regularly varying at zero and tail index $-\theta$ (with $\theta \geq 0$), has a coefficient of *lower* tail dependence equal to $2^{-1/\theta}$. Indeed, by (4.79)

$$\lambda_L = \lim_{u \to 0^+} \frac{\varphi^{-1}(2\varphi(u))}{u} = \lim_{x \to 0^+} \frac{\varphi^{-1}(2x)}{\varphi^{-1}(x)} , \qquad (4.87)$$

and, since $\varphi$ is regularly varying with tail index $-\theta$, $\varphi^{-1}$ is also regularly varying with tail index $-1/\theta$ [57], so that

$$\lambda_L = \lim_{x \to 0^+} \frac{\varphi^{-1}(2x)}{\varphi^{-1}(x)} = 2^{-1/\theta} , \qquad (4.88)$$

Similarly, any Archimedean copula with a generator regularly varying at 1 and with tail index $\theta$ (with $\theta \geq 1$, in order to fulfill the convexity requirement), has a coefficient of *upper* tail dependence equal to $2 - 2^{1/\theta}$.

These results also apply to the frailty model with frailty parameters having a distribution regularly varying at zero with tail index $1/\theta$. Using the properties of the Laplace transform, one can conclude that the copulas generated by such frailty models have generators which are regularly varying at zero with tail index $-\theta$. Therefore, they have a coefficient of lower tail dependence equal to $2^{-1/\theta}$. Similarly, copulas generated by frailty models with frailty parameters with distribution regularly varying at infinity, with tail index $1/\theta$ ($\theta > 1$),

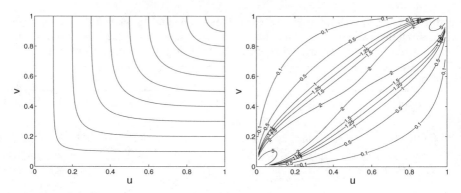

**Fig. 4.7.** Contour plot of the copula with generator (4.89) (*left panel*) and of its density (*right panel*) for the parameters value $\alpha = 1$ and $\beta = 2$

have a coefficient of upper tail dependence equal to $2 - 2^{1/\theta}$, since they lead to generators which are regularly varying at 1, with tail index $\theta$. Finally, to obtain an Archimedean copula with both upper and lower tail dependence, one just has to consider generators which are regularly varying at 0 and 1, or alternatively to have frailty parameters with regular variation at zero and at infinity.

An example is the following generator:

$$\varphi(t) = t^{-\alpha} \cdot (-\ln t)^{\beta}, \qquad (\alpha, \beta) \in [0, \infty) \times [1, \infty), \tag{4.89}$$

with inverse

$$\varphi^{-1}(t) = \exp\left[-\frac{\beta}{\alpha} \cdot \mathrm{W}\left(\frac{\alpha}{\beta} t^{1/\beta}\right)\right], \tag{4.90}$$

where $W(\cdot)$ denotes the Lambert function solution of $W(x) \cdot e^{W(x)} = x$. It allows for upper and lower tail dependence with $\lambda_L = 2^{-1/\alpha}$ and $\lambda_U = 2 - 2^{1/\beta}$. Figure 4.7 shows this copula for $\alpha = 1$ and $\beta = 2$, corresponding to $\lambda_L = 0.5$ and $\lambda_U = 2 - \sqrt{2} \simeq 0.6$.

## Elliptical Copulas

Assuming that $(X, Y)$ are normally distributed with correlation coefficient $\rho$, it can be shown that, for all $\rho \in [-1, 1)$, $\lambda = 0$, while $\bar{\lambda} = \rho$ [149, 229]. This later result expresses, as one can expect, that – despite the absence of tail dependence – extremes appear more likely together for positively correlated variables.

In contrast, if $(X, Y)$ have a Student's copula, one can show that the tail dependence coefficient is

$$\lambda = 2 \cdot \bar{T}_{\nu+1}\left(\sqrt{\nu+1}\sqrt{\frac{1-\rho}{1+\rho}}\right), \tag{4.91}$$

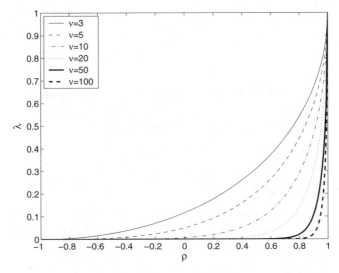

**Fig. 4.8.** Coefficient of upper tail dependence as a function of the correlation co-efficient $\rho$ for various values of the number of degrees of freedom $\nu$ for Student's copula

which is greater than zero for all $\rho > -1$, and thus $\bar{\lambda} = 1$. $T_{\nu+1}$ is the Student distribution with $\nu$ degrees of freedom and the bar denote the complementary distribution. This result $\bar{\lambda} = 1$ proves that extremes appear more likely together whatever the correlation coefficient may be, showing that, in fact, there is no general relationship between the asymptotic dependence and the linear correlation coefficient. Figure 4.8 shows the coefficient of upper tail dependence as a function of the correlation coefficient $\rho$ for various values of the number of degrees of freedom $\nu$.

These distinctive properties of the Gaussian and Student's copulas, characterized by the absence or presence of tail dependence, are illustrated in Fig. 4.9 which shows the realizations of two random variables with identical standard Gaussian marginals, with a Gaussian copula or a Student's copula with three degrees of freedom and the same correlation coefficient $\rho = 0.8$. In the right panel for the Student's copula, the realizations (dots) are found to lie within a diamond-shaped domain with narrower and narrower tips as more extreme values are considered. This phenomenon can be observed, not only for the bottom-left and upper-right quadrants, but also for the upper-left and bottom-right quadrants. This results from the fact that the tail dependence coefficient remains nonzero even for negative correlation coefficients as illustrated in Fig. 4.8.

The Gaussian and Student's distributions are two examples of elliptical distributions. More generally, the following result is known: elliptically distributed random variables present a nonzero tail dependence if and only if they are regularly varying, i.e., their distributions behave asymptotically like

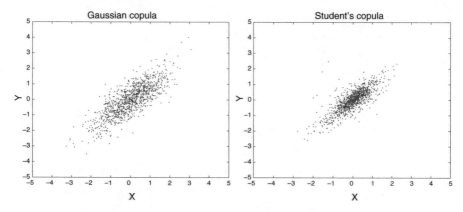

**Fig. 4.9.** Realizations of two random variables with Gaussian marginals and with a Gaussian copula (*left panel*) and a Student's copula with three degrees of freedom (*right panel*) with the same correlation coefficient $\rho = 0.8$

power laws with some exponent $\nu > 0$ [239]. In such a case, for every regularly varying pair of random variables which are elliptically distributed, the coefficient of tail dependence $\lambda$ is given by expression (4.91). This result is natural since the correlation coefficient is an invariant quantity within the class of elliptical distributions and since the coefficient of tail dependence is only determined by the asymptotic behavior of the distribution, so that it does not matter that the distribution is a Student's distribution with $\nu$ degrees of freedom or any other elliptical distribution as long as they have the same asymptotic behavior in the tail.

## Linear Factor Models

Consider the one-factor model

$$X_1 = \beta_1 \cdot Y + \epsilon_1 , \tag{4.92}$$
$$X_2 = \beta_2 \cdot Y + \epsilon_2 , \tag{4.93}$$

where the $\epsilon_i$'s are random variables independent of $Y$ and the $\beta_i$'s are nonrandom *positive* coefficients.

The tail dependence $\lambda$ of $X_1$ and $X_2$ can be simply expressed as the minimum of the tail dependence coefficients $\lambda_1$ and $\lambda_2$ between the two random variables $X_1$ and $Y$, on the one hand and $X_2$ and $Y$, on the other hand [332, 335]:

$$\lambda = \min\{\lambda_1, \lambda_2\} . \tag{4.94}$$

To understand this result, note that the tail dependence between $X_1$ and $X_2$ is created only through the common factor $Y$. It is thus natural that the tail dependence between $X_1$ and $X_2$ is bounded from above by the weakest tail dependence between the $X_i$'s and $Y$ while deriving the equality requires more work. The result (4.94) generalizes to an arbitrary number of random variables and shows that the study of the tail dependence in linear factor models can be reduced to the analysis of the tail dependence between each individual $X_i$ and the common factor $Y$. In the following, we thus omit the subscript $i$ and consider without loss of generality one $X$ linearly regressed on a factor $Y$ according to $X = \beta \cdot Y + \epsilon$.

A general result concerning the tail dependence generated by factor models for any kind of factor and noise distributions is as follows [332, 335]: the coefficient of (upper) tail dependence between $X$ and $Y$ is given by

$$\lambda = \int_{\max\{1, \frac{l}{\beta}\}}^{\infty} dx \ f(x) \ , \tag{4.95}$$

where, provided that they exist,

$$l = \lim_{u \to 1} \frac{F_X^{-1}(u)}{F_Y^{-1}(u)} \ , \tag{4.96}$$

$$f(x) = \lim_{t \to \infty} \frac{t \cdot P_Y(t \cdot x)}{\bar{F}_Y(t)} \ , \tag{4.97}$$

where $F_X$ and $F_Y$ are the marginal distribution functions of $X$ and $Y$ respectively, and $P_Y$ is the density of $Y$.

As a direct consequence, one can show that any rapidly varying factor, which encompasses the Gaussian, the exponential or the gamma distributed factors for instance, leads to a vanishing coefficient of tail dependence, whatever the distribution of the idiosyncratic noise may be. This result is obvious when both the factor and the idiosyncratic noise are normally distributed, since then $X$ and $Y$ follow a bivariate Gaussian distribution, whose tail dependence has been said to be zero.

On the contrary, regularly varying factors, like the Student's distributed factors, lead to a tail dependence, provided that the distribution of the idiosyncratic noise does not become fatter-tailed than the factor distribution. One can thus conclude that, in order to generate tail dependence, the factor must have a sufficiently "wild" distribution. To present an explicit example, let us assume now that the factor $Y$ and the idiosyncratic noise $\epsilon$ have centered Student's distributions with the same number $\nu$ of degrees of freedom and scale factors respectively equal to 1 and $\sigma$. The choice of the scale factor equal to 1 for $Y$ is not restrictive but only provides a convenient normalization for $\sigma$. Appendix 4.A shows that the tail dependence coefficient is given by

$$\lambda = \frac{1}{1 + \left(\frac{\sigma}{\beta}\right)^{\nu}} \ . \tag{4.98}$$

This expression shows that, the larger the typical scale $\sigma$ of the fluctuation of $\epsilon$ and the weaker the coupling coefficient $\beta$, the smaller is the tail dependence, in accordance with intuition.

The linear correlation coefficient $\rho$ for the one-factor model is given by $\rho = (1 + \frac{\sigma^2}{\beta^2})^{-1/2}$, which allows us to rewrite the coefficient of upper tail dependence in terms of $\rho > 0$ and $\nu > 2$:

$$\lambda = \frac{\rho^\nu}{\rho^\nu + (1 - \rho^2)^{\nu/2}} = \frac{1}{1 + \left(\frac{1-\rho^2}{\rho^2}\right)^{\nu/2}} . \tag{4.99}$$

Surprisingly, $\lambda$ does not go to zero for all $\rho$'s as $\nu$ goes to infinity, as could be anticipated from the fact that the Student's distribution converges to the Gaussian distribution which is known to have zero tail dependence. Expression (4.99) predicts that $\lambda \to 0$ when $\nu \to \infty$ for all $\rho$'s smaller than $1/\sqrt{2}$. But, and here lies the surprise, $\lambda \to 1$ for all $\rho$ larger than $1/\sqrt{2}$ when $\nu \to \infty$. This counterintuitive result is due to a non-uniform convergence which makes the order of the two limits non-commutative: taking first the limit $u \to 1$ and then $\nu \to \infty$ is different from taking first the limit $\nu \to \infty$ and then $u \to 1$. In a sense, by taking first the limit $u \to 1$, we always ensure the power law regime even if $\nu$ is later taken to infinity. This is different from first "sitting" on the Gaussian limit $\nu \to \infty$. This paradoxical behavior reveals the sometimes paradoxical consequences of taking the limit $u \to 1$ in the definition of the tail dependence.

As an illustration, Fig. 4.10 presents the coefficient of tail dependence for the Student's factor model as a function of $\rho$ for various values of $\nu$. It is interesting to compare this figure with Fig. 4.8 depicting the coefficient of tail dependence for the Student's copula. Note that $\lambda$ is vanishing for all negative $\rho$'s in the case of the factor model, while $\lambda$ remains nonzero for negative values of the correlation coefficient for bivariate Student's variables.

If $Y$ and $\epsilon$ have different numbers $\nu_Y$ and $\nu_\epsilon$ of degrees of freedom, two cases occur. For $\nu_Y < \nu_\epsilon$, $\epsilon$ is negligible asymptotically and $\lambda = 1$. For $\nu_Y > \nu_\epsilon$, $X$ becomes asymptotically identical to $\epsilon$. Then, $X$ and $Y$ have the same tail-dependence as $\epsilon$ and $Y$, which is zero by construction.

A straightforward generalization of this result can be derived for the multifactor model [72]:

$$X_1 = \beta_{1,1} \cdot Y_1 + \cdots + \beta_{1,n} \cdot Y_n + \epsilon_1 , \tag{4.100}$$

$$X_2 = \beta_{2,1} \cdot Y_1 + \cdots + \beta_{2,n} \cdot Y_n + \epsilon_2 . \tag{4.101}$$

The following generalization of (4.98) gives the coefficient of tail dependence between $X_1$ and one of the $Y_i$ as

$$\lambda_{1,i} = \frac{\beta_{1,i}^\nu}{\sum_{j=1}^n \beta_{1,j}^\nu + \sigma^\nu} , \tag{4.102}$$

provided that the $Y_i$'s remain independent factors. For simplicity, we have assumed that all the factors are standardized, i.e., their scale factors are all equal to one. Generalizing expression (4.94), the coefficient of tail dependence between $X_1$ and $X_2$ is

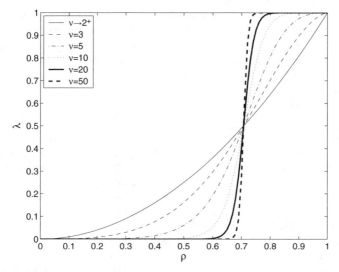

**Fig. 4.10.** Coefficient of upper tail dependence as a function of the correlation coefficient $\rho$ for various values of the number of degrees of freedom $\nu$ for the Student's factor model

$$\lambda = \sum_{i=1}^{n} \mathbf{1}_{\{\beta_{1,i} \cdot \beta_{2,i} > 0\}} \cdot \min\left(\lambda_{1,i}, \lambda_{2,i}\right) . \tag{4.103}$$

These results are of particular interest for portfolio analysis and risk management, as we shall see in the next section.

### 4.5.4 Practical Implications

Let us now give two straightforward applications of the tail dependence for financial purposes. We also refer the reader to [390] for other financial applications.

**Portfolio Tail Risk Management**

Table 4.3 presents the results obtained on the estimations of the upper and lower coefficients of tail dependence between several major stocks and the market represented here by the Standard & Poor's 500, over the last decade. The estimation has been performed under the assumption that (4.92–4.93) hold, in which the factor is represented by the Standard & Poor's 500. Using the market index as the factor is reasonable since, according to standard financial theory, the market's return is well-known to be the most important explanatory factor for the return of each individual asset.[4] The coefficient of

---
[4] In a situation where the common factor cannot be easily identified or estimated, the results for elliptic distributions obtained in Sect. 4.5.3 may provide a convenient alternative.

**Table 4.3.** This table presents the coefficients of lower and of upper tail dependence of the companies traded on the NYSE and listed in the first column with the Standard & Poor's 500. The returns used for the calculations are sampled in the time interval from January 1991 to December 2000. The numbers within the parentheses are the estimated standard deviations of the empirical coefficients of tail dependence. Reproduced from [332]

|                              | $\lambda_L$ |        | $\lambda_U$ |        |
| ---------------------------- | ----------- | ------ | ----------- | ------ |
| Bristol-Myers Squibb Co.     | 0.16        | (0.03) | 0.14        | (0.01) |
| Chevron Corp.                | 0.05        | (0.01) | 0.03        | (0.01) |
| Hewlett-Packard Co.          | 0.13        | (0.01) | 0.12        | (0.01) |
| Coca-Cola Co.                | 0.12        | (0.01) | 0.09        | (0.01) |
| Minnesota Mining & MFG Co.   | 0.07        | (0.01) | 0.06        | (0.01) |
| Philip Morris Cos Inc.       | 0.04        | (0.01) | 0.04        | (0.01) |
| Procter & Gamble Co.         | 0.12        | (0.02) | 0.09        | (0.01) |
| Pharmacia Corp.              | 0.06        | (0.01) | 0.04        | (0.01) |
| Schering-Plough Corp.        | 0.12        | (0.01) | 0.11        | (0.01) |
| Texaco Inc.                  | 0.04        | (0.01) | 0.03        | (0.01) |
| Texas Instruments Inc.       | 0.17        | (0.02) | 0.12        | (0.01) |
| Walgreen Co.                 | 0.11        | (0.01) | 0.09        | (0.01) |

tail dependence between any two assets is then easily derived from (4.94). It is interesting to observe that the coefficients of tail dependence seem almost identical in the lower and the upper tail. Nonetheless, the coefficient of lower tail dependence is always slightly larger than the upper one, showing that large losses are more likely to come together compared with large gain occurrences.

Two clusters of assets clearly stand out: those with a tail dependence of about 10% (or more) and those with a tail dependence of about 5%. Let us exploit this observation and explore some consequences of the existence of stocks with drastically different tail dependence coefficients with the index. These stocks offer the interesting possibility of constructing a prudential portfolio which can be significantly less sensitive to the large market moves. Figure 4.11 compares the daily returns of the Standard & Poor's 500 with those of two portfolios $P_1$ and $P_2$: $P_1$ is made of the four stocks (Chevron Corp., Philip Morris Cos Inc., Pharmacia Corp., and Texaco Inc.,) with the smallest $\lambda$'s while $P_2$ is made of the four stocks (Bristol-Meyer Squibb Co., Hewlett-Packard Co., Schering-Plough Corp., and Texas Instruments Inc.,) with the largest $\lambda$'s. In fact, we have constructed two variants of $P_1$ and two variants of $P_2$. The first variant corresponds to choose the same weight $1/4$ of each asset in each class of assets (with small $\lambda$'s for $P_1$ and large $\lambda$'s for $P_2$). The second variant has asset weights in each class chosen in addition to minimize the variance of the resulting portfolio. We find that the results are almost the same between the equally weighted and minimum-variance portfolios. This makes sense since the tail dependence coefficient of a bivariate random vector does not depend on the variances of the components, which only account for the price moves of moderate amplitudes.

Figure 4.11 presents the results for the equally weighted portfolios generated from the two groups of assets. Observe that only one large drop occurs simultaneously for $P_1$ and for the Standard & Poor's 500 in contrast with $P_2$ for which several large drops are associated with the largest drops of the index and only a few occur desynchronized. The figure clearly shows an almost circular scatter plot for the large moves of $P_1$ and the index compared with a rather narrow ellipse, whose long axis is approximately along the first diagonal, for the large returns of $P_2$ and the index, illustrating that the small tail dependence between the index and the four stocks in $P_1$ automatically implies that their mutual tail dependence is also very small, according to (4.94). As a consequence, $P_1$ offers a better diversification with respect to large drops than $P_2$. This effect already, quite significant for such small portfolios, should be overwhelming for large ones. The most interesting result stressed in Fig. 4.11 is that optimizing for minimum tail dependence automatically diversifies away the large risks.

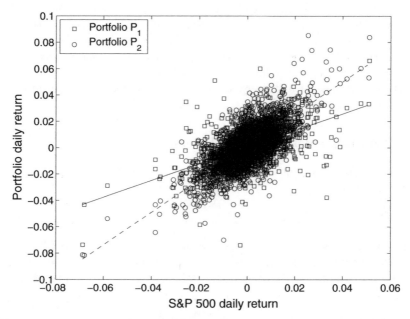

**Fig. 4.11.** Daily returns of two equally weighted portfolios $P_1$ (made of four stocks with small $\lambda \leq 0.06$) and $P_2$ (made of four stocks with large $\lambda \geq 0.12$) as a function of the daily returns of the Standard & Poor's 500 over the period January 1991 to December 2000. The straight (resp. dashed) line represents the regression of portfolio $P_1$ (resp. $P_2$) on the Standard & Poor's 500. Reproduced from [332]

These advantages of portfolio $P_1$ with small tail dependence compared with portfolio $P_2$ with large tail dependence with respect to the Standard & Poor's 500 index come at almost no cost in terms of the daily Sharpe ratio, equal respectively to 0.058 and 0.061 for the equally weighted and minimum variance $P_1$ and to 0.069 and 0.071 for the equally weighted and minimum variance $P_2$.

The straight lines in Fig. 4.11 represent the linear regressions of the returns of the two portfolios on the index returns, and show that there is significantly less linear correlation between $P_1$ and the index (correlation coefficient of 0.52 for both the equally weighted and the minimum variance $P_1$) compared with $P_2$ and the index (correlation coefficient of 0.73 for the equally weighted $P_2$ and of 0.70 for the minimum variance $P_2$). Theoretically, it is possible to construct two random variables with small correlation coefficient and large $\lambda$ and vice-versa. Recall that the correlation coefficient and the tail dependence coefficient are two opposite end-members of dependence measures: the correlation coefficient quantifies the dependence between relatively small moves while the tail dependence coefficient measures the dependence during extreme events. The finding that $P_1$ comes with both the smallest correlation and the smallest tail dependence coefficients suggests that they are not independent properties of assets. This intuition is in fact explained and encompassed by the factor model since the larger $\beta$ is, the larger is the correlation coefficient and the larger is the tail dependence. Diversifying away extreme shocks may provide a useful diversification tool for less extreme dependences, thus improving the potential usefulness of a strategy of portfolio management based on the tail dependence proposed here.

### Impact on Dependent Default Modeling

Consider $N$ obligators with individual default probability $\pi_i$, $i = 1, \ldots, N$ and default indicator

$$D_i = 1 \iff X_i \leq T_i , \tag{4.104}$$

where $(X_1, \ldots, X_N)$ denotes the vector of latent variables and $(T_1, \ldots, T_N)$ the vector of thresholds below which default occurs, Sect. 3.6.4 has shown that the probability that $k$ obligators – labeled $i_1, \ldots, i_k$ – among $N$ default is given by

$$\Pr\left[D_{i_1} = 1, \ldots, D_{i_k} = 1\right] = C\left(\pi_{i_1}, \ldots, \pi_{i_k}\right) , \tag{4.105}$$

where $C$ is the copula of the latent variables $X_i$ under consideration.

This equation emphasizes the key role of the copula in credit risk modeling. Since default probabilities are generally very low, specifically for very high quality obligators, the behavior of the copula in the extreme is crucial. As a consequence, it could seem natural that the presence or the absence of

**Table 4.4.** Ratio of the 99% quantiles of the distribution of defaulting obligators when the latent variables have a Student copula with $\nu$ degrees of freedom, normalized with respect to the Gaussian copula, for portfolios of 10,000 homogeneous credits with default probability $\pi_i$ and correlation $\rho$. The values of $\pi_i$ and $\rho$ are the same as in [186]

| $\pi_i$ | $\rho$ | $\nu = 50$ | $\nu = 10$ | $\nu = 4$ |
|---------|--------|-----------|-----------|----------|
| 0.01% | 2.58% | 2.33 | 5.62 | 6.00 |
| 0.50% | 3.80% | 1.66 | 3.75 | 6.84 |
| 7.50% | 9.21% | 1.09 | 1.39 | 1.78 |

tail dependence between the latent variables $X_i$ would be of particular importance. When latent variables are asymptotically independent – as assumed in traditional models exposed in Sect. 3.6.4 – one can reasonably guess that such models would underestimate the actual occurrence of concomitant defaults.

This view has been advocated by Frey *et al.* [185, 186], among others. Considering large credit portfolios, they investigate the evolution of the total number of defaulting obligators when the dependence structure describing their interaction changes. Table 4.4 gives the ratios of the 99% quantiles of the distribution of defaulting obligators for a Student's copula normalized with respect to the Gaussian copula, for three credit groups of different quality. The ratio of the quantiles increases when the number of degrees of freedom of the Student's copula decreases, since the dependence between extremes becomes stronger. We also observe that the ratio of the quantiles decreases with the quality of the obligator, *i.e.*, when the default probability increases. Indeed, in such a case, the tail dependence has a weaker impact on the portfolio loss, and therefore the exact shape of the copula in the neighborhood of $(0,0)$ or of $(1,1)$ is less important.

Overall, these simulations tend to give substance to the assertion that the choice of the copula is fundamental. However, some recent studies support the opposite point of view. For the practical purpose of pricing credit derivatives, several authors [291, 430] have shown that the choice of the copula has in fact only a weak impact on the value of such contracts. As an example, Laurent and Gregory [291] show that the premium for the first-to-default swap in basket default swaps is almost the same for a Gaussian copula and for a Clayton copula. Such results also hold for CDO[5] tranches. Schloegl and O'Kane confirm these results for the Student's copula [430], for which they find that it does not provide significant improvement with respect to the Gaussian copula.

To sum up, for credit derivative pricing, the choice of the copula does not appear to be crucial. However, taking into account the simulation results in [186], we clearly see that the dependence structure has a real impact on the

---

[5] Collaterized Debt Obligation

loss distribution of the credit portfolio. Therefore, even if the copula is not so important for derivative pricing, it could be really crucial to establish hedging strategies. This point has not yet been really explored to our knowledge, but appears as an important future development of the research on credit derivatives.

# Appendix

## 4.A  Tail Dependence Generated by Student's Factor Model

We consider two random variables $X$ and $Y$, related by

$$X = \beta Y + \epsilon , \tag{4.A.1}$$

where $\epsilon$ is a random variable independent of $Y$ and $\beta$ a nonrandom positive coefficient. Let us assume that $Y$ and $\epsilon$ have a Student's distribution with density:

$$P_Y(y) = \frac{C_\nu}{\left(1 + \frac{y^2}{\nu}\right)^{(\nu+1)/2}} , \tag{4.A.2}$$

$$P_\epsilon(\epsilon) = \frac{C_\nu}{\sigma \left(1 + \frac{\epsilon^2}{\nu\,\sigma^2}\right)^{(\nu+1)/2}} . \tag{4.A.3}$$

**Lemma 4.5.1.** *The probability that $X$ is larger than $F_X^{-1}(u)$ knowing that $Y$ is larger than $F_Y^{-1}(u)$ is given by :*

$$\Pr[X > F_X^{-1}(u)|Y > F_Y^{-1}(u)] = \bar{F}_\epsilon(\eta)$$
$$+ \frac{\beta}{1-u} \int_{F_Y^{-1}(u)}^{\infty} dy \; \bar{F}_Y(y) \cdot P_\epsilon[\beta F_Y^{-1}(u) + \eta - \beta y] , \tag{4.A.4}$$

*with*

$$\eta = F_X^{-1}(u) - \beta F_Y^{-1}(u) . \tag{4.A.5}$$

The proof of this lemma relies on a simple integration by part and a change of variable, which are detailed in Appendix 4.A.1.

Introducing the notation

$$\tilde{Y}_u = F_Y^{-1}(u) , \tag{4.A.6}$$

Appendix 4.A.2 shows that

$$\eta = \beta \left[ \left(1 + \left(\frac{\sigma}{\beta}\right)^\nu\right)^{1/\nu} - 1 \right] \tilde{Y}_u + \mathcal{O}(\tilde{Y}_u^{-1}) . \tag{4.A.7}$$

This allows us to conclude that $\eta \to +\infty$ as $u \to 1$. Thus, $\bar{F}_\epsilon(\eta) \to 0$ as $u \to 1$ and

$$\lambda = \lim_{u \to 1} \frac{\beta}{1 - u} \int_{\tilde{Y}_u}^\infty dy \ \bar{F}_Y(y) \cdot P_\epsilon(\beta \tilde{Y}_u + \eta - \beta y) \ . \tag{4.A.8}$$

Now, using the following result:

**Lemma 4.5.2.** *Assuming $\nu > 0$ and $x_0 > 1$,*

$$\lim_{\epsilon \to 0} \frac{1}{\epsilon} \int_1^\infty dx \ \frac{1}{x^\nu} \frac{C_\nu}{\left[1 + \left(\frac{x - x_0}{\epsilon}\right)^2\right]^{\frac{\nu + 1}{2}}} = \frac{1}{x_0^\nu} \ , \tag{4.A.9}$$

whose proof is given in Appendix 4.A.3, it is straightforward to show that

$$\lambda = \frac{1}{1 + \left(\frac{\sigma}{\beta}\right)^\nu} \ . \tag{4.A.10}$$

The final steps of this derivation are given in Appendix 4.A.4.

### 4.A.1 Proof of Lemma 4.5.1

By definition,

$$\Pr[X > F_X^{-1}(u), Y > F_Y^{-1}(u)] = \int_{F_X^{-1}(u)}^\infty dx \int_{F_Y^{-1}(u)}^\infty dy \ P_Y(y) \cdot P_\epsilon(x - \beta y)$$

$$= \int_{F_Y^{-1}(u)}^\infty dy \ P_Y(y) \cdot \bar{F}_\epsilon[F_X^{-1}(u) - \beta y] \ .$$

Let us perform an integration by part:

$$\Pr[X > F_X^{-1}(u), Y > F_Y^{-1}(u)]$$

$$= \left[-\bar{F}_Y(y) \cdot \bar{F}_\epsilon(F_X^{-1}(u) - \beta y)\right]_{F_Y^{-1}(u)}^\infty + \beta \int_{F_Y^{-1}(u)}^\infty dy \ \bar{F}_Y(y) \cdot P_\epsilon(F_X^{-1}(u) - \beta y)$$

$$= (1 - u)\bar{F}_\epsilon(F_X^{-1}(u) - \beta F_Y^{-1}(u)) + \beta \int_{F_Y^{-1}(u)}^\infty dy \ \bar{F}_Y(y) \cdot P_\epsilon(F_X^{-1}(u) - \beta y) \ .$$

Defining $\eta = F_X^{-1}(u) - \beta F_Y^{-1}(u)$ (see (4.A.5)), and dividing each term by

$$\Pr[Y > F_Y^{-1}(u)] = 1 - u \ , \tag{4.A.11}$$

we obtain the result given in (4.A.4)

## 4.A.2 Derivation of Equation (4.A.7)

The factor $Y$ and the idiosyncratic noise $\epsilon$ are distributed according to the Student's distributions with $\nu$ degrees of freedom given by (4.A.2) and (4.A.3) respectively. It follows that the survival distributions of $Y$ and $\epsilon$ are:

$$\bar{F}_Y(y) = \frac{\nu^{\frac{\nu-1}{2}} C_\nu}{y^\nu} + \mathcal{O}(y^{-(\nu+2)}) , \tag{4.A.12}$$

$$\bar{F}_\epsilon(\epsilon) = \frac{\sigma^\nu \, \nu^{\frac{\nu-1}{2}} C_\nu}{\epsilon^\nu} + \mathcal{O}(\epsilon^{-(\nu+2)}) , \tag{4.A.13}$$

and

$$\bar{F}_X(x) = \frac{(\beta^\nu + \sigma^\nu) \, \nu^{\frac{\nu-1}{2}} C_\nu}{x^\nu} + \mathcal{O}(x^{-(\nu+2)}) . \tag{4.A.14}$$

Using the notation (4.A.6), (4.A.5) can be rewritten as

$$\bar{F}_X(\eta + \beta \tilde{Y}_u) = \bar{F}_Y(\tilde{Y}_u) = 1 - u , \tag{4.A.15}$$

whose solution for large $\tilde{Y}_u$ (or equivalently as u goes to 1) is

$$\eta = \beta \left[ \left( 1 + \left( \frac{\sigma}{\beta} \right)^\nu \right)^{1/\nu} - 1 \right] \tilde{Y}_u + \mathcal{O}(\tilde{Y}_u^{-1}) . \tag{4.A.16}$$

To obtain this equation, we have used the asymptotic expressions of $\bar{F}_X$ and $\bar{F}_Y$ given in (4.A.14) and (4.A.12).

## 4.A.3 Proof of Lemma 4.5.2

The change of variable

$$u = \frac{x - x_0}{\epsilon} , \tag{4.A.17}$$

gives

$$\frac{1}{\epsilon} \int_1^\infty dx \, \frac{1}{x^\nu} \frac{C_\nu}{\left[ 1 + \frac{1}{\nu} \left( \frac{x - x_0}{\epsilon} \right)^2 \right]^{\frac{\nu+1}{2}}} = \int_{\frac{1-x_0}{\epsilon}}^\infty du \, \frac{1}{(\epsilon u + x_0)^\nu} \frac{C_\nu}{(1 + \frac{u^2}{\nu})^{\frac{\nu+1}{2}}}$$

$$= \frac{1}{x_0^\nu} \int_{\frac{1-x_0}{\epsilon}}^\infty du \, \frac{1}{(1 + \frac{\epsilon u}{x_0})^\nu} \frac{C_\nu}{(1 + \frac{u^2}{\nu})^{\frac{\nu+1}{2}}}$$

$$= \frac{1}{x_0^\nu} \int_{\frac{1-x_0}{\epsilon}}^{\frac{x_0}{\epsilon}} du \, \frac{1}{(1 + \frac{\epsilon u}{x_0})^\nu} \frac{C_\nu}{(1 + \frac{u^2}{\nu})^{\frac{\nu+1}{2}}}$$

$$+ \frac{1}{x_0^\nu} \int_{\frac{x_0}{\epsilon}}^\infty du \, \frac{1}{(1 + \frac{\epsilon u}{x_0})^\nu} \frac{C_\nu}{(1 + \frac{u^2}{\nu})^{\frac{\nu+1}{2}}} .$$

Consider the second integral in the right-hand side of the last equality. We have

$$u \geq \frac{x_0}{\epsilon} , \tag{4.A.18}$$

which allows us to write

$$\frac{1}{(1+u^2)^{\frac{\nu+1}{2}}} \leq \frac{\nu^{\frac{\nu+1}{2}} \epsilon^{\nu+1}}{x_0^{\nu+1}} , \tag{4.A.19}$$

so that

$$\left| \int_{\frac{x_0}{\epsilon}}^{\infty} du \frac{1}{(1+\frac{\epsilon u}{x_0})^\nu} \frac{C_\nu}{(1+u^2)^{\frac{\nu+1}{2}}} \right| \leq \frac{\nu^{\frac{\nu+1}{2}} \epsilon^{\nu+1}}{x_0^{\nu+1}} \int_{\frac{x_0}{\epsilon}}^{\infty} du \frac{C_\nu}{(1+\frac{\epsilon u}{x_0})^\nu} \tag{4.A.20}$$

$$= \frac{\nu^{\frac{\nu+1}{2}} \epsilon^\nu}{x_0^\nu} \int_1^\infty dv \frac{C_\nu}{(1+v)^\nu} \tag{4.A.21}$$

$$= \mathcal{O}(\epsilon^\nu). \tag{4.A.22}$$

The next step of the proof is to show that

$$\int_{\frac{1-x_0}{\epsilon}}^{\frac{x_0}{\epsilon}} du \frac{1}{(1+\frac{\epsilon u}{x_0})^\nu} \frac{C_\nu}{(1+\frac{u^2}{\nu})^{\frac{\nu+1}{2}}} \longrightarrow 1 \quad \text{as} \quad \epsilon \longrightarrow 0 . \tag{4.A.23}$$

Let us calculate

$$\left| \int_{\frac{1-x_0}{\epsilon}}^{\frac{x_0}{\epsilon}} du \frac{1}{(1+\frac{\epsilon u}{x_0})^\nu} \frac{C_\nu}{(1+\frac{u^2}{\nu})^{\frac{\nu+1}{2}}} - 1 \right|$$

$$= \left| \int_{\frac{1-x_0}{\epsilon}}^{\frac{x_0}{\epsilon}} du \frac{1}{(1+\frac{\epsilon u}{x_0})^\nu} \frac{C_\nu}{(1+\frac{u^2}{\nu})^{\frac{\nu+1}{2}}} - \int_{-\infty}^{\infty} du \frac{C_\nu}{(1+\frac{u^2}{\nu})^{\frac{\nu+1}{2}}} \right|$$

$$= \left| \int_{\frac{1-x_0}{\epsilon}}^{\frac{x_0}{\epsilon}} du \left[ \frac{1}{(1+\frac{\epsilon u}{x_0})^\nu} - 1 \right] \frac{C_\nu}{(1+\frac{u^2}{\nu})^{\frac{\nu+1}{2}}} \right.$$

$$\left. - \int_{-\infty}^{\frac{1-x_0}{\epsilon}} du \frac{C_\nu}{(1+\frac{u^2}{\nu})^{\frac{\nu+1}{2}}} - \int_{\frac{x_0}{\epsilon}}^{\infty} du \frac{C_\nu}{(1+\frac{u^2}{\nu})^{\frac{\nu+1}{2}}} \right|$$

$$\leq \left| \int_{\frac{1-x_0}{\epsilon}}^{\frac{x_0}{\epsilon}} du \left[ \frac{1}{(1+\frac{\epsilon u}{x_0})^\nu} - 1 \right] \frac{C_\nu}{(1+\frac{u^2}{\nu})^{\frac{\nu+1}{2}}} \right| + \left| \int_{-\infty}^{\frac{1-x_0}{\epsilon}} du \frac{C_\nu}{(1+\frac{u^2}{\nu})^{\frac{\nu+1}{2}}} \right|$$

$$+ \left| \int_{\frac{x_0}{\epsilon}}^{\infty} du \frac{C_\nu}{(1+\frac{u^2}{\nu})^{\frac{\nu+1}{2}}} \right| . \tag{4.A.24}$$

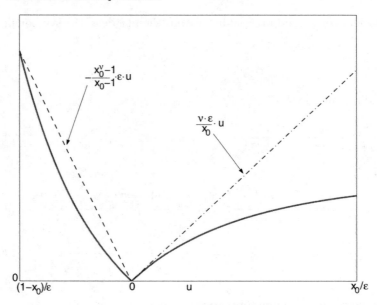

**Fig. 4.12.** The graph of the function (4.A.25) (*thick solid line*), the cord which gives an upper bound of the function within $\left[\frac{1-x_0}{\epsilon}, 0\right]$ (*dashed line*) and the tangent at $0^+$ which gives an upper bound of the function in the interval $\left[0, \frac{x_0}{\epsilon}\right]$ (*dash dotted line*)

The second and third integrals obviously behave like $\mathcal{O}(\epsilon^\nu)$ when $\epsilon$ goes to zero since we have assumed $x_0 > 1$ which ensures that $\frac{1-x_0}{\epsilon} \to -\infty$ and $\frac{x_0}{\epsilon} \to \infty$ when $\epsilon \to 0^+$. For the first integral, we have

$$\left| \int_{\frac{1-x_0}{\epsilon}}^{\frac{x_0}{\epsilon}} du \left[ \frac{1}{(1 + \frac{\epsilon u}{x_0})^\nu} - 1 \right] \frac{C_\nu}{(1 + \frac{u^2}{\nu})^{\frac{\nu+1}{2}}} \right|$$

$$\leq \int_{\frac{1-x_0}{\epsilon}}^{\frac{x_0}{\epsilon}} du \left| \frac{1}{(1 + \frac{\epsilon u}{x_0})^\nu} - 1 \right| \frac{C_\nu}{(1 + \frac{u^2}{\nu})^{\frac{\nu+1}{2}}} .$$

The function

$$\left| \frac{1}{(1 + \frac{\epsilon u}{x_0})^\nu} - 1 \right| \tag{4.A.25}$$

vanishes at $u = 0$, is convex for $u \in [\frac{1-x_0}{\epsilon}, 0)$ and concave for $u \in (0, \frac{x_0}{\epsilon}]$ (see Fig. 4.12), so that there are two constants $A, B > 0$ such that

$$\left| \frac{1}{(1 + \frac{\epsilon u}{x_0})^\nu} - 1 \right| \le - \frac{x_0^\nu - 1}{x_0 - 1} \epsilon \cdot u = -A \cdot \epsilon \cdot u , \quad \forall u \in \left[ \frac{1 - x_0}{\epsilon}, 0 \right] \quad (4.A.26)$$

$$\left| \frac{1}{(1 + \frac{\epsilon u}{x_0})^\nu} - 1 \right| \le \frac{\nu \epsilon}{x_0} u = B \cdot \epsilon \cdot u, \quad \forall u \in \left[ 0, \frac{x_0}{\epsilon} \right] .\quad (4.A.27)$$

We can thus conclude that

$$\left| \int_{\frac{1-x_0}{\epsilon}}^{\frac{x_0}{\epsilon}} du \left[ \frac{1}{(1 + \frac{\epsilon u}{x_0})^\nu} - 1 \right] \frac{C_\nu}{(1 + \frac{u^2}{\nu})^{\frac{\nu+1}{2}}} \right| \le -A \cdot \epsilon \int_{\frac{1-x_0}{\epsilon}}^0 du \frac{u \cdot C_\nu}{(1 + \frac{u^2}{\nu})^{\frac{\nu+1}{2}}}$$

$$+ B \cdot \epsilon \int_0^{\frac{x_0}{\epsilon}} du \frac{u \cdot C_\nu}{(1 + \frac{u^2}{\nu})^{\frac{\nu+1}{2}}}$$

$$= \mathcal{O}(\epsilon^\alpha) , \quad (4.A.28)$$

with $\alpha = \min\{\nu, 1\}$. Indeed, the two integrals can be performed exactly, which shows that they behave as $\mathcal{O}(1)$ if $\nu > 1$ and as $\mathcal{O}(\epsilon^{\nu-1})$ otherwise. Thus, we finally obtain

$$\left| \int_{\frac{1-x_0}{\epsilon}}^{\frac{x_0}{\epsilon}} du \frac{1}{(1 + \frac{\epsilon u}{x_0})^\nu} \frac{C_\nu}{(1 + \frac{u^2}{\nu})^{\frac{\nu+1}{2}}} - 1 \right| = \mathcal{O}(\epsilon^\alpha) . \quad (4.A.29)$$

Putting together (4.A.22) and (4.A.29) gives

$$\left| \frac{1}{\epsilon} \int_1^\infty dx \, \frac{1}{x^\nu} \frac{C_\nu}{\left[ 1 + \frac{1}{\nu} \left( \frac{x-x_0}{\epsilon} \right)^2 \right]^{\frac{\nu+1}{2}}} - \frac{1}{x_0^\nu} \right| = \mathcal{O}(\epsilon^{\min\{\nu,1\}}) , \quad (4.A.30)$$

which concludes the proof.

### 4.A.4 Derivation of Equation (4.A.10)

From (4.A.12), we can deduce

$$\bar{F}_Y(y) = \frac{\nu^{\frac{\nu-1}{2}} C_\nu}{y^\nu} \left( 1 + \mathcal{O}(y^{-2}) \right) . \quad (4.A.31)$$

Using (4.A.3) and (4.A.7), we obtain

$$P_\epsilon(\beta \tilde{Y}_u + \eta - \beta y) = P_\epsilon(\gamma \tilde{Y}_u - \beta y) \cdot \left( 1 + \mathcal{O}(\tilde{Y}_u^{-2}) \right) , \quad (4.A.32)$$

where

$$\gamma = \beta \left( 1 + \left( \frac{\sigma}{\beta} \right)^\nu \right)^{1/\nu} . \quad (4.A.33)$$

Putting together these results yields for the leading order

$$\int_{\tilde{Y}_u}^{\infty} dy \; \bar{F}_Y(y) \cdot P_\epsilon(\beta\tilde{Y}_u + \eta - \beta y)$$

$$= \int_{\tilde{Y}_u}^{\infty} dy \; \frac{\nu^{\frac{\nu-1}{2}} C_\nu}{y^\nu} \cdot \frac{C_\nu}{\sigma \left(1 + \frac{(\gamma\tilde{Y}_u - \beta y)^2}{\nu \, \sigma^2}\right)^{\frac{\nu+1}{2}}}$$

$$= \frac{\nu^{\frac{\nu-1}{2}} C_\nu}{\beta \, \tilde{Y}_u^{\,\nu}} \int_1^{\infty} dx \; \frac{1}{x^\nu} \cdot \frac{C_\nu \frac{\beta\tilde{Y}_u}{\sigma}}{\left(1 + \frac{1}{\nu}\left(\frac{x - \frac{\gamma}{\beta}}{\frac{\sigma}{\beta\tilde{Y}_u}}\right)^2\right)^{\frac{\nu+1}{2}}} \;, \qquad (4.A.34)$$

where the change of variable $x = \frac{y}{\tilde{Y}_u}$ has been performed in the last equation.

We now apply Lemma 4.5.2 with $x_0 = \frac{\gamma}{\beta} > 1$ and $\epsilon = \frac{\sigma}{\beta\tilde{Y}_u}$ which goes to zero as $u \to 1$. This gives

$$\int_{\tilde{Y}_u}^{\infty} dy \; \bar{F}_Y(y) \cdot P_\epsilon(\beta\tilde{Y}_u + \eta - \beta y) \sim_{u\to 1} \frac{\nu^{\frac{\nu-1}{2}} C_\nu}{\beta \, \tilde{Y}_u^\nu} \left(\frac{\beta}{\gamma}\right)^\nu \;, \qquad (4.A.35)$$

which shows that

$$\Pr[X > F_X^{-1}(u), Y > F_Y^{-1}(u)] \sim_{u\to 1} F_Y^{-1}(\tilde{Y}_u) \left(\frac{\beta}{\gamma}\right)^\nu = (1-u) \left(\frac{\beta}{\gamma}\right)^\nu \;.$$

Therefore

$$\Pr[X > F_X^{-1}(u) | Y > F_Y^{-1}(u)] \sim_{u\to 1} \left(\frac{\beta}{\gamma}\right)^\nu \;, \qquad (4.A.36)$$

which finally yields

$$\lambda = \frac{1}{1 + \left(\frac{\sigma}{\beta}\right)^\nu} \;. \qquad (4.A.37)$$

# 5

# Description of Financial Dependences with Copulas

There are two general methods for estimating empirically the copula best describing the dependence structure of a basket of assets, and more generally of a portfolio made of different financial and/or actuarial risks: parametric and nonparametric. The latter class is by far the most general since it does not require the *a priori* specification of a model, and should thus avoid the problem of misspecification (model error). In contrast, the parametric approach has the advantage that, if a model is correctly specified, it leads to a much more precise parametric estimation. In addition, the reduced number of parameters involved in the description of the selected copula can be interpreted as being the relevant meaningful variables that summarize the dependence properties between the assets. Consider for instance the Gaussian representation, or more generally any presentation in terms of elliptical distributions, whose dependence structure is, to large extent (see Chap. 4), summarized by the set of linear coefficients of correlation. These coefficients of correlation thus play a pivotal role and it is tempting to interpret them as the macrovariables (or phenomenological variables) synthesizing all possible microstructural interactions between economic agents leading to the observed dependence. Let us recall that identifying the "correct variables" constitutes the critical first step in model building to obtain the best possible representation of observed phenomena. The usefulness of the parametric estimation is thus obvious from this point of view.

The first section of this chapter reviews the most representative methods to estimate copulas, with an emphasis on the description of parametric approaches. The following section focuses on the problem of model selection and on goodness-of-fit tests. Indeed, the estimation procedure has no sense if the quality and the likelihood of the model are not assessed. Instead of reviewing the many available goodness-of-fit tests, we discuss how to best describe the dependence structure of asset returns and we compare the relative merits of the different models considered in the literature to address this question.

## 5.1 Estimation of Copulas

There is a significant body of literature on the estimation of copulas. This section aims at summarizing some of the most popular techniques which have appeared in the statistical literature and which are now of common use in modeling financial and economic variables as well as actuarial risks.

### 5.1.1 Nonparametric Estimation

#### The Empirical Copula

The very first copula estimation method dates back to the work by De-heuvels [121, 122]. It relies on a simple generalization of the usual estimator of a multivariate distribution. Indeed, considering an $n$-dimensional random vector $\boldsymbol{X} = (X_1, \ldots, X_n)$ whose copula is $C$ and given a sample of size $T$ $\{(x_1(1), x_2(1), \ldots, x_n(1)), \ldots, (x_1(T), x_2(T), \ldots, x_n(T))\}$, a natural idea is to estimate the empirical distribution function $F$ of $\boldsymbol{X}$ as

$$\hat{F}(\boldsymbol{x}) = \frac{1}{T} \sum_{k=1}^{T} \mathbf{1}_{\{x_1(k) \leq x_1, \ldots, x_n(k) \leq x_n\}} , \qquad (5.1)$$

and the empirical marginal distribution functions of the $X_i$'s as

$$\hat{F}_i(x_i) = \frac{1}{T} \sum_{k=1}^{T} \mathbf{1}_{\{x_i(k) \leq x_i\}} . \qquad (5.2)$$

The application of Sklar's theorem would then appear to obtain a nonpara-metric estimation of the copula $C$. Unfortunately, even if the margins of $F$ are continuous, their empirical counterparts are not. Therefore, one cannot determine a *unique* estimated copula $\hat{C}$.[1] Following this approach, one can, however, obtain a unique nonparametric estimator of $C$ defined at the dis-crete points $\left( \frac{i_1}{T}, \frac{i_2}{T}, \ldots, \frac{i_n}{T} \right)$, with $i_k \in \{1, 2, \ldots, T\}$. Inverting the empirical marginal distribution function, we obtain

$$\hat{C}\left( \frac{i_1}{T}, \frac{i_2}{T}, \ldots, \frac{i_n}{T} \right) = \frac{1}{T} \sum_{k=1}^{T} \mathbf{1}_{\{x_1(k) \leq x_1(i_1;T), \ldots, x_n(k) \leq x_n(i_n;T)\}} , \qquad (5.3)$$

where $x_p(k; T)$ denotes the $k^{\text{th}}$ order statistics of the sample $\{x_p(1), \ldots, x_p(T)\}$. Following Deheuvels, one can define an *empirical copula* as any copula which satisfies the relation (5.3).

It is well-known that the empirical distribution function $\hat{F}$ converges, al-most surely, uniformly to the underlying distribution function $F$ from which

---

[1] The same issue arises, of course, for the empirical estimation of marginal as well as multivariate distributions.

the sample is drawn, as the sample size $T$ goes to infinity. This property still holds for the nonparametric estimator defined by the empirical copula

$$\sup_{u \in [0,1]^n} \left| \hat{C}(u) - C(u) \right| \xrightarrow{a.s} 0 . \tag{5.4}$$

Similarly, the empirical copula density $\hat{c}$ can be estimated by

$$\hat{c}\left(\frac{i_1}{T}, \ldots, \frac{i_n}{T}\right) = \begin{cases} \frac{1}{T}, & \text{if } \{x_1(i_1; T), \ldots, x_n(i_n; T)\} \text{ belongs to the sample,} \\ 0, & \text{otherwise.} \end{cases} \tag{5.5}$$

The following relation holds between the empirical copula $\hat{C}$ and the empirical copula density:

$$\hat{c}\left(\frac{i_1}{T}, \ldots, \frac{i_n}{T}\right) = \sum_{k_1=1}^{2} \cdots \sum_{k_n=1}^{2} \left[ (-1)^{k_1 + \cdots + k_n} \times \right.$$
$$\left. \times \hat{C}\left(\frac{i_1 - k_1 + 1}{T}, \ldots, \frac{i_n - k_n + 1}{T}\right) \right] . \tag{5.6}$$

A natural question arises: what is the estimated value of $C(u)$ or $c(u)$ when $u$ does not belong to the lattice defined by the set of points $\left(\frac{i_1}{T}, \frac{i_2}{T}, \ldots, \frac{i_n}{T}\right)$, with $i_k \in \{1, 2, \ldots, T\}$? It would seem that this is nothing but a straightforward interpolation problem, which could be solved by constructing a simple staircase function or applying spline functions, for instance. However, such methods of interpolation do not ensure that the function so obtained fulfills the requirements for a copula, according to Definition 3.2.1; in particular, the function must be $n$-increasing, which requires a multilinear interpolation scheme. In the bivariate case (for simplicity), given any point $(u, v) \in \left[\frac{k_u}{T}, \frac{k_u+1}{T}\right] \times \left[\frac{k_v}{T}, \frac{k_v+1}{T}\right]$, where $k_u, k_v \in \{0, 1, \ldots T - 1\}$ denotes the integer part of $T \cdot u$ and $T \cdot v$ respectively, the following interpolation

$$\tilde{C}(u, v) = \hat{C}\left(\frac{k_u}{T}, \frac{k_v}{T}\right) \cdot (k_u + 1 - T \cdot u)(k_v + 1 - T \cdot v)$$
$$+ \hat{C}\left(\frac{k_u}{T}, \frac{k_v + 1}{T}\right) \cdot (k_u + 1 - T \cdot u)(T \cdot v - k_v)$$
$$+ \hat{C}\left(\frac{k_u + 1}{T}, \frac{k_v}{T}\right) \cdot (T \cdot u - k_u)(k_v + 1 - T \cdot v)$$
$$+ \hat{C}\left(\frac{k_u + 1}{T}, \frac{k_v + 1}{T}\right) \cdot (T \cdot u - k_u)(T \cdot v - k_v) , \tag{5.7}$$

defines a *bona fide* empirical copula. Indeed, by construction $\tilde{C}$ is a copula (see [370, p. 16]) and $\tilde{C}\left(\frac{i}{T}, \frac{j}{T}\right) = \hat{C}\left(\frac{i}{T}, \frac{j}{T}\right)$ for all $i, j \in \{1, \ldots, T\}$.

Li *et al.* [304, 305] have provided some other insightful methods. One of them relies on the use of Bernstein polynomials,

$$P_{i,n}(x) = \binom{n}{i} x^i (1-x)^{n-i}. \tag{5.8}$$

Defining

$$\hat{C}_B(\boldsymbol{u}) = \sum_{i_1=1}^{T} \cdots \sum_{i_n=1}^{T} P_{i_1,T}(u_1) \cdots P_{i_n,T}(u_n) \cdot \hat{C}\left(\frac{i_1}{T}, \frac{i_2}{T}, \ldots, \frac{i_n}{T}\right), \tag{5.9}$$

one obtains a copula which converges uniformly to $C$, almost surely as $T$ goes to infinity (a weak form of the Stone-Weierstrass theorem).[2]

If the method provides a smooth infinitely differentiable copula, it comes however, with two severe drawbacks:

- First, it is easy to show [138] that any differentiable copula in the neighborhood of $(1,1)$ (or of $(0,0)$) has a vanishing coefficient of tail dependence $\lambda$ (see Chap. 4). Indeed, a necessary condition for $\lambda$ not to vanish is that the copula be non-differentiable in the neighborhood of $(1,1)$.[3] Thus, by construction, a nonparametric estimation of copulas using the interpolation method described above automatically forbids a correct estimation of the tail dependence parameter. Such an estimation amounts to project the copula onto the set of copulas with vanishing tail dependence.
- Second, the convergence of the derivatives of $\hat{C}_B$ toward the derivatives of $C$ is not *a priori* ensured. As a consequence, it is not possible to use these estimates to generate simulated data enjoying the same dependence structure as that of the sample (see Sect. 3.5). This is particularly harmful since one often has to resort to Monte Carlo simulations and bootstrap methods to assess portfolio risk or to valuate derivative assets. It is thus necessary to look for nonparametric estimators of both the copula and its derivatives.

## Kernel Copula Estimator

Smooth joint estimates of a copula and of its derivatives can be obtained by using a kernel-based approach [168]. Still considering an $n$-dimensional random vector $\boldsymbol{X}$ with copula $C$, let us call its joint distribution function $F$ and its marginal distribution functions $F_i$ such that $F(X_1, \ldots, X_n) =$

---

[2] According to the Weierstrass approximation theorem, any continuous function defined on an interval $[a,b]$ can be uniformly approximated as closely as desired by a polynomial function. The Stone-Weierstrass theorem generalizes the Weierstrass approximation theorem in two directions by considering an arbitrary compact Hausdorff space instead of a compact interval $[a,b]$, and approximations with elements from more general sets than polynomials.

[3] Note that this condition is necessary but not sufficient as shown for instance by the Gaussian copula which is not differentiable at $(1,1)$ but nevertheless has vanishing tail dependence.

$C\left(F_1\left(X_1\right),\ldots,F_n\left(X_n\right)\right)$. The most commonly used kernel is probably the Gaussian kernel,

$$\varphi(x) = \frac{1}{\sqrt{2\pi}}e^{-\frac{1}{2}x^2} \ . \tag{5.10}$$

We present the general procedure detailed in [168] on this particular example. Let us first estimate the joint distribution of $\boldsymbol{X}$. Given the sample of size $T \ \{(x_1(1),x_2(1), \ldots,x_n(1)),\ldots,(x_1(T),x_2(T),\ldots,x_n(T))\}$, the kernel estimates of $F_i(x)$ and $F(x)$ are

$$\hat{F}_i(x_i) = \frac{1}{T}\sum_{t=1}^{T}\Phi\left(\frac{x_i - x_i(t)}{h_i}\right) \ , \tag{5.11}$$

and

$$\hat{F}(x) = \frac{1}{T}\sum_{t=1}^{T}\prod_{i=1}^{n}\Phi\left(\frac{x_i - x_i(t)}{h_i}\right) \ , \tag{5.12}$$

where

$$\Phi(x) = \int_{-\infty}^{x}\varphi(t)\ dt \tag{5.13}$$

and $(h_1,\ldots,h_n)$ is the bandwidth, a function of $T$ with value in $\mathbb{R}^n$ and satisfying

$$h_i(T) > 0, \quad \forall T, i \in \{1,\ldots,n\} \ , \tag{5.14}$$

$$\prod_{i=1}^{n}h_i(T) + \left[T \cdot \prod_{i=1}^{n}h_i(T)\right]^{-1} \longrightarrow 0 \ , \quad \text{as } T \longrightarrow \infty \ . \tag{5.15}$$

In practice, one usually chooses $h_i = \hat{\sigma}_i \cdot (4/3T)^{1/5}$, where $\hat{\sigma}_i$ denotes the sample standard deviation of $\{x_i(1),\ldots,x_i(T)\}$.

Defining $\hat{\boldsymbol{q}}$, the vector whose $i^{\text{th}}$ component is the $u_i$-quantile of $\hat{F}_i$,

$$\hat{q}_i = \inf_{x\in\mathbb{R}}\{x : \hat{F}_i(x) \geq u_i\}, \quad u_i \in (0,1) \ , \tag{5.16}$$

the kernel estimator of the copula $C$ is simply given by

$$\hat{C}(u_1,\ldots,u_n) = \hat{F}(\hat{\boldsymbol{q}}) \ . \tag{5.17}$$

Under mild regularity conditions, this kernel estimator is asymptotically Gaussian. From Proposition 1 in [168], one can show that

$$\left(T \cdot \prod_{i=1}^{n}h_i\right)^{1/2} \cdot \left(\hat{C}(\boldsymbol{u}) - C(\boldsymbol{u})\right) \longrightarrow \mathcal{N}\left(0, C(\boldsymbol{u})\right) \ . \tag{5.18}$$

This asymptotic behavior holds even when the sample is not iid, provided that the underlying process satisfies some strong mixing conditions, roughly speaking (see [168] for details). Therefore, this method can be applied to financial asset returns, which are known to exhibit volatility clustering among other time dependence patterns.

By construction of the kernel estimator, it can be differentiated with respect to the $u_i$'s. It is thus easy to obtain an estimator of a partial derivative of the copula with respect to one (or more) of the variables. For instance, the kernel estimator of the first order partial derivative of $C$ with respect to $u_i$ is

$$\frac{\widehat{\partial C(\boldsymbol{u})}}{\partial u_i} = \frac{\partial \hat{C}(\boldsymbol{u})}{\partial u_i} = \frac{1}{\hat{f}_i\left(\hat{q}_i\left(u_i\right)\right)} \cdot \partial_i \hat{F}\left(\hat{\boldsymbol{q}}(\boldsymbol{u})\right) \tag{5.19}$$

where $\hat{f}_i$ denotes the kernel estimate of the marginal density of $X_i$,

$$\hat{f}_i(x_i) = \frac{1}{T \cdot h_i} \sum_{t=1}^{T} \varphi\left(\frac{x_i - x_i(t)}{h_i}\right) , \tag{5.20}$$

and $\partial_i \hat{F}$ is the partial derivative of $\hat{F}$ with respect to its $i$th variable.

Again, under mild regularity conditions, it can be shown that this estimator is asymptotically Gaussian, so that

$$\left(T \cdot \prod_{i=1}^{n} h_i\right)^{1/2} \cdot \left(\frac{\partial \hat{C}(\boldsymbol{u})}{\partial u_i} - \frac{\partial C(\boldsymbol{u})}{\partial u_i}\right) \longrightarrow \mathcal{N}\left(0, \frac{1}{f_i(q_i(u_i))} \cdot \frac{\partial C(\boldsymbol{u})}{\partial u_i}\right). \tag{5.21}$$

Applying the same kind of arguments, one can estimate the higher order partial derivatives of the copula $C$:

$$\frac{\widehat{\partial C(\boldsymbol{u})}}{\partial u_{i_1} \cdots \partial u_{i_k}} = \frac{\partial \hat{C}(\boldsymbol{u})}{\partial u_{i_1} \cdots \partial u_{i_k}} = \frac{\partial_{i_1, \dots, i_k} \hat{F}\left(\hat{\boldsymbol{q}}(\boldsymbol{u})\right)}{\hat{f}_{i_1}\left(\hat{q}_{i_1}\left(u_{i_1}\right)\right) \cdots \hat{f}_{i_k}\left(\hat{q}_{i_k}\left(u_{i_k}\right)\right)} , \tag{5.22}$$

where all the $i_j$'s are assumed different and $k \le n$. As a consequence, it becomes possible to simulate random variables with copula $\hat{C}$, by using the algorithm detailed in Sect. 3.5.2.

When $k = n$, one obtains the kernel estimator of the copula density:

$$\hat{c}(\boldsymbol{u}) = \frac{\hat{f}\left(\hat{\boldsymbol{q}}(\boldsymbol{u})\right)}{\hat{f}_1\left(\hat{q}_1\left(u_1\right)\right) \cdots \hat{f}_n\left(\hat{q}_n\left(u_n\right)\right)} , \tag{5.23}$$

where $\hat{f}$ denotes the kernel estimate of the joint density of $\boldsymbol{X}$,

$$\hat{f}(\boldsymbol{x}) = \frac{1}{T \cdot \prod_i h_i} \sum_{t=1}^{T} \prod_{i=1}^{n} \varphi\left(\frac{x_i - x_i(t)}{h_i}\right). \tag{5.24}$$

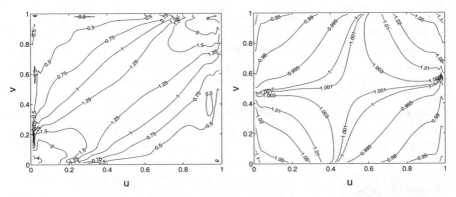

**Fig. 5.1.** Contour plot of the copula density estimated by the kernel method for the daily returns of the couple constituted of the German Mark ($u$ variable) and the Japanese Yen ($v$ variable) over the time interval from 1 May, 1973 to 2 November, 2001 (*left panel*) and for the couple made of General Motors ($u$ variable) and Procter & Gamble ($v$ variable) over the time period from 3 July, 1962 to 29 December, 2000 (*right panel*). The German Mark data has been reconstructed from the Euro data after 31 December, 1999

Two examples of copula densities estimated by the kernel method are shown in Fig. 5.1. Observe that the level curves of the left panel are rather similar to those of Fig. 3.3, which depicts the contour plot of a Student copula. This is suggestive of the relevance of a Student's copula with a moderate number of degrees of freedom as a possible candidate for modeling dependencies between the returns of foreign exchange rates.[4] For stock returns, the situation is less clear, even if one could surmise that a Student copula with a large number of degrees of freedom could be a reasonable model.

To sum up this paragraph on kernel estimators, let us stress that, notwithstanding their seeming attractiveness, they have a severe drawback as they require a very large amount of data. As an illustration, in order to obtain the two pictures of Fig. 5.1, we used between 7,000 and 10,000 data points. With less than 2,500–5,000 points, one obtains unreliable estimates in most cases, showing that the kernel estimators behave badly for small samples. Therefore, with daily returns, an accurate non-parametric estimate of the copula requires between 30 and 40 years of data. Over such a long time period, it is far from given that the dependence structure remains stationary, thus possibly blowing up the whole effort.

### 5.1.2 Semiparametric Estimation

When the number of observations is not large enough and/or when one has a sufficiently accurate idea of the true model, it is in general more profitable

---

[4] Of course, this statement should be formally tested by using rigorous statistical techniques. See the following sections.

to apply a parametric or semiparametric estimation method. By parametric, we mean a method based on an entirely parametric model: in such a case, we assume that the true model belongs to a given family of *multivariate* distributions, *i.e.*, a family of copula plus a family of univariate distribution for each individual marginal law. Such a modeling approach requires a very accurate knowledge of the true distribution and can lead to bad estimations of the copula parameters if the marginals are misspecified. Thus, when in doubt concerning the univariate marginal distributions of the data (see in this respect the cautionary study presented in Chap. 2), a semiparametric approach may be preferable. Indeed, in contrast with fully parametric methods, semiparametric techniques use a parametric representation only for the copula. No assumption is made concerning the marginal distributions, which may either be estimated nonparametrically or not even come into play at all, as we shall see now.

**Estimation Based on Concordance Measures**

Basically, two kinds of semiparametric methods exist. The simplest one is based upon the nonparametric estimation of parameters which only depend on the copula. Concordance measures, such as Kendall's tau and Spearman's rho for instance, provide good examples. They can be easily estimated and, once a parametric family of copulas has been retained, one just has to express the parameters of the copula as functions of these estimated quantities. It is the stance taken by Oakes [375] to estimate the parameter $\theta$ of a Clayton copula (see (3.43)). Table 4.1 gives the following relation between the parameter $\theta$ and Kendall's tau:

$$\theta = \frac{2\tau}{1 - \tau} \ . \tag{5.25}$$

Therefore, a natural estimator of $\theta$ is:

$$\hat{\theta}_T = \frac{2\hat{\tau}_T}{1 - \hat{\tau}_T} \ , \tag{5.26}$$

where $\hat{\tau}_T$ denotes the sample version of Kendall's tau, based on the bivariate sample of size $T$: $\{(x_1, y_1), \ldots, (x_T, y_T)\}$. Let us recall that

$$\hat{\tau}_T = 2 \cdot \frac{C - D}{T \cdot (T - 1)} \ , \tag{5.27}$$

where $C$ (resp. $D$) denotes the number of concordant (resp. discordant) pairs, *i.e.*, such that $(x_i - x_j) \cdot (y_i - y_j) > 0$ (resp. $< 0$).

Based on relation (4.37), a similar approach can be applied to estimate the shape parameter $\rho$ of an elliptical copula. One then obtains

$$\hat{\rho}_T = \sin\left(\frac{\pi}{2}\hat{\tau}_T\right) . \tag{5.28}$$

As an illustration, let us consider again the two samples of the daily returns of the German Mark and the Japanese Yen on the one hand, and of General Motors and Procter & Gamble on the other hand. Assuming that their copula belongs to the class of elliptical copulas, we can apply this method to infer the value of the shape parameter $\rho$ for each pair of assets. For the first one (German Mark/Japanese Yen), we obtain: $\hat{\tau}_T = 0.37$ so that $\hat{\rho}_T = 0.54$, while for the second one: $\hat{\tau}_T = 0.18$ and therefore $\hat{\rho}_T = 0.29$. This shows that the dependence is stronger between the pair of currencies than between the pair of stocks.

This method is particularly attractive due to its simplicity but is a bit naive. While it provides very simple and robust estimators, these estimators are not always very accurate. This justifies turning to more elaborated methods, such as that developed by Genest $et$ $al.$ [197], which relies on the maximization of a pseudo likelihood.

### Pseudo Maximum Likelihood Estimation

Let us still consider a sample of size $T$ $\{(x_1(1), x_2(1), \ldots, x_n(1)), \ldots, (x_1(T),$ $x_2(T), \ldots, x_n(T))\}$, drawn from a common distribution $F$ with copula $C$ and margins $F_i$. By definition of the copula, the random vector $\boldsymbol{U}$ whose $i$th component is given by $U_i = F_i(X_i)$ has a distribution function equal to $C$. Assuming that the copula $C = C(\cdot; \boldsymbol{\theta}^0)$ belongs to the family $\{C(u_1, \ldots, u_n; \boldsymbol{\theta}); \boldsymbol{\theta} \in \Theta \subset \mathbb{R}^p\}$, where $\boldsymbol{\theta}$ denotes the vector parameterizing the copula, the function

$$\ln \mathcal{L} = \sum_{i=1}^{T} \ln c \left( F_1 \left( x_1(i) \right), \ldots, F_n \left( x_n(i) \right); \boldsymbol{\theta} \right) , \tag{5.29}$$

where $c(\cdot; \boldsymbol{\theta})$ denotes the density of $C(\cdot; \boldsymbol{\theta})$, provides the likelihood of the sequence $\{(u_1(k) = F_1(x_1(k)), \ldots, u_N(k) = F_N(x_N(k)))\}_{k=1}^{T}$. Note that the sequence $\{(u_1(k) = F_1(x_1(k)), \ldots, u_N(k) = F_N(x_N(k)))\}_{k=1}^{T}$ is independently and identically distributed provided that the $x_i(k)$'s are independent and identically distributed realizations.

Since the marginal distributions are generally unknown and when no parametric model seems available, it is reasonable to use the empirical marginal distribution functions $\hat{F}_i$ defined by (5.2) to obtain an estimator of $\boldsymbol{U}$,

$$\hat{U} = \left( \hat{F}_1(X_1), \ldots, \hat{F}_n(X_n) \right) . \tag{5.30}$$

Then, one derives the $pseudo$-sample $\{(\hat{u}_1(k), \ldots, \hat{u}_n(k))\}_{k=1}^{T}$, where $\hat{u}_i(k) = \hat{F}_i(x_i(k))$, which is not iid even if the $x_i(k)$'s are. Hence, substituting the $u(k)$'s for the $\hat{u}(k)$'s in the log-likelihood function (5.29), one obtains the $peudo$ log-likelihood of the model, based on the sample $\{(x_1(1), \ldots, x_1(T)), \ldots, (x_n(1), \ldots, x_n(T))\}$:

$$\ln \tilde{\mathcal{L}} = \sum_{i=1}^{T} \ln c \left( \hat{F}_1 \left( x_1(i) \right), \ldots, \hat{F}_n \left( x_n(i) \right) ; \boldsymbol{\theta} \right). \tag{5.31}$$

Finally, the parameter vector $\boldsymbol{\theta}$ is estimated by maximizing the pseudo log-likelihood, so that

$$\hat{\boldsymbol{\theta}}_T = \arg \max_{\boldsymbol{\theta}} \ \ln \tilde{\mathcal{L}} \left( \{ x_1(i), \ldots, x_n(i) \}; \boldsymbol{\theta} \right). \tag{5.32}$$

Under the usual technical regularity conditions ensuring the consistency and asymptotic normality of maximum likelihood estimators, Genest et al. [197] have shown that $\hat{\boldsymbol{\theta}}_T$ is a consistent estimator of $\boldsymbol{\theta}^0$ and that it is asymptotically Gaussian (see also Appendix 5.B):

$$\sqrt{n} \left( \hat{\boldsymbol{\theta}}_T - \boldsymbol{\theta}^0 \right) \xrightarrow{law} \mathcal{N} \left( 0, \boldsymbol{\Sigma}^2 \right) \tag{5.33}$$

with $\boldsymbol{\Sigma}^2 = \boldsymbol{I} \left( \boldsymbol{\theta}^0 \right)^{-1} + \boldsymbol{I} \left( \boldsymbol{\theta}^0 \right)^{-1} \boldsymbol{\Omega} \boldsymbol{I} \left( \boldsymbol{\theta}^0 \right)^{-1}$, where $\boldsymbol{I} \left( \boldsymbol{\theta}^0 \right)$ denotes Fisher's information matrix at $\boldsymbol{\theta}^0$:

$$\left[ \boldsymbol{I} \left( \boldsymbol{\theta}^0 \right) \right]_{ij} = \mathrm{E} \left[ \frac{\partial c(\boldsymbol{U}; \boldsymbol{\theta})}{\partial \theta_i} \cdot \frac{\partial c(\boldsymbol{U}; \boldsymbol{\theta})}{\partial \theta_j} \right]_{\boldsymbol{\theta} = \boldsymbol{\theta}^0}, \tag{5.34}$$

and, with $p = \dim \boldsymbol{\theta}$,

$$\Omega_{ij} = \mathrm{Cov} \left[ \sum_{k=1}^{p} W_{ki}(U_k), \sum_{k=1}^{p} W_{kj}(U_k) \right], \tag{5.35}$$

where

$$W_{ki}(U_k) = \int_{\boldsymbol{u} \in [0,1]^p} \mathbf{1}_{\{ U_k \leq \boldsymbol{u}_k \}} \left. \frac{\partial^2 \ln c(\boldsymbol{u}; \boldsymbol{\theta})}{\partial \theta_i \partial u_i} \right|_{\boldsymbol{\theta} = \boldsymbol{\theta}^0} dC \left( \boldsymbol{u}; \boldsymbol{\theta}^0 \right). \tag{5.36}$$

These results rely on a straightforward application of the consistency and asymptotic normality of functionals of multivariate rank statistics derived by Ruymgaart *et al.* [423, 424] and Rüschendorf [422]. Since $\boldsymbol{\Omega}$ is a positive definite matrix, the covariance matrix of the estimator $\hat{\boldsymbol{\theta}}_T$ is larger[5] than it would be, were the marginal distributions $F_i$ perfectly known. Indeed, in such a case, the covariance matrix of the estimator would be nothing but the inverse of Fisher's information matrix $\boldsymbol{I} \left( \boldsymbol{\theta}^0 \right)^{-1}$.

As an illustration, let us fit the two samples considered in Sect. 5.1.1, namely the daily returns of the FX rate of the German Mark and of the

---

[5] We say that a matrix $A$ is larger than a matrix $B$ if their difference $A - B$ is a positive definite matrix. In particular, it implies that the diagonal terms of $A$ are larger one to one than the diagonal terms of $B$. In the present case, this means that the variance of each component of the pseudo maximum likelihood estimator is larger than the variance of each component of the *actual* maximum likelihood estimator, yielding less accurate estimates of the parameter $\theta_0$.

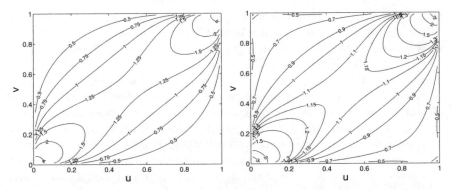

**Fig. 5.2.** Contour plot of the Student copula maximizing the pseudo likelihood (5.31) for the daily returns of the couple German Mark/Japanese Yen over the time interval from 1 May, 1973 to 2 November, 2001 (*left panel*) and for the couple General Motors/Procter & Gamble over the time period from 3 July, 1962 to 29 December, 2000 (*right panel*)

Japanese Yen over the time interval from 1 May, 1973 to 2 November, 2001 and the daily returns of the couple of stocks (General Motors; Procter & Gamble) over the time period from 3 July, 1962 to 29 December, 2000. As aforementioned, the kernel estimates (see Fig. 5.1) of the copulas of these two couples suggest that the Student copula could provide a reasonable description of their dependence structure, at least for the pair of currencies. The pseudo log-likelihood of these samples for a Student copula with $\nu$ degrees of freedom and shape matrix $\rho$ can be straightforwardly derived from (3.37) p. 110. No closed form for $\hat{\rho}$ and $\hat{\nu}$ can be obtained. One has to maximize the pseudo log-likelihood with a numerical procedure. Figure 5.2 depicts the contour plot of the Student copula maximizing the pseudo likelihood for each sample. For the sample (German Mark; Japanese Yen), we obtain the following estimates for the shape parameter and the number of degrees of freedom respectively: $\hat{\rho} = 0.54$ and $\hat{\nu} = 5.82$. Comparing the left panels of Figs. 5.1 and 5.2, we observe that the Student copula estimated from the data seems reasonably close to the kernel estimate of the copula, suggesting that the copula model is realistic in this case. For the couple (General Motors; Procter & Gamble), we find $\hat{\rho} = 0.29$ and $\hat{\nu} = 5.92$. However, when comparing the right panels of Figs. 5.1 and 5.2, one can observe a clear discrepancy between the two models and it is doubtful that the Student copula provides a good representation of the dependence in this case. Settling this question requires to qualify the goodness-of-fit of the model, which will be discussed in Sect. 5.2.

In the mean time, let us stress two important points concerning the practical implementation of the pseudo maximum likelihood estimation method.

- It is convenient to replace the empirical distribution function $\hat{F}_i(\cdot)$, defined by (5.2), by $\frac{T}{T+1}\hat{F}_i(\cdot)$. These two quantities are asymptotically equivalent but the use of the latter allows to prevent potential unboundedness of the pseudo log-likelihood when one of the $\hat{u}_i$'s tends to one.

- Any maximization algorithm requires an initialization. The choice of the starting point is not innocuous since the performance of the algorithm can depend, for a large part, on it. For any elliptical copula, assessing the Kendall's tau and applying relation (5.28) allows one to obtain a good starting point. In fact, the estimation of $\rho_{ij}$ from $\tau_{ij}$ often provides such a good starting point that the pseudo maximum likelihood estimate of $\rho$ does not significantly improve on it [350]. Our examples confirm this point: with both methods, we have obtained the same values (within their confidence interval). In addition, the first estimation method is much faster than the second one. These remarks are specially important when one deals with large portfolios for which the numerical maximization of the pseudo likelihood becomes particularly tricky (and time consuming). Therefore, in such a case, the non-parametric estimation of $\rho$ by relation (5.28) is probably the best method. Then, one can obtain an accurate estimate of the number $\nu$ of degrees of freedom by maximization of the pseudo likelihood with respect to this single parameter only,

$$\hat{\nu}_T = \arg \max_{\nu} \; \tilde{\mathcal{L}}\left(\{x\}; \hat{\rho}_T, \nu\right) \; , \tag{5.37}$$

where $\hat{\rho}_T$ denotes the non-parametric estimator of $\rho$ obtained from (5.28).

To conclude on the two semiparametric estimation methods that we have presented, both methods have their pros and cons. For low dimensional problems, the pseudo maximum likelihood estimator is probably the best. Its variance is usually lower: Genest $et$ $al.$ [197] report that the variance of this estimator is smaller than the variance of the estimator based on Kendall's tau by 10–40% for Clayton's copula (depending on the value of parameter $\theta$). In contrast, when the dimension of the problem is large, the pseudo maximum likelihood method becomes time consuming and less efficient.

## 5.1.3 Parametric Estimation

While many procedures exist, we will only focus on maximum likelihood methods. Among those, two main approaches can be distinguished: the *one*-step maximum likelihood estimation and the *two*-step maximum likelihood estimation.

Given a multivariate distribution function $F(\boldsymbol{x}; \boldsymbol{\theta})$ depending on the vector of parameters $\boldsymbol{\theta} \in \Theta \subset \mathbb{R}^p$, which can be represented as $F(\boldsymbol{x}; \boldsymbol{\theta}) = C\left(F_1(x_1; \boldsymbol{\theta}), \ldots, F_n(x_n; \boldsymbol{\theta}); \boldsymbol{\theta}\right)$, and given a sample of size $T$ $\{(x_1(1), x_2(1), \ldots, x_n(1)), \ldots, (x_1(T), x_2(T), \ldots, x_n(T))\}$, the log-likelihood of the model is

$$\ln \mathcal{L}\left(\{x\}; \boldsymbol{\theta}\right) = \sum_{i=1}^{T} \ln c\left(F_1(x_1(i); \boldsymbol{\theta}), \dots, F_n(x_n(i); \boldsymbol{\theta}); \boldsymbol{\theta}\right)$$

$$+ \sum_{i=1}^{T} \ln f_1\left(x_1(i); \boldsymbol{\theta}\right) + \cdots + \sum_{i=1}^{T} \ln f_n\left(x_n(i); \boldsymbol{\theta}\right), \qquad (5.38)$$

where, as usual, $c(\cdot; \boldsymbol{\theta})$ denotes the density of $C(\cdot; \boldsymbol{\theta})$ and the $f_i(\cdot; \boldsymbol{\theta})$'s are the densities of the marginal distribution function $F_i(\cdot; \boldsymbol{\theta})$'s. The one-step maximum likelihood estimator of $\boldsymbol{\theta}$ is then

$$\hat{\boldsymbol{\theta}}_T = \arg\max_{\boldsymbol{\theta}} \ \ln \mathcal{L}\left(\{x\}; \boldsymbol{\theta}\right). \qquad (5.39)$$

Under the usual regularity conditions, it enjoys the properties of consistency and asymptotic normality, with its asymptotic covariance matrix given by the inverse of Fisher's information matrix.

Consider the dependence structure for a sample, supplied by the Insurance Service Office, of indemnity claims of insurance companies consisting of indemnity payment (or loss) and allocated loss adjustment expenses (ALAE). Applying the one-step maximum likelihood method leading to (5.39), Frees and Valdez [183] and Klugman and Parsa [272] have shown that the dependence of this sample can be reasonably modeled by Gumbel's or Frank's copula. We should stress that, for this procedure to work properly, the choice of the marginals is crucial. It is thus appropriate to model each marginal distribution function and perform a first maximum likelihood estimation of their corresponding parameters. Then, together with the choice of a suitable copula, these preliminary estimates of the parameters of the marginal distributions provide useful starting points to globally maximize (5.38) numerically.

Pushing further this reasoning, consider the situation where one can split the parameter vector $\boldsymbol{\theta}$ under the form $\boldsymbol{\theta} = (\boldsymbol{\alpha}, \boldsymbol{\beta}_1, \dots, \boldsymbol{\beta}_n)$ so that

$$F(\boldsymbol{x}; \boldsymbol{\theta}) = C\left(F_1(x_1; \boldsymbol{\beta}_1), \dots, F_n(x_n; \boldsymbol{\beta}_n); \boldsymbol{\alpha}, \boldsymbol{\beta}_1, \dots, \boldsymbol{\beta}_n\right), \qquad (5.40)$$

i.e., the marginal distributions are functions of independent sets of parameters. As a consequence, the log-likelihood reads

$$\ln \mathcal{L}\left(\{x\}; \boldsymbol{\alpha}, \boldsymbol{\beta}_1, \dots, \boldsymbol{\beta}_n\right) = \sum_{i=1}^{T} \ln c\left(F_1(x_1(i); \boldsymbol{\beta}_1), \dots, F_n(x_n(i); \boldsymbol{\beta}_n); \boldsymbol{\theta}\right)$$

$$+ \sum_{i=1}^{T} \ln f_1\left(x_1(i); \boldsymbol{\beta}_1\right)$$

$$\vdots$$

$$+ \sum_{i=1}^{T} \ln f_n\left(x_n(i); \boldsymbol{\beta}_n\right). \qquad (5.41)$$

Thus, instead of looking for the global maximum

$$\left(\hat{\alpha}_T, \hat{\boldsymbol{\beta}}_{1,T}, \dots, \hat{\boldsymbol{\beta}}_{n,T}\right) = \arg \max_{\boldsymbol{\alpha}, \boldsymbol{\beta}_1, \dots, \boldsymbol{\beta}_n} \ln \mathcal{L}\left(\{\boldsymbol{x}\}; \boldsymbol{\alpha}, \boldsymbol{\beta}_1, \cdots, \boldsymbol{\beta}_n\right),$$

(5.42)

over $(\boldsymbol{\alpha}, \boldsymbol{\beta}_1, \dots, \boldsymbol{\beta}_n)$, one can perform a two-steps–in fact $(n + 1)$-steps– maximization of the likelihood:

$$\hat{\boldsymbol{\beta}}_{1,T} = \arg \max_{\boldsymbol{\beta}_1} \sum_{i=1}^{T} \ln f_1\left(x_1(i); \boldsymbol{\beta}_1\right)$$

(5.43)

$$\vdots$$

$$\hat{\boldsymbol{\beta}}_{n,T} = \arg \max_{\boldsymbol{\beta}_n} \sum_{i=1}^{T} \ln f_n\left(x_n(i); \boldsymbol{\beta}_n\right)$$

(5.44)

$$\hat{\boldsymbol{\alpha}}_T \;\; = \arg \max_{\boldsymbol{\alpha}} \sum_{i=1}^{T} \ln c\left(F_1\left(x_1(i); \hat{\boldsymbol{\beta}}_{1,T}\right), \dots\right.$$

$$\left. \dots, F_n\left(x_n(i); \hat{\boldsymbol{\beta}}_{n,T}\right); \boldsymbol{\alpha}, \hat{\boldsymbol{\beta}}_{1,T}, \dots, \hat{\boldsymbol{\beta}}_{n,T}\right).$$

(5.45)

One can prove that the two-step estimator $\hat{\boldsymbol{\theta}}_T = \left(\hat{\alpha}_T, \hat{\boldsymbol{\beta}}_{1,T}, \dots, \hat{\boldsymbol{\beta}}_{n,T}\right)$ is consistent and asymptotically Gaussian [248, 372, 492],

$$\left(\hat{\boldsymbol{\theta}}_T - \boldsymbol{\theta}_0\right) \xrightarrow{law} \mathcal{N}\left(0, \boldsymbol{A}^{-1} \boldsymbol{B} \boldsymbol{A}^{-1^t}\right),$$

(5.46)

where $\boldsymbol{A}^{-1} \boldsymbol{B} \boldsymbol{A}^{-1^t}$ is the inverse of Godambe's information matrix,[6] with

$$\boldsymbol{A} = \begin{pmatrix} \mathrm{E}\left[\partial_{\beta_1, \beta_1} \ln f_1|_{\beta_1^0}\right] & 0 & 0 & 0 \\ 0 & \ddots & 0 & 0 \\ 0 & 0 & \mathrm{E}\left[\partial_{\beta_n, \beta_n} \ln f_n|_{\beta_n^0}\right] & 0 \\ \mathrm{E}\left[\partial_{\beta_1, \alpha} \ln c|_{\theta^0}\right] & \cdots & \mathrm{E}\left[\partial_{\beta_n, \alpha} \ln c|_{\theta^0}\right] & \mathrm{E}\left[\partial_{\alpha, \alpha} \ln c|_{\theta^0}\right] \end{pmatrix},$$

(5.47)

and

$$\boldsymbol{B} = \mathrm{Cov}\left[\left(\partial_{\beta_1} \ln f_1|_{\beta_1^0}, \cdots, \partial_{\beta_n} \ln f_n|_{\beta_n^0}, \partial_{\alpha} \ln c|_{\theta^0}\right)\right].$$

(5.48)

While asymptotically less efficient than the one-step estimator, this approach has the obvious advantage of reducing the dimensionality of the problem, which is particularly useful when one has to resort to a numerical maximization.

In practice, one has often to deal with samples of different lengths. This may occur for instance when considering simultaneously mature and emerging

---

[6] Godambe's information matrix has been introduced in the context of inference functions (or estimating equations) [205, 354].

markets with different lifespans, or market returns together with the returns of a company which has only been recently introduced in the stock exchange or which has defaulted, or foreign exchange rates where one of the currencies of interest has only a short history, such as the Euro. In such a case, the two-step method is much better than the one-step method. The latter requires using a data set which is the intersection of all the marginal samples, leading often to a significant loss of efficiency in the estimation of the parameters of marginal distributions. In contrast, the two-step method uses the whole set of samples for the estimation of marginal parameters and restricts to the intersection of the marginal samples only for the estimation of the parameters of the copula. This two-step estimator is still consistent and asymptotically Gaussian. Its asymptotic variance can be derived from (5.47–5.48), by accounting for the different lengths of the marginal samples (see Patton [380]). While the one-step estimator still remains asymptotically more efficient than the two-step estimator, Patton reports that the accuracy of the two-step estimator is much better than that of the one-step estimator, when the size of the intersection of the marginal samples is small.

### 5.1.4 Goodness-of-Fit Tests

Many different tests have been developed to check the goodness of fit of a copula. Basically, the simplest approach uses the property that, under the null hypothesis that $C(u_1, \ldots, u_n)$ is the right copula, the set of random variables

$$C_n(U_n|U_1, \ldots, U_{n-1}), \ldots, C_2(U_2|U_1), U_1 , \tag{5.49}$$

with

$$C_k(u_k|u_1, \ldots, u_{k-1}) = \frac{\partial_{u_1} \cdots \partial_{u_{k-1}} C_k(u_1, \ldots, u_k)}{\partial_{u_1} \cdots \partial_{u_{k-1}} C_{k-1}(u_1, \ldots, u_{k-1})} , \tag{5.50}$$

are identically, uniformly, and independently distributed. This property has already been used in Sect. 3.5.2 to provide an algorithm for the generation of random variables with a given copula $C$. Thus, testing the null hypothesis is equivalent to testing that the sample of $T$ vectors

$$\{C_n(\hat{u}_n(t)|\hat{u}_1(t), \cdots, \hat{u}_{n-1}(t)), \cdots, C_2(\hat{u}_2(t)|\hat{u}_1(t)), \hat{u}_1(t)\}_{t=1}^{T} , \tag{5.51}$$

is drawn from a population of uniform random vectors with independent components. Such tests date back to [121, 122]. In the same vein, the more recent Bhattacharya-Matusita-Hellinger dependence metric discussed in Chap. 4 can also be used [214]. It allows in particular to account for censored data [272], which is particularly useful when dealing with insurance data. One can also focus on a restricted area of the unit hypercube by using hit tests [381, 380].

Other alternatives consist in testing the significance of the distance in $L^p$, for some $p$, between the null copula $C$ and the estimated copula $\hat{C}$:

$$D_p = \int_{\boldsymbol{u} \in [0,1]^n} \left| C(\boldsymbol{u}) - \hat{C}(\boldsymbol{u}) \right|^p d\boldsymbol{u} , \tag{5.52}$$

but the statistical properties of such tests are rather poor [169]. A simpler approach focuses on the discrepancy between the fitted copula and the null copula on the main diagonal only by use of the $K$ function:

$$K(z) = \Pr \left[ C \left( U_1, \ldots, U_n \right) \leq z \right] . \tag{5.53}$$

Since $K$ is a univariate distribution function, Kolmogorov or Anderson-Darling tests can be applied.

In fact, this last approach is particularly interesting when one deals with Archimedean copulas since $K$ can be shown to admit the simple closed-form expression [35]

$$K(z) = z + \sum_{k=1}^{n-1} (-1)^k \frac{\varphi^k(z)}{k!} \cdot \chi_{k-1}(z) , \tag{5.54}$$

where $\varphi(\cdot)$ denotes the generator of the copula and:

$$\chi_k(z) = \frac{\partial_z \chi_{k-1}(z)}{\partial_z \varphi(z)}, \text{ with } \chi_0(z) = [\partial_z \varphi(z)]^{-1} . \tag{5.55}$$

## 5.2 Description of Financial Data in Terms of Gaussian Copulas

Section 3.6 has discussed the importance of the Gaussian copula for financial modeling. We now review the empirical tests of the hypothesis, denoted $H_0$, that the Gaussian copula is the correct description of the dependence between financial assets. After summarizing the testing procedure developed in [334], we describe the results.

### 5.2.1 Test Statistics and Testing Procedure

Let us first derive the test statistics which will allow us to reject or not reject the null hypothesis $H_0$. The following proposition, whose proof is given in Appendix 5.A, can be stated.

**Proposition 5.2.1.** *Assuming that the $N$-dimensional random vector $\boldsymbol{X} = (X_1, \ldots, X_N)$ with joint distribution function $F$ and marginals $F_i$, satisfies the null hypothesis $H_0$, then, the variable*

$$Z^2 = \sum_{j,i=1}^{N} \Phi^{-1}(F_i(X_i)) \, (\boldsymbol{\rho}^{-1})_{ij} \, \Phi^{-1}(F_j(X_j)) , \tag{5.56}$$

*where the matrix $\boldsymbol{\rho}$ is*

$$\rho_{ij} = \mathrm{Cov}[\Phi^{-1}(F_i(X_i)), \Phi^{-1}(F_j(X_j))] \,, \tag{5.57}$$

*follows a $\chi^2$-distribution with $N$ degrees of freedom.*

The testing procedure based on this result is now described for $N = 2$ assets. This case $N = 2$ is not restrictive as it would appear at first sight since, for portfolio analysis and risk management purposes, larger baskets of assets should be considered. The testing procedure described here can indeed be applied to any number of assets, and it is only for the sake of simplicity of the exposition that the presentation is restricted to the bivariate case.

Let us consider two financial time series of size $T$: $\{x_1(1), \ldots, x_1(t), \ldots, x_1(T)\}$ and $\{x_2(1), \ldots, x_2(t), \ldots, x_2(T)\}$. We assume that the vectors $\boldsymbol{x}(t) = (x_1(t), x_2(t))$, $t \in \{1, \ldots, T\}$ are independent and identically distributed with distribution $F$, which implies that the variables $x_1(t)$ (respectively $x_2(t)$), $t \in \{1, \ldots, T\}$, are also independent and identically distributed, with distribution $F_1$ (respectively $F_2$). We immediately note that this assumption of independently distributed data is not very realistic. It is well-known that daily returns are uncorrelated but that their volatility exhibits long-range dependence. A natural approach would then be to filter the data with an ARCH or GARCH process and then apply the testing procedure to the residuals. This approach will be discussed in Sect. 5.3.4 and we do not pursue this further here.

The empirical cumulative distribution $\hat{F}_i$ of each variable $X_i$ is given by

$$\hat{F}_i(x_i) = \frac{1}{T} \sum_{k=1}^{T} \mathbf{1}_{\{x_i(k) \le x_i\}} \,. \tag{5.58}$$

We use these estimated cumulative distributions to obtain the nearly Gaussian variables $\hat{y}_i$ as

$$\hat{y}_i(k) = \Phi^{-1}\left(\hat{F}_i(x_i(k))\right) \quad k \in \{1, \ldots, T\} \,. \tag{5.59}$$

The sample covariance matrix $\hat{\boldsymbol{\rho}}$ is estimated by the expression

$$\hat{\boldsymbol{\rho}} = \frac{1}{T} \sum_{i=1}^{T} \hat{\boldsymbol{y}}(i) \cdot \hat{\boldsymbol{y}}(i)^t \tag{5.60}$$

which allows us to calculate the variable

$$\hat{z}^2(k) = \sum_{i,j=1}^{2} \hat{y}_i(k) \, (\boldsymbol{\rho}^{-1})_{ij} \, \hat{y}_j(k) \,, \tag{5.61}$$

as defined in (5.A.6) for $k \in \{1, \ldots, T\}$. This variable $\hat{z}^2(k)$ should be distributed according to a $\chi^2$-distribution if the Gaussian copula hypothesis is correct.

As recalled in Chap. 2, a standard way for comparing an empirical with a theoretical distribution is to measure the distance between these two distributions and to perform the Kolmogorov test or the Anderson-Darling test (for a better accuracy in the tails of the distribution). The Kolmogorov distance is the maximum local distance among all quantiles, which is most often realized in the bulk of the distribution, while the Anderson-Darling distance puts the emphasis on the tails of the two distributions by a suitable normalization (which is nothing but the local standard deviation of the fluctuations of the distance). These two distances can be complemented by two additional measures which are defined as averages of the Kolmogorov distance and of the Anderson-Darling distance respectively,

Kolmogorov: $\qquad\qquad d_1 = \max_z |F_{z^2}(z^2) - F_{\chi^2}(z^2)|$ $\qquad\qquad$ (5.62)

average Kolmogorov: $d_2 = \int |F_{z^2}(z^2) - F_{\chi^2}(z^2)| \, dF_{\chi^2}(z^2)$ $\qquad$ (5.63)

Anderson-Darling: $\qquad\quad d_3 = \max_z \dfrac{|F_{z^2}(z^2) - F_{\chi^2}(z^2)|}{\sqrt{F_{\chi^2}(z^2)[1 - F_{\chi^2}(z^2)]}}$ $\qquad$ (5.64)

average Anderson-Darling: $d_4 = \int \dfrac{|F_{z^2}(z^2) - F_{\chi^2}(z^2)|}{\sqrt{F_{\chi^2}(z^2)[1 - F_{\chi^2}(z^2)]}} dF_{\chi^2}(z^2)$ $\quad$ (5.65)

The Kolmogorov distance $d_1$ and its average $d_2$ are more sensitive to the deviations occurring in the bulk of the distributions. In contrast, the Anderson-Darling distance $d_3$ and its average $d_4$ are more accurate in the tails of the distributions. Considering statistical tests for these four distances is important in order to be as complete as possible with respect to the different sensitivity of the tests. The averaging introduced in the distances $d_2$ and $d_4$ (which are simply the average of $d_1$ and $d_3$ respectively) provides important information. Indeed, the distances $d_1$ and $d_3$ are mainly controlled by the point that maximizes the argument within the max($\cdot$) function. They can thus be quite sensitive to the presence of an outlier. By averaging, $d_2$ and $d_4$ become less sensitive to outliers, since the weight of such points is only of order $1/T$ (where $T$ is the size of the sample) while it equals one for $d_1$ and $d_3$.

For the usual Kolmogorov and Anderson-Darling distance, the law of the empirical counterpart of $d_1$ and $d_2$ is known, at least asymptotically. In addition, it is free from the underlying distribution. However, for such a result to hold, one needs to know the exact value of the covariance matrix $\rho$ and the exact expression of the marginal distribution functions $F_i$. In the present case, the variables $\hat{z}^2(k)$, given by (5.61), are only pseudo-observations since their assessment requires the preliminary estimation of the covariance matrix $\hat{\rho}$ and the marginal distribution functions $\hat{F}_i$. And, as outlined in [200, 201], when one considers the empirical process constructed from the pseudo-sample $\{\hat{z}^2(k)\}_{t=1}^T$, the limiting behavior is not the same as in the case where one would *actually* observe $z^2(k)$, because there are two extra terms: one due to the fact that $F_i$ is replaced by $\hat{F}_i$ and another one due to the fact that $\rho$

is replaced by $\hat{\boldsymbol{\rho}}$. Therefore, one cannot directly use the asymptotic results known for these standard statistical tests.

As a very simple remedy, one can use a bootstrap method [143], whose accuracy has been proved to be at least as good as that given by asymptotic methods used to derive the theoretical distributions [97]. For the assessment of the asymptotic laws of $d_2$ and $d_4$, such a numerical study is compulsory since, even for the true observations $z^2(k)$, one does not know the expression of the asymptotic laws. Putting all this together, a possible implementation of this testing procedure is the following:

1. Given the original sample $\{\boldsymbol{x}(t)\}_{t=1}^{T}$, generate the pseudo-Gaussian variables $\hat{\boldsymbol{y}}(t)$, $t \in \{1,\ldots,T\}$ defined by (5.59).
2. Then, estimate the covariance matrix $\hat{\boldsymbol{\rho}}$ of the pseudo-Gaussian variables $\hat{\boldsymbol{y}}$, which allows one to compute the variables $\hat{z}^2$ and then measure the distance of its estimated distribution to the $\chi^2$-distribution.
3. Given this covariance matrix $\hat{\boldsymbol{\rho}}$, generate numerically a sample of $T$ bivariate Gaussian random vectors with the same covariance matrix $\hat{\boldsymbol{\rho}}$.
4. For the sample of Gaussian vectors synthetically generated with covariance matrix $\hat{\boldsymbol{\rho}}$, estimate its sample covariance matrix $\tilde{\boldsymbol{\rho}}$ and its marginal distribution functions $\tilde{F}_i$.
5. To each of the $T$ vectors of the synthetic Gaussian sample, associate the corresponding realization of the random variable $z^2$, called $\tilde{z}^2(t)$.
6. Construct the empirical distribution for the variable $\tilde{z}^2$ and measure the distance between this distribution and the $\chi^2$-distribution.
7. Repeat 10,000 times (for instance) the steps 3 to 6, and then obtain an accurate estimate of the cumulative distribution of distances between the distribution of the synthetic Gaussian variables and the theoretical $\chi^2$-distribution. This cumulative distribution represents the test statistic, which will allow you to reject or not the null hypothesis $H_0$ at a given significance level.
8. The significance of the distance obtained at step 2 for the true variables – *i.e.*, the probability to observe, at random and under $H_0$, a distance larger than the empirically estimated distance – is finally obtained by a simple reading on the complementary cumulative distribution estimated at step 7.

## 5.2.2 Empirical Results

Empirical tests implementing the previous procedure have been performed on securities, exchange rates, and commodities (metals) in [334]. Focusing on securities and exchange rates, we summarize some of the most striking features concerning the results obtained in this study.

## Currencies

The Federal Reserve Board provides access to a large set of historical quotes of spot foreign exchange rates. Following [334], let us focus on the Swiss Franc, the German Mark, the Japanese Yen, the Malaysian Ringit, the Thai Baht and the British Pound during the time interval of ten years from 25 January, 1989 to 31 December, 1998. All these exchange rates are expressed against the US dollar.

At the 95% significance level, one observes that only 40% (according to $d_1$ and $d_3$) but 60% (according to $d_2$ and $d_4$) of the tested pairs of currencies are compatible with the Gaussian copula hypothesis over the entire time interval. During the first half-period from 25 January, 1989 to 11 January, 1994, 47% (according to $d_3$) and up to about 75% (according to $d_2$ and $d_4$) of the tested currency pairs are compatible with the assumption of the Gaussian copula, while during the second subperiod from 12 January, 1994 to 31 December, 1998, between 66% (according to $d_1$) and about 75% (according to $d_2$, $d_3$ and $d_4$) of the currency pairs remain compatible with the Gaussian copula hypothesis. These results raise several comments both from a statistical and from an economic point of view.

We first have to stress that the most significant rejection of the Gaussian copula hypothesis is obtained for the distance $d_3$, which is indeed the most sensitive to the events in the tail of the distributions. The test statistics given by this distance can indeed be very sensitive to the presence of a single large event in the sample, so much so that the Gaussian copula hypothesis can be rejected only because of the presence of this single event (outlier). The difference between the results given by $d_3$ and $d_4$ (the averaged $d_3$) are very significant in this respect. The case of the German Mark and the Swiss Franc provides a particularly startling example. Indeed, during the time interval from 12 January, 1994 to 31 December, 1998, the probability $p(d)$ of non-rejection is rather high according to $d_1$, $d_2$ and $d_4$ ($p(d) \geq 31\%$) while it is very low according to $d_3$: $p(d) = 0.05\%$, which should lead to the rejection of the Gaussian copula hypothesis on the basis of the distance $d_3$ alone. This discrepancy between the different distances suggests the presence of an outlier in the sample.

To check this hypothesis, we show in the upper panel of Fig. 5.3 the function

$$f_3(t) = \frac{|F_{z^2}(z^2(t)) - F_{\chi^2}(\chi^2(t))|}{\sqrt{F_{\chi^2}(\chi^2)[1 - F_{\chi^2}(\chi^2)]}} , \tag{5.66}$$

used in the definition of the Anderson-Darling distance $d_3 = \max_z f_3(z)$ (see definition (5.64)), expressed in terms of time $t$ rather than $z^2$. The functions have been computed over the two time subintervals separately.

Apart from three extreme peaks occurring on 20 June, 1989, 19 August, 1991 and 16 September, 1992 within the first time subinterval and one extreme peak on 10 September, 1997 within the second time subinterval, the

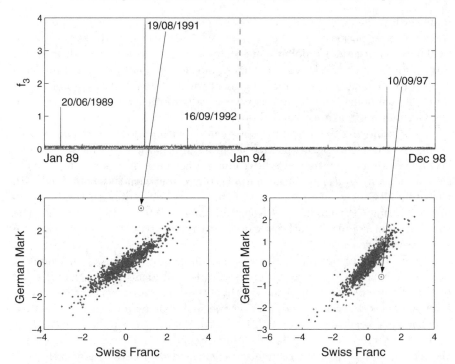

**Fig. 5.3.** The upper panel represents the graph of the function $f_3(t)$ defined in (5.66) used in the definition of the distance $d_3$ for the couple Swiss Franc/German Mark as a function of time $t$, over the time intervals from 25 January, 1989 to 11 January, 1994 and from 12 January, 1994 to 31 December, 1998. The two lower panels represent the scatter plot of the return of the German Mark versus the return of the Swiss Franc during the two previous time periods. The circled dot, in each figure, shows the pair of returns responsible for the largest deviation of $f_3$ during the considered time interval. Reproduced from [332]

statistical fluctuations measured by $f_3(t)$ remain small and of the same order. Removing the contribution of these outlier events in the determination of $d_3$, the new statistical significance derived according to $d_3$ becomes similar to that obtained with $d_1$, $d_2$ and $d_4$ on each subinterval. From the upper panel of Fig. 5.3, it is clear that the Anderson-Darling distance $d_3$ is equal to the height of the largest peak corresponding to the event on 19 August, 1991 for the first period and to the event on 10 September, 1997 for the second period. These events are depicted by a circled dot in the two lower panels of Fig. 5.3, which represent the return of the German Mark versus the return of the Swiss Franc over the two considered time periods.

The event on 19 August, 1991 is associated with the coup against Gorbachev in Moscow: the German mark (respectively the Swiss Franc) lost 3.37% (respectively 0.74%) against the US dollar. The 3.37% drop of the German Mark is the largest daily move of this currency against the US dollar over the whole first period. On 10 September, 1997, the German Mark appreciated by 0.60% against the US dollar while the Swiss Franc lost 0.79%, which represents a moderate move for each currency, but a large joint move. This event is related to the contradictory announcements of the Swiss National Bank about its monetary policy, which put an end to a rally of the Swiss Franc along with the German mark against the US dollar.

Thus, removing the large moves associated with major historical events or events associated with unexpected incoming information[7] – which cannot be accounted for in a statistical study, unless one relies on a stress-test analysis – we obtain, for $d_3$, significance levels compatible with those obtained with the other distances. We can thus conclude that, according to the four distances, during the time interval from 12 January, 1994 to 31 December, 1998 the Gaussian copula hypothesis cannot be rejected for the couple German Mark/Swiss Franc.

From an economic point of view, the impact of regulatory mechanisms between currencies or monetary crises can be well identified by the rejection or the absence of rejection of the null hypothesis. Indeed, consider the couple German Mark/British Pound. During the first half period, their correlation coefficient is very high ($\rho = 0.82$) and the Gaussian copula hypothesis is strongly rejected according to the four distances. On the contrary, during the second half period, the correlation coefficient decreases significantly ($\rho = 0.56$) and none of the four distances allows us to reject the null hypothesis. Such non-stationarity can be easily explained. Indeed, on 1 January, 1990, the British Pound entered the European Monetary System (EMS), so that the exchange rate between the German Mark and the British Pound was not allowed to fluctuate beyond a margin of 2.25%. However, due to a strong speculative attack, the British Pound was devaluated in September 1992 and had to leave the EMS. Thus, between January 1990 and September 1992, the exchange rate of the German Mark and the British Pound was confined within a narrow spread, incompatible with the Gaussian copula description. After 1992, the British Pound exchange rate floated with respect to the German Mark, and the dependence between the two currencies decreased, as shown by their correlation coefficient. In this latter regime, one can no more reject the Gaussian copula hypothesis.

The impact of major crises on the copula can also be clearly identified. An example is given by the Malaysian Ringit/Thai Baht couple. During the period from January 1989 to January 1994, these two currencies have only undergone moderate and weakly correlated fluctuations ($\rho = 0.29$), so that the null hypothesis cannot be rejected at the 95% significance level. In contrast, during the period from January 1994 to October 1998, the Gaussian copula

---

[7] Modeling the volatility by a mean reverting stochastic process with long memory (the multifractal random walk (MRW)), Sornette et al. [456] have demonstrated the outlier nature of the event on 19 August, 1991.

hypothesis is strongly rejected. This rejection is obviously due to the persistent and dependent ($\rho = 0.44$) shocks incurred by the Asian financial and monetary markets during the 7 months of the Asian Crisis from July 1997 to January 1998 [29, 262].

These two cases show that the Gaussian copula hypothesis can be considered reasonable for currencies in the absence of regulatory mechanisms and of strong and persistent crises. They also provide an understanding of why the results of the test over the entire sample are so much weaker than the results obtained for the two subintervals: the time series are strongly nonstationary.

## Stocks

Let us now turn to the description of the dependence properties of the distributions of daily returns for a diversified set of stocks among the largest companies quoted on the New York Stock Exchange. We report the results presented in [334] concerning Appl. Materials, AT&T, Citigroup, Coca Cola, EMC, Exxon-Mobil, Ford, General Electric, General Motors, Hewlett Packard, IBM, Intel, MCI WorldCom, Medtronic, Merck, Microsoft, Pfizer, Procter & Gamble, SBC Communication, Sun Microsystem, Texas Instruments, and Wal Mart.

The dataset covers the time interval from 8 February, 1991 to 29 December, 2000. At the 95% significance level, 75% of the pairs of stocks are compatible with the Gaussian copula hypothesis. Over the time subinterval from February 1991 to January 1996, this percentage becomes larger than 99% for $d_1$, $d_2$ and $d_4$ while it equals 94% according to $d_3$. Over the time subinterval from February 1996 to December 2000, 92% of the pairs of stocks are compatible with the Gaussian copula hypothesis according to $d_1$, $d_2$ and $d_4$ and more than 79% according to $d_3$. Therefore, the Gaussian copula assumption is much more widely accepted for stocks than it was for the currencies reported above. In addition, the nonstationarity observed for currencies does not seem very prominent for stocks.

For the sake of completeness, let us add a word concerning the results of the tests performed for five stocks belonging to the computer sector : Hewlett Packard, IBM, Intel, Microsoft, and Sun Microsystem. During the first half period (from Feb. 1991 to Jan. 1996), all the pairs of stocks qualify the Gaussian copula hypothesis at the 95% significance level. The results are rather different for the second half period (from Feb. 1996 to Dec. 2000) since about 40% of the pairs of stocks reject the Gaussian copula hypothesis according to $d_1$, $d_2$ and $d_3$. This can certainly be ascribed to the existence of a few shocks, notably associated with the crash of the "new economy" in March–April 2000 [450]. However, on the whole, it appears that there is no systematic rejection of the Gaussian copula hypothesis for stocks within the same industrial sector, notwithstanding the fact that one can expect correlations stronger than the average between such stocks.

## 5.3 Limits of the Description in Terms of the Gaussian Copula

### 5.3.1 Limits of the Tests

A severe limitation of existing tests applied to Gaussian copulas [334, 350] is their inability to clearly distinguish between Gaussian and some relatively close alternative models such as the Student's copulas when these latter copulas have a sufficiently large number of degrees of freedom, typically of the order of or larger than 10–20. As recalled in Chap. 3, the Student copula becomes very close to the Gaussian copula in its bulk when it has a large number of degrees of freedom. In contrast, these two copulas still differ significantly in the corners of the unit square (see Figs. 3.2 and 3.3). This difference has no serious consequences for "normal" events but leads to important implications for extremes. Indeed, as discussed in Sect. 4.5.3, an alternative model to the Gaussian copula, such as the Student's copula, presents a significant tail dependence, even for moderately large numbers of degrees of freedom, while the Gaussian copula has absolutely no asymptotic tail dependence; these tail dependences are controlled mathematically by the behavior of the copulas in the corners of the unit square. Therefore, if the tests previously described are unable to distinguish between a Student's and a Gaussian copula, Occam's razor (simplicity and parsimony) suggests choosing the Gaussian copula and, as a consequence, one may underestimate severely the dependence between extreme events if the correct description turns out to be the Student's copula. This may have catastrophic consequences in risk assessment and portfolio management.

Figure 4.8 p. 173 provides a quantification of the dangers incurred by mistaking a Student copula for a Gaussian one. Consider the case of a Student copula with $\nu = 20$ degrees of freedom with a correlation coefficient $\rho$ lower than $0.3 \sim 0.4$; its tail dependence $\lambda_\nu(\rho)$ turns out to be less than 0.7%, *i.e.*, the probability that one variable becomes extreme knowing that the other one is extreme is less than 0.7%. In this case, the Gaussian copula with a zero probability of simultaneous extreme events is not a bad approximation of the Student's copula. In contrast, consider a Student copula with a correlation $\rho$ larger than 0.7–0.8, corresponding to a tail dependence larger than 10%, which is a nonnegligible probability for simultaneous extreme events. The effect of tail dependence becomes of course much stronger as the number $\nu$ of degrees of freedom decreases.

These examples stress the importance of determining whether the previous testing procedure distinguishes between a Student copula with $\nu = 20$ (or less) degrees of freedom and a given correlation coefficient of the order of $\rho = 0.5$, for instance, and a Gaussian copula with an appropriate correlation coefficient $\rho'$. Due to the strikingly different behavior of these two models in the extremes, the non-rejection of the Gaussian copula hypothesis previously found for most assets can lead to an underestimation of the extreme risks, as a result of the

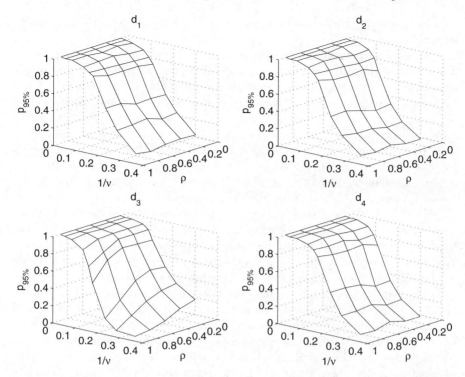

**Fig. 5.4.** Probability of non-rejection of the Gaussian copula hypothesis when the true copula is given by a Student copula with $\nu$ degrees of freedom and a correlation coefficient equal to $\rho$ (error of type II: "false positive"), when the error of type I ("false negative") of the test is set equal to 5%, for the four distances $d_1$–$d_4$

weak sensitivity of the test in the extreme regions of the copula. It is therefore important to discuss the sensitivity of the test presented in Sect. 5.2.1 and to review the other alternatives proposed in the literature.

### 5.3.2 Sensitivity of the Method

The previous section has found that the Gaussian copula provides a reasonably good model, in the sense that it cannot be rejected by a statistical test at the 95% significance level. However, could this be due to the lack of power of the statistical test rather than to the goodness of the Gaussian copula?

Let us denote by $H_{\nu,\rho}$ the hypothesis that the true copula of the data is the Student copula with $\nu$ degrees of freedom with the correlation coefficient $\rho$. Considering the alternative hypothesis $H_{\nu,\rho}$, one needs to know what is the probability that one cannot reject the null hypothesis $H_0$ when the true model is $H_{\nu,\rho}$. A complementary information is: what is the minimum $p$-value (significance level) of the test allowing us to reject the Gaussian copula hypothesis for instance 95 times out of 100 when the true copula is the Student copula. Answering these questions on the power of the test require a numerical study.

Figure 5.4 shows the minimum $p$-value of the test, denoted by $p_{95\%}$, as a function of the (inverse of the) number of degrees of freedom $\nu$ and of the correlation coefficient $\rho$ of the true Student copula. Overall, the four tests associated with the four different distances $d_1$–$d_4$ behave similarly. As expected, for large $\nu$, namely $\nu \geq 10 - 20$ ($1/\nu \leq 0.05 - 0.1$), a very high $p$-value is required to reject the Gaussian hypothesis. In such a case, it is almost impossible to distinguish between the Gaussian hypothesis and a Student copula for most realizations. If one leaves out distance $d_3$, the power of the tests is almost independent of the value of the correlation coefficient. For $d_3$, the power is clearly weaker for the smallest correlations.

In the light of these results on the performance of the tests, the previous conclusion on the relevance of the Gaussian copula for the modeling of the dependence between financial risks must be reconsidered. Concerning currencies, the non-rejection of the Gaussian copula hypothesis does not exclude at the 95% significance level that the dependence of the currency pairs may be described by a Student copula with adequate values of $\nu$ and $\rho$. For the German Mark/Swiss Franc pair, a Student copula with about five degrees of freedom was found to obtain the same $p$-values [334]. For the correlation coefficient $\rho = 0.92$ of the German Mark/Swiss Franc pair, Student's copula with five degrees of freedom predicts a tail dependence coefficient $\lambda_5(0.92) = 63\%$, in constrast with a zero value for the Gaussian copula. Such a large value of $\lambda_5(0.92)$ implies that, when an extreme event occurs for the German Mark, it also occurs for the Swiss Franc with a frequency of 63%. Therefore, a stress scenario based on the assumption of a Gaussian copula would fail to account for such coupled extreme events, which may represent as many as two-third of all extreme events, if it would turn out that the true copula was Student's copula with five degrees of freedom. Note that, with such a large value of the correlation coefficient, the tail dependence remains high even if the number of degrees of freedom is as large as 20 or more (see Fig. 4.8).

The Swiss Franc and Malaysian Ringit pair offers a very different case. For instance, during the time period from January 1994 to December 1998, the test statistics are so high that the description of the dependence with Student's copula would require it to have at least 7–10 degrees of freedom. In addition, the correlation coefficient of the two currencies is only $\rho = 0.16$, so that, even in the most pessimistic situation $\nu = 7$, the choice of the Gaussian copula would amount to neglecting the tail dependence coefficient $\lambda_5(0.16) = 4\%$ predicted by Student's copula. In this case, stress scenarios based on the Gaussian copula would predict uncoupled extreme events, which would be wrong only once in 25 times.

These two examples highlight the fact that, as much as the number of degrees of freedom of Student's copula which is necessary to describe the data, the correlation coefficient remains an important parameter.

### 5.3.3 The Student Copula: An Alternative?

The tests performed in [83, 350] show that the Student copula can provide a significantly better description of the data than the Gaussian copula, particularly for foreign exchange (FX) rates, for which the number of degrees of freedom of the Student copula is about 5–6 for daily returns. In both cases, the testing procedure is based on the pseudo likelihood estimation method detailed in Sect. 5.1.2.

Using the Akaike information criterion defined by the following formula:

$$\mathrm{AIC} = -2\ln\tilde{\mathcal{L}}\left(\{x_1(i),\cdots,x_n(i)\};\hat{\boldsymbol{\theta}}\right) + 2\dim\boldsymbol{\theta} \ . \tag{5.67}$$

Breymann and his co-authors [83] have shown that the dependence structure of the German Mark/Japanese Yen couple is better described by Student's copula with about six degrees of freedom (for daily returns) than with a Gaussian copula, the latter being the second best copula among a set of five copulas comprising Clayton's, Gumbel's, and Frank's copulas. This result refines those obtained in [334] and is in line with the results obtained by non-parametric and semiparametric estimation shown in Figs. 5.1–5.2. In addition, Student's copula is found to provide an even better description when one considers FX returns calculated at smaller time scales [83]. Indeed, the Student copula seems to provide a reliable model for FX returns calculated for time scales larger than 2 hours. The number of degrees of freedom is found to increase with the time scale: it increases from 4 at the 2 hours time scale to 6 at the daily time scale. Such a result is expected since, under time aggregation, the distribution of returns should converge to the Gaussian distribution according to the central limit theorem, therefore the dependence structure of the returns is expected to also converge toward the Gaussian copula at a large time scale. At time scales smaller than 2 hours, the study by Breymann *et al.* shows that neither the Gaussian nor the Student copulas are sufficient to describe the dependence structure of the distributions of FX returns. At these small time scales, microstructural effects probably come into play and require more elaborated copulas to model the dependences observed at very high frequencies.

In addition, for all time scales, the copula of the bivariate excess returns for high (or low) threshold appears to be best described by Clayton's (or by the survival Clayton) copula. This result can lead us to the following interpretation on the existence of concomitant extremes. Assume that, conditional on a frailty random variable representing the information flow, the assets returns are independent. The copula of the returns of the 2 assets then exhibit the behavior reported in the study by Breymann *et al.* [83] if one assumes that the random variable representing the information flow has a regularly varying distribution – which means that pieces of information with great impact on asset returns arrive relatively often.

In contrast with the case of foreign exchange rates, the estimated number of degrees of freedom of the Student copula best fitting the dependence between stocks is rather high, so that the probability of concomitant extreme risks remains weak for their usual level of correlation. In the study by Mashal and Zeevi [350], the dependence between stocks is claimed to be significantly better accounted for by a Student copula with 11–12 degrees of freedom than by a Gaussian copula. Their conclusions are drawn from a generalization of the log-likelihood ratio (or Wilks) test, based on the fact that the Gaussian copula is nothing but a Student copula with an infinite number of degrees of freedom. This allows them to compare directly the relevance of the Student copula with respect to the Gaussian copula. Indeed, given two nested copulas $C_1$ and $C_2$, i.e., two copulas such that the space of parameter vectors $\boldsymbol{\theta}_1$ of $C_1$ is a subspace of the space of parameter vectors $\boldsymbol{\theta}_2$ of $C_2$, then the statistic

$$\Lambda_T = -2\ln \frac{\tilde{\mathcal{L}}_1(\hat{\boldsymbol{\theta}}_1)}{\tilde{\mathcal{L}}_2(\hat{\boldsymbol{\theta}}_2)} \tag{5.68}$$

is asymptotically distributed as a $\chi_1^2$ with one degree of freedom if $\dim \boldsymbol{\theta}_1 = \dim \boldsymbol{\theta}_2 - 1$, up to a scale factor $1 + \gamma$ larger than one due to the use of a pseudo maximum-likelihood instead of the true maximum likelihood:

$$\Lambda_T \xrightarrow{law} (1 + \gamma)\chi_1^2, \qquad \text{as } T \longrightarrow \infty. \tag{5.69}$$

The positive parameter $\gamma$ depends on the choice of the model and can be determined by numerical simulations. In more general cases where $\dim \boldsymbol{\theta}_2 - \dim \boldsymbol{\theta}_1 = m > 1$, $\Lambda_T$ does not follows an asymptotic $\chi_m^2$-distribution with $m$ degrees of freedom as in standard tests of nested hypotheses. This results from the fact that the log-likelihood ratio statistic does not converge to a $\chi^2$ distribution when the model is misspecified, which is the relevant situation when using the pseudo likelihood instead of the true likelihood (see Appendix 5.B). In such a case, the Wald or Lagrange multiplier tests are more appropriate [209, 372].

While these results improve somewhat on the initial study [334], in contrast with the case of currencies, one can question the existence of a real improvement brought by the Student copula to describe the dependence between stocks. Indeed, correlation coefficients between two stocks are hardly greater than 0.4–0.5, so that the tail dependence of a Student copula with 11–12 degrees of freedom is about 2.5% or less. In view of all the different sources of uncertainty during the estimation process in addition to the possible non-stationarity of the data, one can doubt that such a description eventually leads to concrete improvements for practical purposes. To highlight this point, let us consider several portfolios made of 50% of the Standard & Poor's 500 index and 50% of one stock (whose name is indicated in the first column of Table 5.1). Let us then estimate the probability $P_r$ that this portfolio incurs a loss larger than $n$ times its standard deviation ($n = 2, \ldots, 5$). For the same portfolio, let us estimate the probability $P_g$ (resp. $P_s$) that it incurs the same loss ( i.e., $n$ times its standard deviation) when the dependence between the index and the stock is given by a Gaussian copula (resp. a Student copula with ten degrees of freedom). The row named $P_r/P_{g/s}$ gives the average values of $P_r/P_g$ and $P_r/P_s$ over the 20 portfolios. For shocks of two- and three-standard deviations, the values of $P_r/P_g$ close to 1 indicate that the dependence structure is correctly captured by a Gaussian copula. For shocks of four- and five-standard deviations, $P_r/P_g$ becomes larger than 1, showing that large shocks are more probable than predicted by the Gaussian dependence, and all the more so, the larger the amplitude of the shocks. This occurs notwithstanding the use of marginals with heavy tails, suggesting the effect of a non-zero tail dependence in the true data. In contrast, the values of $P_r/P_s$ are significantly smaller than 1 showing that the Student copula overestimates the frequency of large shocks. In addition, this overestimation is surprisingly worse for larger shocks (by as much as a Factor 2.5) in the range in which the Gaussian copula becomes less adequate. This suggests that the tail dependence of the Student copula is too large to describe this data set. This simple exercise illustrates that neither the Gaussian copula nor a Student copula with a reasonable number of degrees of freedom provide an accurate description of the dependence between stock returns.[8] The discrepancies between these two models and the real dependence structure becomes all the more important, the more extreme is the amplitude of the shock. And in fact, the situation is worse for the Student copula. This suggests that, for practical applications, Student's copula may not provide a real improvement with respect to the Gaussian copula for traditional portfolio management.

### 5.3.4 Accounting for Heteroscedasticity

The aforementioned studies have not taken into account, or only partially, the well-known volatility clustering phenomenon, which certainly impacts on the dependence properties of assets returns. This issue has been addressed by Patton [380], who has shown that the two-step maximum likelihood estimation can be extended to conditional copulas to account for the time-varying nature of financial time series. Filtering marginal data by a GARCH process, Patton has shown that the conditional dependence structure between exchange rates (Japanese Yen against Euro) is better described by Clayton's copula than by the Gaussian copula. We also note that Muzy *et al.* [366] have constructed a multivariate "multifractal" process to account for both volatility clustering and the dependence between assets. In this case, the conditional copula is (nearly) Gaussian.

The main limitation of Patton's approach comes from the fact that filtering the data does not leave the dependence structure, *i.e.*, the copula, unchanged. Thus, the copula of the residuals is not the same as the copula of the raw returns. Moreover, the copula of the residuals changes with

---

[8] This point confirms the doubts raised by the comparison of the nonparametric and the semiparametric estimates of the density of the copula of the daily returns of General Motors and Procter & Gamble, represented in Figs. 5.1–5.2.

**Table 5.1.** Portfolios made of 50% of the Standard & Poors 500 index and 50% of one stock (whose name is indicated in the first column) are considered

| | $100 \times \Pr[R \leq -n \cdot \sigma]$ | | | | | | | | | | | |
| | $n=2$ | | | $n=3$ | | | $n=4$ | | | $n=5$ | | |
| | $P_r$ | $P_g$ | $P_s$ | $P_r$ | $P_g$ | $P_s$ | $P_r$ | $P_g$ | $P_s$ | $P_r$ | $P_g$ | $P_s$ |
|---|---|---|---|---|---|---|---|---|---|---|---|---|
| Abbott Labs | 2.07 | 2.06 | 2.62 | 0.58 | 0.51 | 0.84 | 0.21 | 0.15 | 0.39 | 0.09 | 0.07 | 0.26 |
| American Home Products Corp. | 1.98 | 2.07 | 2.72 | 0.51 | 0.56 | 0.98 | 0.30 | 0.24 | 0.43 | 0.17 | 0.13 | 0.22 |
| Boeing Co. | 2.03 | 1.96 | 2.50 | 0.53 | 0.51 | 0.95 | 0.21 | 0.18 | 0.44 | 0.13 | 0.09 | 0.19 |
| Bristol-Myers Squibb Co. | 1.56 | 1.81 | 2.33 | 0.55 | 0.48 | 0.98 | 0.26 | 0.22 | 0.81 | 0.11 | 0.1 | 0.42 |
| Chevron Corp. | 1.94 | 1.99 | 2.26 | 0.40 | 0.42 | 0.88 | 0.13 | 0.15 | 0.55 | 0.08 | 0.07 | 0.30 |
| Du Pont (E.I.) de Nemours & Co. | 2.13 | 2.02 | 2.59 | 0.51 | 0.47 | 0.87 | 0.21 | 0.19 | 0.58 | 0.09 | 0.07 | 0.32 |
| Disney (Walt) Co. | 1.83 | 1.87 | 2.40 | 0.47 | 0.53 | 1.28 | 0.24 | 0.22 | 0.73 | 0.15 | 0.12 | 0.43 |
| General Motors Corp. | 1.73 | 1.95 | 2.12 | 0.45 | 0.42 | 0.76 | 0.21 | 0.13 | 0.59 | 0.08 | 0.06 | 0.36 |
| Hewlett-Packard Co. | 1.77 | 2.08 | 2.54 | 0.53 | 0.51 | 0.99 | 0.21 | 0.19 | 0.44 | 0.08 | 0.09 | 0.15 |
| Coca-Cola Co. | 1.60 | 1.83 | 2.13 | 0.45 | 0.5 | 0.77 | 0.19 | 0.18 | 0.58 | 0.09 | 0.07 | 0.46 |
| Minnesota Mining & MFG Co. | 1.85 | 2.01 | 2.23 | 0.57 | 0.49 | 0.80 | 0.19 | 0.19 | 0.60 | 0.08 | 0.09 | 0.52 |
| Philip Morris Cos Inc. | 2.00 | 2.07 | 2.33 | 0.45 | 0.5 | 1.10 | 0.21 | 0.19 | 0.65 | 0.13 | 0.12 | 0.34 |
| Pepsico Inc. | 1.92 | 2.08 | 2.50 | 0.51 | 0.49 | 0.83 | 0.15 | 0.18 | 0.39 | 0.15 | 0.07 | 0.22 |
| Procter & Gamble Co. | 1.51 | 1.67 | 2.05 | 0.45 | 0.48 | 0.95 | 0.24 | 0.21 | 0.82 | 0.13 | 0.09 | 0.67 |
| Pharmacia Corp. | 1.81 | 1.94 | 2.69 | 0.53 | 0.54 | 1.06 | 0.23 | 0.25 | 0.80 | 0.11 | 0.12 | 0.45 |
| Schering-Plough Corp. | 1.85 | 1.94 | 2.01 | 0.49 | 0.44 | 0.73 | 0.11 | 0.14 | 0.58 | 0.08 | 0.06 | 0.31 |
| Texaco Inc. | 1.90 | 1.94 | 2.77 | 0.55 | 0.55 | 1.01 | 0.28 | 0.23 | 0.41 | 0.11 | 0.11 | 0.21 |
| Texas Instruments Inc. | 1.87 | 2.02 | 2.09 | 0.49 | 0.5 | 0.89 | 0.21 | 0.15 | 0.66 | 0.06 | 0.07 | 0.16 |
| United Technologies Corp | 2.17 | 2.1 | 2.28 | 0.47 | 0.45 | 0.78 | 0.17 | 0.14 | 0.47 | 0.11 | 0.06 | 0.30 |
| Walgreen Co. | 1.81 | 1.96 | 2.28 | 0.47 | 0.41 | 0.92 | 0.23 | 0.14 | 0.40 | 0.09 | 0.08 | 0.21 |
| $P_r/P_{g/s}$ | 0.95 | | 0.79 | 1.02 | | 0.55 | 1.15 | | 0.39 | 1.24 | | 0.38 |

We estimate the probability $P_r$ that each portfolio incurs a loss larger than $n$ times its standard deviation ($n = 2, \ldots, 5$). For each portfolio, we also estimate the probability $P_g$ (resp. $P_s$) that it incurs the same loss (*i.e.*, $n$ times its standard deviation) when the dependence between the index and the stock is given by a Gaussian copula (resp. a Student copula with ten degrees of freedom). The row named $P_r/P_{g/s}$ gives the average values of $P_r/P_g$ and $P_r/P_s$ over the 20 portfolios.

the chosen filter. Residuals are not the same when one filters the data with an ARCH, a GARCH or a Multifractal Random Walk. In addition, for an arbitrage-free market, the (multivariate) log-price process can be expressed as a time changed multivariate Brownian motion[9] [264], so that conditional on the (realized) volatility [8, 38], the log-price process is nothing but a multivariate Brownian motion. As a consequence, conditional on the volatility, the multivariate distribution of returns should be Gaussian, and, therefore, the copula of conditional returns should also be the Gaussian copula. Thus, the estimation of the conditional copula does not really bring new insights. *In fine*, the discrepancy between the Gaussian copula and the conditional copula provided by some other model mainly highlights the weakness of the model under consideration. This raises the question whether performing a model-free analysis (without any pre-filtering process) is not a more satisfying alternative. Obviously, the price to pay for such a model-free approach is a weakening of the power of the statistical test due to the presence of (temporal) dependence between data. There is no free lunch, neither on financial markets, nor in statistics.

## 5.4 Summary

The Gaussian paradigm has had a long life in finance. While it is now clear that marginal distributions cannot be described by Gaussian laws, especially in their tails (see Chap. 2), the dependence structure between two or more assets is much less known and nothing suggests to reject *a priori* the Gaussian copula as a correct description of the observed dependence structure. In addition, the Gaussian copula can be derived in a very natural way from a principle of maximum entropy [265, 453].[10] The Gaussian copula has also the advantage of being the simplest possible one in the class of elliptical copulas, since it is entirely specified by the knowledge of the correlation coefficients while, for instance, Student's copula requires in addition the specification of the number of degrees of freedom. This has led to taking the Gaussian copula as a logical starting point for the study of the dependence structure between financial assets.

However, as recalled in Chap. 3, if the Gaussian and Student copulas are very similar in their bulk, they become significantly different in their tails.

---

[9] More precisely, in an arbitrage-free market, any $n$-dimensional square-integrable log-price process $\ln p(t)$, with continuous sample path, satisfies

$$r_\tau(t) = \ln p(t + \tau) - \ln p(t) = \int_t^{t+\tau} \mu(s) \, ds + \int_t^{t+\tau} \sigma(s) \, dW(s) \,,$$

where $\mu$ is a predictable $n$-dimensional vector and $\sigma$ is an $n$-by-$n$ matrix. $W$ denotes an $n$-dimensional standard Brownian motion.

[10] For other examples of the determination of distributions using the principle of maximum entropy, see [410].

Concretely, the essential difference between the Gaussian and Student copulas is that the former has independent extremes (in the sense of the asymptotic tail dependence; see Chap. 4), while the latter generates concomitant extremes with a non-zero probability which is all the larger, the smaller is the number of degrees of freedom and the larger is the correlation coefficient. Thus, by providing a slight departure from the Gaussian copula in the bulk of the distributions, Student's copula could also be a good candidate to model financial dependencies. It turns out that it is indeed a good model for foreign exchange rates. The situation is not so clear of stock returns, as Student's copula does not seem to perform significantly better than the Gaussian copula, both being apparently approximations of the true copula. From a practical point of view, there have been several efforts to find better copulas, but the obtained gains are not clear. From an economic point of view, the reasons explaining the difference between the dependence structure of the FX rate and the stock returns remain to be found. The differences between stock markets and FX markets organizations can be seen as an obvious reason, but direct links between markets organization and returns distribution or copula have not yet been clearly articulated.

One of the motivations in introducing the tail dependence coefficient $\lambda$ is to quantify the potential risks incurred in modeling the dependence structure between assets with Gaussian copulas, for which $\lambda = 0$. Indeed, for assets with large correlation coefficients, it may be dangerous to use Gaussian copulas as long as one does not have a better idea of the value of the tail dependence coefficient. Parametric models do not provide readily this information since they fix the tail dependence coefficient and therefore do not provide an independent test of whether $\lambda$ is small (and undistinguishable from 0) or large. To get further insight, nonparametric methods could thus be useful.

Nonparametric models have the advantage of being much more general since, by construction, they do not assume a specific copula and might thus allow for an independent determination of the tail dependence coefficient. Some of these methods have the advantage of leading to estimated copulas which are smooth and differentiable everywhere, which is convenient for the generation of random variables having the estimated copula, for sensitivity analysis and for the generation of synthetic scenarios [149]. However, this advantage comes with the main drawback that the tail dependence coefficient vanishes by construction. In sum, all methods mentioned until now suffer from the same problem of neglecting concomitant extremes. It thus seems that the use of copulas is not the easiest path to calibrate extreme events. We address this problem in the next chapter, in particular by describing direct methods for estimating extreme concomitant events.

# Appendix

## 5.A  Proof of the Existence of a $\chi^2$-Statistic for Testing Gaussian Copulas

To prove proposition 5.2.1, we first consider an $n$-dimensional random vector $\boldsymbol{X} = (X_1, \ldots, X_n)$. Let us denote by $F$ its distribution function and by $F_i$ the marginal distribution of each $X_i$. Let us now assume that the distribution function $F$ satisfies $H_0$, so that $F$ has a Gaussian copula with correlation matrix $\rho$ while the $F_i$'s can be any distribution functions. According to Theorem 3.2.1, the distribution $F$ can be represented as :

$$F(x_1, \ldots, x_n) = \Phi_{\rho,n}(\Phi^{-1}(F_1(x_1)), \ldots, \Phi^{-1}(F_N(x_n,))) . \qquad (5.A.1)$$

Let us now transform the $X_i$'s into Normal random variables $Y_i$'s:

$$Y_i = \Phi^{-1}(F_i(X_i)) . \qquad (5.A.2)$$

Since the mapping $\Phi^{-1}(F_i(\cdot))$ is increasing, the invariance Theorem 3.2.2 allows us to conclude that the copula of the variables $Y_i$'s is identical to the copula of the variables $X_i$'s. Therefore, the variables $Y_i$'s have Normal marginal distributions and a Gaussian copula with correlation matrix $\rho$. Thus, by definition, the multivariate distribution of the $Y_i$'s is the multivariate Gaussian distribution with correlation matrix $\rho$:

$$G(\mathbf{y}) = \Phi_{\rho,n}(\Phi^{-1}(F_1(x_1)), \ldots, \Phi^{-1}(F_n(x_n))) \qquad (5.A.3)$$

$$= \Phi_{\rho,n}(y_1, \ldots, y_n) , \qquad (5.A.4)$$

and $\boldsymbol{Y}$ is a Gaussian random vector. From (5.A.3–5.A.4), we have

$$\rho_{ij} = \mathrm{Cov}[\Phi^{-1}(F_i(X_i)), \Phi^{-1}(F_j(X_j))] . \qquad (5.A.5)$$

Consider now the random variable

$$Z^2 = \boldsymbol{Y}^t \boldsymbol{\rho}^{-1} \boldsymbol{Y} = \sum_{i,j=1}^{n} Y_i \, (\boldsymbol{\rho}^{-1})_{ij} \, Y_j , \qquad (5.A.6)$$

where $\cdot^t$ denotes the transpose operator. It is well known that the variable $Z^2$ follows a $\chi^2$-distribution with $n$ degrees of freedom. Indeed, since $\boldsymbol{Y}$ is a Gaussian random vector with covariance matrix[11] $\rho$, it follows that the components of the vector

$$\tilde{\boldsymbol{Y}} = \boldsymbol{A}\boldsymbol{Y} , \qquad (5.A.7)$$

are *independent* Normal random variables. Here, $\boldsymbol{A}$ denotes the square root of the matrix $\rho^{-1}$, obtained by the Cholevsky decomposition, so that $\boldsymbol{A}^t \boldsymbol{A} = \rho^{-1}$. Thus, the sum $\tilde{\boldsymbol{Y}}^t \tilde{\boldsymbol{Y}} = Z^2$ is the sum of the squares of $n$ independent Normal random variables, which follows a $\chi^2$-distribution with $n$ degrees of freedom.

---

[11] Up to now, the matrix $\rho$ was named *correlation matrix*. But in fact, since the variables $Y_i$'s have unit variance, their correlation matrix is also their *covariance matrix*.

## 5.B Hypothesis Testing with Pseudo Likelihood

Let us consider the iid sample $\{(x_1(1), x_2(1), \ldots, x_n(1)), \ldots, (x_1(T), x_2(T), \ldots, x_n(T))\}$ drawn from the $n$-dimensional distribution $F$ with copula $C$ and margins $F_i$. We aim at estimating the unknown copula $C$ by use of the semi-parametric method presented in Sect. 5.1.2. Its pseudo likelihood reads

$$\ln \tilde{\mathcal{L}}_T = \sum_{i=1}^{T} \ln c \left(\hat{u}_1(i), \ldots, \hat{u}_n(i); \boldsymbol{\theta}\right) , \qquad (5.B.8)$$

with $\hat{u}_k(i) = \hat{F}_k\left(x_k(i)\right)$, where the $\hat{F}_i$'s are the empirical estimates of the marginal distribution functions $F_i$'s, and $c(\cdot; \boldsymbol{\theta})$ denotes the copula density $C_{\boldsymbol{\theta}}$, $\boldsymbol{\theta} \in \Theta \subset \mathbb{R}^p$. The parameter vector $\boldsymbol{\theta}$ can be estimated by maximization of this pseudo log-likelihood, so that

$$\hat{\boldsymbol{\theta}}_T = \arg\max_{\boldsymbol{\theta}} \; \ln \tilde{\mathcal{L}}\left(\{\hat{u}_1(i), \ldots, \hat{u}_n(i)\}; \boldsymbol{\theta}\right) . \qquad (5.B.9)$$

Under usual regularity conditions, it can be shown that $\hat{\boldsymbol{\theta}}_T$ is a consistent estimator of $\boldsymbol{\theta}^0$, which is asymptotically Gaussian [197],

$$\sqrt{T}\left(\hat{\boldsymbol{\theta}}_T - \boldsymbol{\theta}^0\right) \xrightarrow{law} \mathcal{N}\left(0, \boldsymbol{\Sigma}^2\right) \qquad (5.B.10)$$

with $\boldsymbol{\Sigma}^2 = \boldsymbol{I}\left(\boldsymbol{\theta}^0\right)^{-1} + \boldsymbol{I}\left(\boldsymbol{\theta}^0\right)^{-1} \boldsymbol{\Omega} \boldsymbol{I}\left(\boldsymbol{\theta}^0\right)^{-1}$, where $\boldsymbol{I}\left(\boldsymbol{\theta}^0\right)$ represents Fisher's information matrix at $\boldsymbol{\theta}^0$,

$$\left[I\left(\boldsymbol{\theta}^0\right)\right]_{ij} = \mathrm{E}\left[\frac{\partial c(\boldsymbol{U}; \boldsymbol{\theta})}{\partial \theta_i} \cdot \frac{\partial c(\boldsymbol{U}; \boldsymbol{\theta})}{\partial \theta_j}\right]_{\boldsymbol{\theta} = \boldsymbol{\theta}^0} , \qquad (5.B.11)$$

and $\boldsymbol{U}$ denotes an $n$-dimensional random vector with distribution function $C$ and with

$$\Omega_{ij} = \mathrm{Cov}\left[\sum_{k=1}^{\dim \boldsymbol{\theta}} W_{ki}(U_k), \sum_{k=1}^{\dim \boldsymbol{\theta}} W_{kj}(U_k)\right] , \qquad (5.B.12)$$

where

$$W_{ki}(U_k) = \int_{\boldsymbol{u} \in [0,1]^{\dim \boldsymbol{\theta}}} \mathbf{1}_{\{U_k \leq u_k\}} \left.\frac{\partial^2 \ln c\left(\boldsymbol{u}; \boldsymbol{\theta}\right)}{\partial \theta_i \partial u_i}\right|_{\boldsymbol{\theta} = \boldsymbol{\theta}^0} dC\left(\boldsymbol{u}; \boldsymbol{\theta}^0\right) . \qquad (5.B.13)$$

These results come from a straightforward application of the consistency and asymptotic normality of functionals of multivariate rank statistics derived by Ruymgaart *et al.* [423, 424] and Rüschendorf [422]. Indeed, concerning asymptotic normality, the maximum pseudo likelihood estimator $\hat{\boldsymbol{\theta}}_T$ satisfies:

$$h_T\left(\hat{\boldsymbol{\theta}}_T\right) = 0 \, , \tag{5.B.14}$$

where

$$h_T(\boldsymbol{\theta}) = \frac{1}{T} \sum_{i=1}^{T} \nabla_{\boldsymbol{\theta}} \ln c\left(\hat{\boldsymbol{u}}(i); \, \boldsymbol{\theta}\right) \, . \tag{5.B.15}$$

Now, expanding $h_T$ around $\boldsymbol{\theta}^0$, we have

$$h_T(\boldsymbol{\theta}) = h_T(\boldsymbol{\theta}^0) + \tilde{\boldsymbol{A}}_T(\boldsymbol{\theta}^0)\left(\boldsymbol{\theta} - \boldsymbol{\theta}^0\right) + \cdots \tag{5.B.16}$$

where $\tilde{\boldsymbol{A}}_T(\boldsymbol{\theta})$ is the Hessian matrix of $h_T(\boldsymbol{\theta})$,

$$\left(\tilde{\boldsymbol{A}}_T(\boldsymbol{\theta})\right)_{ij} = \frac{1}{T} \sum_{k=1}^{T} \partial^2_{\theta_i \theta_j} \ln c\left(\hat{\boldsymbol{u}}(k); \, \boldsymbol{\theta}\right) \, . \tag{5.B.17}$$

Proposition A.1 in [197] provides a generalized form of the law of large numbers for functionals of rank statistics, so that

$$\tilde{\boldsymbol{A}}_T(\boldsymbol{\theta}) \xrightarrow{a.s} \mathrm{E}\left[\partial^2_{\theta_i \theta_j} \ln c\left(U; \, \boldsymbol{\theta}\right)\right]_{\boldsymbol{\theta} = \boldsymbol{\theta}^0} = -\boldsymbol{I}\left(\boldsymbol{\theta}^0\right) \, , \tag{5.B.18}$$

where $\boldsymbol{I}\left(\boldsymbol{\theta}^0\right)$ denotes Fisher's information matrix (5.B.11). Evaluating (5.B.16) at $\boldsymbol{\theta} = \hat{\boldsymbol{\theta}}_T$, one finally obtains

$$\sqrt{T} \cdot h_T(\boldsymbol{\theta}^0) = \sqrt{T} \cdot I\left(\boldsymbol{\theta}^0\right)\left(\hat{\boldsymbol{\theta}}_T - \boldsymbol{\theta}^0\right) + o_p(1) \, , \tag{5.B.19}$$

as usual.

Proposition A.1 in [197] also states a generalized form of the central limit theorem for functionals of rank statistics, which allows one to write

$$\sqrt{T} \cdot h_T(\boldsymbol{\theta}^0) \longrightarrow \mathcal{N}\left(0, \boldsymbol{\Gamma}\left(\boldsymbol{\theta}^0\right)\right) \, , \tag{5.B.20}$$

where $\boldsymbol{\Gamma}\left(\boldsymbol{\theta}^0\right) = \boldsymbol{I}\left(\boldsymbol{\theta}^0\right) + \boldsymbol{\Omega}$. Then, (5.B.19–5.B.20) allow us to conclude that

$$\sqrt{T} \cdot \left(\hat{\boldsymbol{\theta}}_T - \boldsymbol{\theta}^0\right) \longrightarrow \mathcal{N}\left(0, \boldsymbol{\Sigma}^2\right) \, , \tag{5.B.21}$$

where $\boldsymbol{\Sigma}^2$ stands for $\boldsymbol{I}\left(\boldsymbol{\theta}^0\right)^{-1} + \boldsymbol{I}\left(\boldsymbol{\theta}^0\right)^{-1} \boldsymbol{\Omega} \boldsymbol{I}\left(\boldsymbol{\theta}^0\right)^{-1}$.

Since $\boldsymbol{\Omega}$ is a positive definite matrix, the variance of the estimator $\hat{\boldsymbol{\theta}}_T$ is larger than it would be, were the marginal distributions $F_i$ perfectly known. Indeed, in such a case, the variance of the estimator would be nothing but the inverse of Fisher's information matrix $\boldsymbol{I}\left(\boldsymbol{\theta}^0\right)^{-1}$.

Now, let us write the vector $\boldsymbol{\theta}$ of parameters as follows:

$$\boldsymbol{\theta} = \begin{pmatrix} \boldsymbol{\theta}_1 \\ \boldsymbol{\theta}_2 \end{pmatrix} \tag{5.B.22}$$

with $\dim \boldsymbol{\theta}_1 = d$ and $\dim \boldsymbol{\theta}_2 = p - d$. We would like to test the null hypothesis according to which $\boldsymbol{\theta}_1 = \boldsymbol{\theta}_1^0$, i.e, $H_0 = \{\boldsymbol{\theta} \in \Theta, \boldsymbol{\theta}_1 = \boldsymbol{\theta}_1^0\}$. In Mashal and Zeevi's approach [350], this amounts to test $H_0 = \{(\nu, \boldsymbol{\Sigma}^2) \, ; \, \nu = \infty\}$, where $\nu$ denotes the number of degrees of freedom of the Student's copula and $\boldsymbol{\Sigma}^2$ its shape matrix.

If the likelihood $\tilde{\mathcal{L}}$ was the actual likelihood, and not a pseudo likelihood, the log-likelihood ratio test would allow us to test such a null hypothesis. Indeed, under the null, Wilks' theorem would hold and one would have

$$\Lambda_T = 2 \cdot \left[\tilde{\mathcal{L}}_T\left(\hat{\boldsymbol{\theta}}_T\right) - \sup_{\boldsymbol{\theta} \in H_0} \tilde{\mathcal{L}}(\boldsymbol{\theta})\right] \longrightarrow \chi_d^2 \,, \tag{5.B.23}$$

where $\chi_d^2$ denotes the $\chi^2$ distribution with $d$ degrees of freedom (see Chap. 2, Sect. 2.4.4).

Unfortunately, this test does not apply with the pseudo likelihood, as previously assumed [209, 372, 491]. Actually, expanding the pseudo log-likelihood (5.B.8) around $\boldsymbol{\theta}^0$ and accounting for (5.B.19), we obtain

$$\tilde{\mathcal{L}}_T\left(\hat{\boldsymbol{\theta}}_T\right) = \tilde{\mathcal{L}}_T\left(\boldsymbol{\theta}^0\right) + \frac{T}{2}\left(\hat{\boldsymbol{\theta}}_T - \boldsymbol{\theta}^0\right)^t \boldsymbol{I}\left(\boldsymbol{\theta}^0\right)\left(\hat{\boldsymbol{\theta}}_T - \boldsymbol{\theta}^0\right) + o_p(1) \,. \tag{5.B.24}$$

Denoting by $\hat{\boldsymbol{\theta}}_T^0$ the pseudo maximum likelihood estimator under the null hypothesis (*i.e.*, assuming $\boldsymbol{\theta}_1 = \boldsymbol{\theta}_1^0$):

$$\hat{\boldsymbol{\theta}}_T^0 = \arg\max_{\boldsymbol{\theta} \in H_0} \sum_{i=1}^{T} \ln c\left(\hat{\boldsymbol{u}}(i); \, \boldsymbol{\theta}\right) \,, \tag{5.B.25}$$

and expanding $\boldsymbol{h}_T = T^{-1}\nabla_{\boldsymbol{\theta}}\tilde{\mathcal{L}}_T$ around $\boldsymbol{\theta}^0$, which yields

$$\frac{1}{\sqrt{T}} \cdot \begin{pmatrix} \nabla_{\boldsymbol{\theta}_1}\tilde{\mathcal{L}}_T\left(\hat{\boldsymbol{\theta}}_T^0\right) \\ 0 \end{pmatrix} = \sqrt{T} \cdot \boldsymbol{h}_T\left(\boldsymbol{\theta}^0\right) - \sqrt{T} \cdot \boldsymbol{I}(\boldsymbol{\theta}^0) \begin{pmatrix} 0 \\ \hat{\boldsymbol{\theta}}_{2,T}^0 - \boldsymbol{\theta}_2^0 \end{pmatrix} + o_p(1) \,, \tag{5.B.26}$$

the expansion of the pseudo likelihood around $\boldsymbol{\theta}^0$ reads

$$\tilde{\mathcal{L}}_T\left(\hat{\boldsymbol{\theta}}_T^0\right) = \tilde{\mathcal{L}}_T\left(\boldsymbol{\theta}^0\right) + \frac{T}{2}\left(\hat{\boldsymbol{\theta}}_T^0 - \boldsymbol{\theta}^0\right)^t \boldsymbol{I}\left(\boldsymbol{\theta}^0\right)\left(\hat{\boldsymbol{\theta}}_T^0 - \boldsymbol{\theta}^0\right) + o_p(1) \,. \tag{5.B.27}$$

The notation

$$\hat{\boldsymbol{\theta}}_T^0 = \begin{pmatrix} \boldsymbol{\theta}_1^0 \\ \hat{\boldsymbol{\theta}}_{2,T}^0 \end{pmatrix} \,, \tag{5.B.28}$$

has been used in (5.B.26).

$\varLambda_T$, defined in (5.B.23), is now obtained by taking the difference between (5.B.24) and (5.B.27):

$$\varLambda_T = T\left(\hat{\boldsymbol{\theta}}_T - \boldsymbol{\theta}^0\right)^t \boldsymbol{I}\left(\boldsymbol{\theta}^0\right)\left(\hat{\boldsymbol{\theta}}_T - \boldsymbol{\theta}^0\right) - T\left(\hat{\boldsymbol{\theta}}_T^0 - \boldsymbol{\theta}^0\right)^t \boldsymbol{I}(\boldsymbol{\theta}^0)\left(\hat{\boldsymbol{\theta}}_T^0 - \boldsymbol{\theta}^0\right) + o_p(1),$$

$$=T\left(\hat{\boldsymbol{\theta}}_T - \hat{\boldsymbol{\theta}}_T^0\right)^t \boldsymbol{I}\left(\boldsymbol{\theta}^0\right)\left(\hat{\boldsymbol{\theta}}_T - \hat{\boldsymbol{\theta}}_T^0\right) + 2T\left(\hat{\boldsymbol{\theta}}_T^0 - \boldsymbol{\theta}^0\right)^t \boldsymbol{I}\left(\boldsymbol{\theta}^0\right)\left(\hat{\boldsymbol{\theta}}_T - \hat{\boldsymbol{\theta}}_T^0\right) + o_p(1)\ ,$$

where the last equality uses the fact that each term is a scalar and is thus equal to its transpose.

Substituting (5.B.19) in (5.B.26) yields

$$\frac{1}{\sqrt{T}} \cdot \begin{pmatrix} \nabla_{\boldsymbol{\theta}_1} \tilde{\mathcal{L}}_T\left(\hat{\boldsymbol{\theta}}_T^0\right) \\ 0 \end{pmatrix} = \sqrt{T} \cdot \boldsymbol{I}(\boldsymbol{\theta}^0)\left(\hat{\boldsymbol{\theta}}_T - \hat{\boldsymbol{\theta}}_n^0\right) + o_p(1) \qquad (5.B.29)$$

and, left-multiplying by $\sqrt{T} \cdot \left(\hat{\boldsymbol{\theta}}_T^0 - \boldsymbol{\theta}^0\right)^t = \sqrt{T} \cdot \begin{pmatrix} 0 \\ \hat{\boldsymbol{\theta}}_{2,T}^0 - \boldsymbol{\theta}_2^0 \end{pmatrix}^t$ shows that

$$T\left(\hat{\boldsymbol{\theta}}_T^0 - \boldsymbol{\theta}^0\right)^t \boldsymbol{I}\left(\boldsymbol{\theta}^0\right)\left(\hat{\boldsymbol{\theta}}_T - \hat{\boldsymbol{\theta}}_T^0\right) = o_p(1)\ , \qquad (5.B.30)$$

which allows us to conclude that

$$\varLambda_T = T\left(\hat{\boldsymbol{\theta}}_T - \hat{\boldsymbol{\theta}}_T^0\right)^t \boldsymbol{I}\left(\boldsymbol{\theta}^0\right)\left(\hat{\boldsymbol{\theta}}_T - \hat{\boldsymbol{\theta}}_T^0\right) + o_p(1)\ . \qquad (5.B.31)$$

Now, since (5.B.29) is equivalent to

$$\sqrt{T} \cdot \left(\hat{\boldsymbol{\theta}}_T - \hat{\boldsymbol{\theta}}_T^0\right) = \frac{1}{\sqrt{T}} \cdot \boldsymbol{I}\left(\boldsymbol{\theta}^0\right)^{-1} \begin{pmatrix} \nabla_{\boldsymbol{\theta}_1} \tilde{\mathcal{L}}_T\left(\hat{\boldsymbol{\theta}}_T^0\right) \\ 0 \end{pmatrix} + o_p(1)\ , \quad (5.B.32)$$

we have:

$$\varLambda_T = \frac{1}{T} \begin{pmatrix} \nabla_{\boldsymbol{\theta}_1} \tilde{\mathcal{L}}_T\left(\hat{\boldsymbol{\theta}}_T^0\right) \\ 0 \end{pmatrix}^t \boldsymbol{I}\left(\boldsymbol{\theta}^0\right)^{-1} \begin{pmatrix} \nabla_{\boldsymbol{\theta}_1} \tilde{\mathcal{L}}_T\left(\hat{\boldsymbol{\theta}}_T^0\right) \\ 0 \end{pmatrix} + o_p(1)\ , \quad (5.B.33)$$

$$= T^{-1} \cdot \nabla_{\boldsymbol{\theta}_1} \tilde{\mathcal{L}}_T\left(\hat{\boldsymbol{\theta}}_T^0\right)^t \left[\boldsymbol{I}^{-1}\right]_{11} \nabla_{\boldsymbol{\theta}_1} \tilde{\mathcal{L}}_T\left(\hat{\boldsymbol{\theta}}_T^0\right) + o_p(1)\ , \qquad (5.B.34)$$

where $\left[\boldsymbol{I}^{-1}\right]_{11}$ denotes the $p \times p$ submatrix of the $p$ first rows and columns of the inverse of $\boldsymbol{I}(\boldsymbol{\theta}^0)$.

From (5.B.29) again, we have

$$\frac{1}{\sqrt{T}} \cdot \left[\boldsymbol{I}^{-1}\right]_{11} \nabla_{\boldsymbol{\theta}_1} \tilde{\mathcal{L}}_T\left(\hat{\boldsymbol{\theta}}_T^0\right) = \sqrt{T}\left(\hat{\boldsymbol{\theta}}_{1,T} - \boldsymbol{\theta}_1^0\right) + o_p(1)\ , \qquad (5.B.35)$$

so that

$$\Lambda_T = T^{-1} \cdot \nabla_{\boldsymbol{\theta}_1} \tilde{\mathcal{L}}_T \left(\hat{\boldsymbol{\theta}}_T^0\right)^t \left[\boldsymbol{I}^{-1}\right]_{11} \nabla_{\boldsymbol{\theta}_1} \tilde{\mathcal{L}}_T \left(\hat{\boldsymbol{\theta}}_n^0\right)$$

$$= T \cdot \left(\hat{\boldsymbol{\theta}}_{1,T} - \boldsymbol{\theta}_1^0\right)^t \left\{\left[\boldsymbol{I}^{-1}\right]_{11}\right\}^{-1} \left(\hat{\boldsymbol{\theta}}_{1,T} - \boldsymbol{\theta}_1^0\right) + o_p(1) \ . \qquad (5.\text{B}.36)$$

Now, by (5.B.21), we have

$$\sqrt{T} \cdot \left(\hat{\boldsymbol{\theta}}_{1,T} - \boldsymbol{\theta}_1^0\right) \longrightarrow \mathcal{N}\left(0, \left[\boldsymbol{\Sigma}^2\right]_{11}\right) . \qquad (5.\text{B}.37)$$

Denoting by $B$ a symmetric positive definite matrix such that

$$\boldsymbol{B} \cdot \boldsymbol{B} = \left[\boldsymbol{\Sigma}^2\right]_{11} \qquad (5.\text{B}.38)$$

and by $\boldsymbol{\xi}_d$ a $d$-dimensional standard Gaussian vector, we obtain:

$$\Lambda_T = \boldsymbol{\xi}_d^{\,t} \cdot \boldsymbol{B} \left\{\left[\boldsymbol{I}^{-1}\right]_{11}\right\}^{-1} \boldsymbol{B} \cdot \boldsymbol{\xi}_d + o_p(1) \ . \qquad (5.\text{B}.39)$$

As a consequence,

$$\Lambda_T \nrightarrow \chi_d^2, \qquad \text{as } T \to \infty \qquad (5.\text{B}.40)$$

unless

$$\boldsymbol{B} \left\{\left[\boldsymbol{I}^{-1}\right]_{11}\right\}^{-1} \boldsymbol{B} = \text{Id}_d \ , \qquad (5.\text{B}.41)$$

which holds when $\boldsymbol{\Omega} = 0$, for instance. Therefore, when one resorts to the pseudo likelihood instead of the actual likelihood, the asymptotic distribution of $\Lambda_n$ is not a simple $\chi^2$ distribution and the log-likelihood ratio test becomes impracticable. In the particular case where dim $\boldsymbol{\theta}_1 = 1$, as in [350], $\boldsymbol{B} \left\{\left[\boldsymbol{I}^{-1}\right]_{11}\right\}^{-1} \boldsymbol{B}$ is a scalar so that $\Lambda_n$ follows a $\chi^2$ distribution with one degree of freedom, up the scale factor.

# 6

# Measuring Extreme Dependences

In this chapter, we investigate the relative information content of several measures of dependence between two random variables $X$ and $Y$ in various models of financial series. We consider measures of dependence especially defined for large and extreme events. These measures of dependence are of two types: (i) unconditional such as with the coefficient of tail dependence already introduced in Chap. 4 and (ii) conditional such as with the correlation coefficient conditional over a given threshold. The introduction of conditioning over values of one or both variables reaching above some threshold is a natural approach to discriminate the dependence in the tails. It explodes the concept of dependence into a multidimensional set of measures, each adapted to certain ranges spanned by the random variables. We present explicit analytical formulas as well as numerical and empirical estimations for these measures of dependence. The main overall insight is that conditional measures of dependence may be very different from the unconditional ones and can often lead to paradoxical interpretations, whose origins are explained in detail.

When the dependence properties are studied as a function of time, one can often observe that conditional measures vary with time. Such time variation has initiated a vigorous discussion in the literature on its possible economic meaning. We review the mechanism by which conditioning provides a straightforward and general mechanism for explaining changes of correlations based on changes of volatility or of trends: for a given conditional threshold, if the volatility of one or both time series changes in some time interval, then the corresponding quantiles sampled in the conditional measure will also change; as a result, the conditional measure will not sample the same part of the tails of the distributions, effectively changing the definition of the conditional measure. In this explanation, the variation with time of conditional measures of dependence results solely from a change of volatility but does not reflect a genuine change of dependence. In other words, a constant dependence structure together with time-varying volatility may give rise to changing conditional measures of dependence, which would be incorrectly interpreted as reflecting genuine changes of dependence. Thus, tools based upon conditional quantities should be used with caution since conditioning alone induces a change in the dependence structure which has nothing to do with a genuine change of unconditional dependence. In this respect, for its stability, the coefficient of tail dependence should be preferred to the conditional correlations. Moreover, the various measures of dependence exhibit different and sometimes opposite behaviors, showing that extreme dependence properties possess a multidimensional character that can be revealed in various ways.

As an illustration, the theoretical results and their interpretation presented below are applied to the controversial contagion problem across Latin American markets during the turmoil periods associated with the Mexican crisis in 1994 and with the Argentinean crisis that started in 2001. The analysis of several measures of dependence between the Argentinean, Brazilian, Chilean and Mexican markets shows that the above conditioning effect does not fully explain the behavior of the Latin American stock indexes, confirming the existence of a possible genuine contagion. Our analysis below suggests that the 1994 Mexican crisis has spread over to Argentina and Brazil through contagion mechanisms and to Chile only through co-movements. Concerning the recent Argentinean crisis that started in 2001, no evidence of contagion to the other Latin American countries (except perhaps in the direction of Brazil) can be found but significant co-movements are identified.

The chapter is organized as follows. Sect. 6.1 motivates the whole chapter by presenting a number of historically important cases which suggested to previous authors that, "during major market events, correlations change dramatically" [71]. This section then offers a review of the different existing view points on conditional dependences.

Section 6.2 describes three conditional correlation coefficients:

- the correlation $\rho_v^+$ conditioned on signed exceedance of one variable,
- or on both variables ($\rho_u$) and
- the correlation $\rho_v^s$ conditioned on the exceedance of the absolute value of one variable (amounting to a conditioning on large values of the volatility).

Boyer *et al.* [78] have provided the general expression of $\rho_v^+$ and $\rho_v^s$ for the Gaussian bivariate model, which we use to derive their $v$ dependence for large thresholds $v$. This analysis shows that, for a given distribution, the conditional correlation coefficient changes even if the unconditional correlation is left unchanged, and the nature of this change depends on the conditioning set. We then give the general expression of $\rho_v^+$ and $\rho_v^s$ for the Student's bivariate model with $\nu$ degrees of freedom and for the factor model $X = \beta Y + \epsilon$, for arbitrary distributions of $Y$ and $\epsilon$. By comparison with the Gaussian model, these expressions exemplify that, for a fixed conditioning set, the behavior of the conditional correlation change dramatically from one distribution to another one. Conditioning on both variables, we give the asymptotic dependence of $\rho_u$ for the bivariate Gaussian model and show that it essentially behaves like $\rho_v^+$. Applying these results to the Latin American stock indexes, we find that one cannot entirely explain the behavior of the conditional correlation coefficient for these markets by the conditioning effect, suggesting the existence of a possible genuine contagion as mentioned above.

In Sect. 6.3, to account for several deficiencies of the correlation coefficient, we study an alternative measure of dependence, the conditional rank correlation (Spearman's rho) which, in its unconditional form, is related to the probability of concordance and discordance of several events drawn from the same probability distribution, as recalled in Chap. 4. This measure provides an important improvement with respect to the correlation coefficient since it only takes into account the dependence structure of the variable and is not sensitive to the marginal behavior of each variable. Numerical computations allow us to derive the behavior of the conditional Spearman's rho, denoted by $\rho_s(v)$. This allow us to prove that there is no direct relation between the Spearman's rho conditioned on large values and the correlation coefficient conditioned on the same values. Therefore, each of these coefficients quantifies a different kind of extreme dependence. Then, calibrating the models on the Latin American market data confirms that the conditional effect cannot fully explain the observed dependence and that contagion can therefore be invoked. These results are much clearer for the conditional Spearman's rho than for the condition (linear) correlation coefficient, due to the greater impact of large statistical fluctuations in the later.

Section 6.4 discusses the tail-dependence parameters $\lambda$ and $\bar{\lambda}$, introduced in Chap. 4. Applying the procedure of [390], we estimate nonparametrically the tail dependence coefficients. We find them significant and thus conclude that, with or without contagion mechanism, extreme co-movements must naturally occur on the various Latin American markets as soon as one of them undergoes a crisis.

Section 6.5 provides a comparison between these different results and a synthesis. A first important message is that there is no unique measure of extreme dependence. Each of the coefficients of extreme dependence that we have presented provides a specific quantification that is sensitive to a certain combination of the marginals and of the copula of the two random variables. Similarly to risks whose adequate characterization requires an extension beyond the restricted one-dimensional measure in terms of the variance (volatility) to include the knowledge of the full distribution, tail-dependence has also a multidimensional character. A second important message is that the increase of some of the conditional coefficients of extreme dependence when weighting more and more the extreme tail range does not necessarily signal a genuine increase of the unconditional correlation or dependence between the two variables. The calculations presented here firmly confirm that this increase is a general and unavoidable result of the statistical properties of many multivariate models of dependence. From the standpoint of the contagion across Latin American markets, the theoretical and empirical results suggest an asymmetric contagion phenomenon from Chile and Mexico towards Argentina and Brazil: large moves of the Chilean and Mexican markets tend to propagate to Argentina and Brazil through contagion mechanisms, *i.e.*, with a change in the dependence structure, while the converse does not hold. As a consequence, this seems to prove that the 1994 Mexican crisis had spread over to Argentina and Brazil through contagion mechanisms and to Chile only through co-movements. Concerning the more recent Argentinean crisis starting in 2001, no evidence of contagion to the other Latin American countries is found (except perhaps in the direction of Brazil) and only co-movements can be identified.

## 6.1 Motivations

### 6.1.1 Suggestive Historical Examples

The 19 October, 1987, stock-market crash stunned Wall Street professionals, hacked about $1 trillion off the value of all U.S. stocks, and elicited predictions of another Great Depression. On "Black Monday," the Dow Jones industrial average plummeted 508 points, or 22.6 percent, to 1,738.74. Contrary to common belief, the US was not the first to decline sharply. Non-Japanese Asian markets began a severe decline on 19 October, 1987, their time, and this decline was echoed first on a number of European markets, then in North American, and finally in Japan. However, most of the same markets had experienced significant but less severe declines in the latter part of the previous week. With the exception of the US and Canada, other markets continued downward through the end of October, and some of these declines were as large as the great crash on 19 October.

On 19 December, 1994, the Mexican government, facing a solvency crisis, chose to devaluate the peso and abandoned its exchange rate parity with the dollar. This devaluation plunged the country into a major financial crisis which quickly propagated to the rest of the Latin American countries.

From July 1997 to December 1997, several East Asian markets crashed, starting with the Thai market on 2 July , 1997 and ending with the Hong Kong market on 17 October, 1997. After this regional event, the turmoil spread over to the American and European markets.

The "slow" crash and in particular the turbulent behavior of the stock markets worldwide starting mid-August 1998 are widely associated with and even attributed to the plunge of the Russian financial markets, the devaluation of its currency and the default of the government on its debt obligations.

The Nasdaq Composite index dropped precipitously with a low of 3227 on 17 April, 2000, corresponding to a cumulative loss of 37% counted from its all-time high of 5133 reached on 10 March, 2000. The drop was mostly driven by the so-called "New Economy" stocks which have risen nearly four-fold over 1998 and 1999 compared to a gain of only 50% for the Standard & Poor's 500 index. And without technology, this benchmark would be flat.

All these events epitomize the observation often reported by market professionals that, "during major market events, correlations change dramatically"

[71], as mentioned above. The possible existence of changes of correlation, or more precisely of changes of dependence, between assets and between markets in different market phases has obvious implications in risk assessment, portfolio management and in the way policy and regulation should be formulated. Concerning portfolio management, these questions related to state-varying-dependence are important for practical applications since in such a case the optimal portfolio will also become state-dependent. Neglecting this effect can lead to very inefficient asset allocations [14, 15]. In this spirit, the Argentinean crisis in 2001 has triggered fears of a contagion to other Latin American markets. Also, the Enron financial scandal at the end of 2001 seems to have opened a flux of similar bankruptcies in other "new economy" companies.

### 6.1.2 Review of Different Perspectives

In the academic world, all these manifestations of propagating crises have given birth to an intense activity concerning the notion of contagion (see [102] for a review). According to the most commonly accepted definition, contagion is characterized by as an increase in the correlation (or, more generally, dependence) across markets during periods of turmoil. In fact, as we shall see, there are two distinct classes of mechanisms for understanding "changes of correlations," not necessarily mutually exclusive.

- It is possible that there are genuine changes with time of the unconditional (with respect to amplitudes) correlations and thus of the underlying structure of the dynamical processes, as observed by identifying shifts in ARMA-ARCH/GARCH processes [440], in regime-switching models [14, 15] or in contagion models [395, 396]. Many workers (see for instance [314, 477]) have shown that the hypothesis of a constant conditional correlation for stock returns or international equity returns must be rejected. In fact, there is strong evidence that the correlations are not only time-dependent but also state-dependent. Indeed, as shown in [271, 397], the correlations increase in periods of large volatility. Moreover, Longin and Solnik [315] have proved that the correlations across international equity markets are also trend-dependent.
- A second class of explanation is that correlations between two variables conditioned on signed exceedance (one-sided) or on absolute value (volatility) exceedance of one or both variables may deviate significantly from the unconditional correlation [78, 316, 317]. In other words, with a fixed unconditional correlation $\rho$, the measured correlation conditional of a given bullish trend, bearish trend, high or low market volatility, may in general differ from $\rho$ and can be viewed as a function of the specific market phase. According to this explanation, changes of correlation may be only a fallacious appearance that stems from a change of volatility or a change of trend of the market and not from a real change of unconditional correlation or dependence.

The existence of the second class of explanation is appealing by its parsimony, as it posits that observed "changes of correlation" may simply result from the way the measure of dependence is performed. This approach has been followed by several authors but is often open to misinterpretation, as stressed in [178]. In addition, it may also be misleading since it does not provide a signature or procedure for identifying the existence of a genuine contagion phenomenon, if any. Therefore, in order to clarify the situation and eventually develop more adequate tools for probing the dependences between assets and between markets, it is highly desirable to characterize the different possible ways with which higher or lower conditional dependence can occur in models with constant unconditional dependence. In order to make progress, it is necessary to first distinguish between the different measures of dependence between two variables for large or extreme events that have been introduced in the literature. This is because the conclusions that one can draw about the variability of dependence are sensitive to the choice of its measure. These measures include the following.

1. The correlation conditioned on signed exceedance of one or both variables [101, 78, 316, 317] that we call respectively $\rho_v^+$ and $\rho_u$, where $u$ and $v$ denote the thresholds above which the exceedances are calculated.
2. The correlation conditioned on absolute value exceedance (or large volatility), above the threshold $v$, of one or both variables [101, 78, 316, 317] that we call $\rho_v^s$ (for a condition of exceedance on one variable).
3. The local correlation (whose definition is given in Sect. 4.1.2), which is immune to the biases associated with the two aforementioned conditional correlation coefficients. Bradley and Taqqu have used it to introduce a new diagnostic of contagion: contagion from market $X$ to market $Y$ is qualified if there is more dependence between $X$ and $Y$ when $X$ is doing badly than when $X$ exhibits typical performance, that is, if there is more dependence at the loss tail distribution of $X$ than at its center [80, 81, 82].
4. The tail-dependence parameter $\lambda$, which has a simple analytical expression when using copulas [149, 147] such as the Gumbel copula [315], and whose estimation provides useful information about the occurrence of extreme co-movements [260, 334, 390].
5. The spectral measure associated with the tail index (assumed to be the same for all assets) of extreme value multivariate distributions [43, 224, 462].
6. Tail indices of extremal correlations defined as the upper or lower correlation of exceedances of ordered log-values [395].
7. Confidence weighted forecast correlations [53] or algorithmic complexity measures [342].

The contribution of this chapter is both methodological and empirical. On the methodological front, first of all, we review the existing tools available for probing the dependence between large or extreme events for several models of interest for financial time series; second, we provide explicit analytical expressions for these measures of dependence between two variables; third, this allows us to quantify the misleading interpretations of certain conditional coefficients commonly used for exploring the evolution of the dependence associated with a change in the market conditions (an increase of the volatility, for instance). On the empirical front, the theoretical results are applied to the controversial problem of the occurrence or absence of a contagion phenomenon across Latin American markets during the turmoil period associated with the Mexican crisis in 1994 or with the recent Argentinean crisis in 2001. For this purpose, the novel insight derived from the analysis of several measures of dependence is applied to the question of a possible evolution of the dependence between the Argentinean, Brazilian, Chilean and Mexican markets with respect to the market conditions.

The dependence measures discussed below are the conditional correlation coefficients $\rho_v^+$, $\rho_v^s$, $\rho_u$, the conditional Spearman's rho $\rho_s(v)$ and the tail dependence coefficients $\lambda$ and $\bar{\lambda}$, whose properties have been summarized in Chap. 4, for several models among which are the bivariate Gaussian distribution, the bivariate Student's distribution, and the one factor model for various distributions of the factor. A priori, one could hope for the existence of logical links between some of these measures, such as a vanishing tail-dependence parameter $\lambda$ implies vanishing asymptotic conditional correlation coefficients. In fact, this turns out to be wrong and one can construct simple examples for which all possible combinations occur. Therefore, each of these measures probe a different quality of the dependence between two variables for large or extreme events. In addition, even if the conditional correlation coefficients are asymptotically zero, they decay in general extremely slowly, as inverse powers of the value of the threshold, and may thus remain significant for most practical applications. These results will allow us to assert that, somewhat similarly to risks whose adequate characterization requires an extension beyond the restricted one-dimensional measure in terms of the variance to include all higher order cumulants or more generally the knowledge of the full distribution [453, 6], these results suggest that large and/or extreme dependences have also a multidimensional character.

## 6.2 Conditional Correlation Coefficient

In this section, we discuss the properties of the correlation coefficient conditioned on one variable. We study the difference between conditioning on the signed values or on absolute values of the variable (conditioning on the absolute value of the variable of interest is only meaningful when its distribution is symmetric). This allows us to conclude that conditioning on signed values generally provides more information than conditioning on absolute values. Moreover, as already underlined for instance by Boyer *et al.* [78], the conditional correlation coefficient is shown to suffer from a bias which forbids its use as a measure of change in the correlation between two assets when the volatility increases (many papers on contagion unfortunately use the conditional correlation coefficient as a probe to detect changes of dependence). We then present an empirical illustration of the evolution of the correlation between several stock indexes of Latin American markets.

## 6.2.1 Definition

Let us consider the correlation coefficient $\rho_{\mathcal{A}}$ of two real random variables $X$ and $Y$ conditioned on $Y \in \mathcal{A}$, where $\mathcal{A}$ is a subset of $\mathbb{R}$ such that $\Pr\{Y \in \mathcal{A}\} > 0$. By definition, the conditional correlation coefficient $\rho_{\mathcal{A}}$ is given by

$$\rho_{\mathcal{A}} = \frac{\mathrm{Cov}(X, Y \mid Y \in \mathcal{A})}{\sqrt{\mathrm{Var}(X \mid Y \in \mathcal{A}) \cdot \mathrm{Var}(Y \mid Y \in \mathcal{A})}} \ . \tag{6.1}$$

This general expression of the conditional correlation coefficient can be transformed into closed formula for several standard distributions and models. This will allow us to investigate the influence of the conditioning set and the underlying model on the behavior of $\rho_{\mathcal{A}}$.

## 6.2.2 Influence of the Conditioning Set

Let the variables $X$ and $Y$ have a multivariate Gaussian distribution with (unconditional) correlation coefficient $\rho$. The following result has been proved [78]:

$$\rho_{\mathcal{A}} = \frac{\rho}{\sqrt{\rho^2 + (1 - \rho^2) \frac{\mathrm{Var}(Y)}{\mathrm{Var}(Y \mid Y \in \mathcal{A})}}} \ . \tag{6.2}$$

Note that $\rho$ and $\rho_{\mathcal{A}}$ have the same sign, that $\rho_{\mathcal{A}} = 0$ if and only if $\rho = 0$ and that $\rho_{\mathcal{A}}$ does not depend directly on $\mathrm{Var}(X)$. Note also that $\rho_{\mathcal{A}}$ can be either greater or smaller than $\rho$ since $\mathrm{Var}(Y \mid Y \in \mathcal{A})$ can be either greater or smaller than $\mathrm{Var}(Y)$. Let us illustrate this property in the two following examples, with a conditioning on large positive (or negative) returns and a conditioning on large volatility. The difference comes from the fact that in the first case, one accounts for the trend while one neglects this information in the second case.

These two simple examples will show that, in the case of two Gaussian random variables, the two conditional correlation coefficients $\rho_v^+$ and $\rho_v^s$ exhibit opposite behaviors since the conditional correlation coefficient $\rho_v^+$ is a decreasing function of the conditioning threshold $v$ (and goes to zero as $v \to +\infty$) while the conditional correlation coefficient $\rho_v^s$ is an increasing function of $v$ and goes to one as $v \to \infty$. These opposite behaviors seem very general and do not depend on the particular choice of the joint distribution of $X$ and $Y$, namely the Gaussian distribution studied until now, as it will be seen in the sequel.

This result underlines the importance of the choice of the conditioning set with the following two caveats that we stress again. First, as already stressed by many authors, the conditional correlation $\rho_v^+$ or $\rho_v^s$ change with the value of the threshold $v$ even if the unconditional correlation $\rho$ remains unchanged. Thus, the observation of a change in the conditional correlation does not provide a reliable signature of a change in the true (unconditional) correlation. Second, the conditional correlations can exhibit really opposite behaviors depending on the conditioning sets. Specifically, accounting for a signed trend or only for its amplitude may yield a decrease or an increase of the conditional correlation with respect to the unconditional one, so that these changes cannot be interpreted as a strengthening or a weakening of the correlations.

### Example 1: Conditioning on Large (Positive) Returns

Let us first consider the conditioning set $\mathcal{A} = [v, +\infty)$, with $v \in \mathbb{R}_+$. Thus $\rho_{\mathcal{A}}$ is the correlation coefficient conditioned on the returns $Y$ larger than a given positive threshold $v$. It will be denoted by $\rho_v^+$ in the sequel. Assuming for simplicity, but without loss of generality that $\mathrm{Var}(Y) = 1$, an exact calculation given below shows that, for large $v$,

$$\rho_v^+ \sim_{v\to\infty} \frac{\rho}{\sqrt{1-\rho^2}} \cdot \frac{1}{|v|} \,, \tag{6.3}$$

which slowly goes to *zero* as $v$ goes to infinity. Obviously, by symmetry, the conditional correlation coefficient $\rho_v^-$, conditioned on $Y$ smaller than $v$, obeys the same formula.

*Proof.* We start with the calculation of the first and the second moments of $Y$ conditioned on $Y$ larger than $v$:

$$E(Y \mid Y > v) = \frac{\sqrt{2}}{\sqrt{\pi} e^{\frac{v^2}{2}} \mathrm{erfc}\left(\frac{v}{\sqrt{2}}\right)} = v + \frac{1}{v} - \frac{2}{v^3} + \mathcal{O}\left(\frac{1}{v^5}\right), \tag{6.4}$$

$$E(Y^2 \mid Y > v) = 1 + \frac{\sqrt{2}v}{\sqrt{\pi} e^{\frac{v^2}{2}} \mathrm{erfc}\left(\frac{v}{\sqrt{2}}\right)} = v^2 + 2 - \frac{2}{v^2} + \mathcal{O}\left(\frac{1}{v^4}\right), \tag{6.5}$$

which allows us to obtain the variance of $Y$ conditioned on $Y$ larger than $v$:

$$\mathrm{Var}(Y \mid Y > v) = 1 + \frac{\sqrt{2}v}{\sqrt{\pi} e^{\frac{v^2}{2}} \mathrm{erfc}\left(\frac{v}{\sqrt{2}}\right)} - \left(\frac{\sqrt{2}}{\sqrt{\pi} e^{\frac{v^2}{2}} \mathrm{erfc}\left(\frac{v}{\sqrt{2}}\right)}\right)^2$$

$$= \frac{1}{v^2} + \mathcal{O}\left(\frac{1}{v^4}\right), \tag{6.6}$$

which together with (6.2) yields (6.3) for large $v$.

## Example 2: Conditioning on Large Volatilities

Let now the conditioning set be $\mathcal{A} = (-\infty, -v] \cup [v, +\infty)$, with $v \in \mathbb{R}_+$. Thus $\rho_{\mathcal{A}}$ is the correlation coefficient conditioned on $|Y|$ larger than $v$, i.e., it is conditioned on a large volatility of $Y$. Still assuming $\mathrm{Var}(Y) = 1$, this correlation coefficient is denoted by $\rho_v^s$ and, for large $v$

$$\rho_v^s \sim_{v \to \infty} \frac{\rho}{\sqrt{\rho^2 + \frac{1-\rho^2}{2+v^2}}} \sim_{v \to \infty} \mathrm{sgn}(\rho) \cdot \left(1 - \frac{1}{2} \frac{1-\rho^2}{\rho^2} \frac{1}{v^2}\right) , \qquad (6.7)$$

which goes to (plus or minus) 1 as $v$ goes to infinity according to $1 - |\rho_v^s| \sim_{v \to \infty} \frac{1-\rho^2}{2\rho^2} v^{-2}$.

*Proof.* The correlation coefficient conditioned on $|Y|$ larger than $v$ can be written

$$\rho_v^s = \frac{\rho}{\sqrt{\rho^2 + \frac{1-\rho^2}{\mathrm{Var}(Y \mid |Y| > v)}}} . \qquad (6.8)$$

The first and second moment of $Y$ conditioned on $|Y|$ larger than $v$ can be easily calculated:

$$E(Y \mid |Y| > v) = 0 , \qquad (6.9)$$

$$E(Y^2 \mid |Y| > v) = 1 + \frac{\sqrt{2}v}{\sqrt{\pi}e^{\frac{v^2}{2}}\mathrm{erfc}\left(\frac{v}{\sqrt{2}}\right)} = v^2 + 2 - \frac{2}{v^2} + \mathcal{O}\left(\frac{1}{v^4}\right) . \qquad (6.10)$$

Expression (6.10) is the same as (6.6) as it should. This gives the following conditional variance:

$$\mathrm{Var}(Y \mid |Y| > v) = 1 + \frac{\sqrt{2}v}{\sqrt{\pi}e^{\frac{v^2}{2}}\mathrm{erfc}\left(\frac{v}{\sqrt{2}}\right)} = v^2 + 2 + \mathcal{O}\left(\frac{1}{v^2}\right) , \qquad (6.11)$$

and finally yields (6.7), for large $v$.

## Intuitive Meaning

Let us provide an intuitive explanation (see also [315]). As seen from (6.2), $\rho_v^+$ is controlled by $\mathrm{Var}(Y \mid Y > v) \propto 1/v^2$ derived in the example 1. In contrast, as seen from (6.8), $\rho_v^s$ is controlled by $\mathrm{Var}(Y \mid |Y| > v) \propto v^2$ given in the example 2. The difference between $\rho_v^+$ and $\rho_v^s$ can thus be traced back to that between $\mathrm{Var}(Y \mid Y > v) \propto 1/v^2$ and $\mathrm{Var}(Y \mid |Y| > v) \propto v^2$ for large $v$.

This results from the following effect. For $Y > v$, one can picture the possible realizations of $Y$ as those of a random particle on the line, which is strongly attracted to the origin by a spring (the Gaussian distribution that prevents $Y$ from performing significant fluctuations beyond a few standard deviations) while being forced to be on the right to a wall at $Y = v$. It is clear that the fluctuations of the position of this particle are very small as it is strongly glued to the impenetrable wall by the restoring spring, hence the result $\mathrm{Var}(Y \mid Y > v) \propto 1/v^2$. In contrast, for the condition $|Y| > v$, by the same argument, the fluctuations of the particle are constrained to be very close to $|Y| = v$, i.e., very close to $Y = +v$ or $Y = -v$. Thus, the fluctuations of $Y$ typically flip from $-v$ to $+v$ and vice-versa. It is thus not surprising to find $\mathrm{Var}(Y \mid |Y| > v) \propto v^2$.

This argument makes intuitive the results $\mathrm{Var}(Y \mid Y > v) \propto 1/v^2$ and $\mathrm{Var}(Y \mid |Y| > v) \propto v^2$ for large $v$ and thus the results for $\rho_v^+$ and for $\rho_v^s$ if we use (6.2) and (6.8). We now attempt to justify $\rho_v^+ \sim_{v\to\infty} \frac{1}{v}$ and $1 - \rho_v^s \sim_{v\to\infty} 1/v^2$ directly by the following intuitive argument. Using the picture of particles, $X$ and $Y$ can be visualized as the positions of two particles which fluctuate randomly. Their joint bivariate Gaussian distribution with nonzero unconditional correlation amounts to the existence of a spring that ties them together. Their Gaussian marginals also exert a spring-like force attaching them to the origin. When $Y > v$, the $X$-particle is teared off between two extremes, between $0$ and $v$. When the unconditional correlation $\rho$ is less than $1$, the spring attracting to the origin is stronger than the spring attracting to the wall at $v$. The particle $X$ thus undergoes tiny fluctuations around the origin that are relatively less and less attracted by the $Y$-particle, hence the result $\rho_v^+ \sim_{v\to\infty} \frac{1}{v} \to 0$. In contrast, for $|Y| > v$, notwithstanding the still strong attraction of the $X$-particle to the origin, it can follow the sign of the $Y$-particle without paying too much cost in matching its amplitude $|v|$. Relatively tiny fluctuation of the $X$-particle but of the same sign as $Y \approx \pm v$ will result in a strong $\rho_v^s$, thus justifying that $\rho_v^s \to 1$ for $v \to +\infty$.

### 6.2.3 Influence of the Underlying Distribution for a Given Conditioning Set

For a fixed conditioning set defining a specific conditional correlation coefficient like $\rho_v^+$ or $\rho_v^s$, the behavior of these coefficients can be dramatically different from a pair of random variables to another one, depending on their underlying joint distribution. As an example, let the variables $X$ and $Y$ have a multivariate Student's distribution with $\nu$ degrees of freedom and an (unconditional) correlation coefficient $\rho$. According to the proposition stated in Appendix 6.B.1, we have the exact formula

$$\rho_{\mathcal{A}} = \frac{\rho}{\sqrt{\rho^2 + \frac{\mathrm{E}[\mathrm{E}(X^2 \mid Y) - \rho^2 Y^2 \mid Y \in \mathcal{A}]}{\mathrm{Var}(Y \mid Y \in \mathcal{A})}}} . \tag{6.12}$$

Explicit formulas for $\mathrm{E}[\mathrm{E}(X^2 \mid Y) - \rho^2 Y^2 \mid Y \in \mathcal{A}]$ and $\mathrm{Var}(Y \mid Y \in \mathcal{A})$ are also given in Appendix 6.B.1. The proof of (6.12) is presented in Appendix 6.B.2.

Expression (6.12) is the analog for a Student bivariate distribution to (6.2) derived above for the Gaussian bivariate distribution. Again, $\rho$ and $\rho_A$ share the following properties: they have the same sign, $\rho_A$ equals zero if and only if $\rho$ equals zero and $\rho_A$ can be either greater or smaller than $\rho$. Applying this general formula (6.12) to the calculus of $\rho_v^+$ and $\rho_v^s$, we find (see Appendices 6.B.3 and 6.B.4) that, conditioning on large returns,

$$\rho_v^+ \xrightarrow{\ v \to +\infty\ } \frac{\rho}{\sqrt{\rho^2 + (\nu - 1)}\sqrt{\frac{\nu-2}{\nu}}\,(1 - \rho^2)}\,, \tag{6.13}$$

while when conditioning on large volatility,

$$\rho_v^s \xrightarrow{\ v \to +\infty\ } \frac{\rho}{\sqrt{\rho^2 + \frac{1}{(\nu-1)}}\sqrt{\frac{\nu-2}{\nu}}\,(1 - \rho^2)}\,. \tag{6.14}$$

$\rho_v^+$ and $\rho_v^s$ converge both, at infinity, to nonvanishing constants (excepted for $\rho = 0$). Moreover, for $\nu$ larger than $\nu_c \simeq 2.839$, this constant is smaller than the unconditional correlation coefficient $\rho$, for all value of $\rho$, in the case of $\rho_v^+$, while for $\rho_v^s$ it is always larger than $\rho$, whatever $\nu$ (larger than two) may be.

These results show that, conditioned on large returns, $\rho_v^+$ is a decreasing function of the threshold $v$ (at least when $\nu \geq 2.839$), while, conditioned on large volatilities, $\rho_v^s$ is an increasing function of $v$.

To give another example, let us now assume that $X$ and $Y$ are two random variables following the equation:

$$X = \beta Y + \epsilon\,, \tag{6.15}$$

where $\alpha$ is a nonrandom real coefficient and $\epsilon$ an idiosyncratic noise independent of $Y$, whose distribution admits a centered moment of second order $\sigma_\epsilon^2$. Let us also denote by $\sigma_y^2$ the second centered moment of the variable $Y$. This relation between $X$ and $Y$ corresponds to the so-called *one factor model*. This one factor model with independence between $Y$ and $\epsilon$ is of course naive for concrete applications, as it neglects the potential influence of other factors in the determination of $X$. However, it has been argued to be a useful model in the context of contagion, and several studies have been based upon it (see [29] or [178], for instance). Moreover, it provides a simple illustrative model with rich and somewhat surprising results.

One can straightforwardly show that the conditional correlation coefficient of $X$ and $Y$ is

$$\rho_A = \frac{\rho}{\sqrt{\rho^2 + (1 - \rho^2)\frac{\mathrm{Var}(y)}{\mathrm{Var}(y \mid y \in A)}}}\,, \tag{6.16}$$

where

$$\rho = \frac{\beta \cdot \sigma_y}{\sqrt{\beta^2 \cdot \sigma_y^2 + \sigma_\epsilon^2}} \tag{6.17}$$

denotes the unconditional correlation coefficient of $X$ and $Y$. Note that the term $\sigma_\epsilon^2$ in the expression (6.17) of $\rho$ is the only place where the influence of the idiosyncratic noise is felt.

Expression (6.16) is the same as (6.2) for the bivariate Gaussian situation studied in Sect. 6.2.2. This is not surprising since, in the case where $Y$ and $\epsilon$ have univariate Gaussian distributions, the joint distribution of $X$ and $Y$ is a bivariate Gaussian distribution. The new fact is that this expression (6.16) remains true whatever the distribution of $Y$ and $\epsilon$, provided that their second moments exist.

We now present the asymptotic expression of $\rho_A$ for $Y$ with a Gaussian or a Student's distribution. Note that the expression of $\rho_A$ is simple enough to allow for exact calculations for a larger class of distributions, but for illustration, these two simple cases will be sufficient.

Assuming that $Y$ has a Gaussian distribution, while the distribution of $\epsilon$ can be everything (provided that $E[\epsilon^2] < \infty$), allows us to show that the same results as those given by (6.3) and (6.7) still hold, so that $\rho_v^+$ goes to zero, while $\rho_v^s$ goes to one.

In contrast, assuming that $Y$ has a Student's distribution yields for both $\rho_v^+$ and $\rho_v^s$:

$$\rho_v^{+,s} \sim \frac{\mathrm{sgn}(\beta)}{\sqrt{1 + \frac{K}{v^2}}} \; , \tag{6.18}$$

where $K$ is a positive constant. $\rho_v^{+,s}$ thus goes to $\pm 1$ as $v$ goes to infinity with $1 - |\rho_v^{+,s}| \propto 1/v^2$, which shows that they can have similar behaviors.

### 6.2.4 Conditional Correlation Coefficient on Both Variables

Since the exploration of the behavior of the correlation coefficient conditioned on only one variable clearly indicates that it can exhibit any kind of behavior, it is natural to look for the effect of a more constraining conditioning. To this aim, let us consider two random variables $X$ and $Y$ and define their conditional correlation coefficient $\rho_{A,B}$, conditioned upon $X \in A$ and $Y \in B$, where $A$ and $B$ are two subsets of $\mathbb{R}$ such that $\Pr\{X \in A, Y \in B\} > 0$, by

$$\rho_{A,B} = \frac{\mathrm{Cov}(X, Y \mid X \in A, Y \in B)}{\sqrt{\mathrm{Var}(X \mid X \in A, Y \in B) \cdot \mathrm{Var}(Y \mid X \in A, Y \in B)}} \; . \tag{6.19}$$

In this case, it is much more difficult to obtain general results for any specified class of distributions compared with the previous case of conditioning on a single variable. Here, we give the asymptotic behavior for a Gaussian distribution in the situation detailed below, using the expressions in [252, page 113], or proposition A.1 of [15].

Let us assume that the pair of random variables (X,Y) has a Normal distribution with unit unconditional variance and unconditional correlation coefficient $\rho$. The subsets $\mathcal{A}$ and $\mathcal{B}$ are both chosen equal to $[u, +\infty)$, with $u \in \mathbb{R}_+$, so that we focus on the correlation coefficient conditional on the returns of both $X$ and $Y$ larger than the threshold $u$. Denoting by $\rho_u$ the correlation coefficient conditional on this particular choice for the subsets $\mathcal{A}$ and $\mathcal{B}$, Appendix 6.A shows that, for large $u$,

$$\rho_u \sim_{u \to \infty} \rho \, \frac{1+\rho}{1-\rho} \cdot \frac{1}{u^2} \, , \tag{6.20}$$

which goes to zero. This decay is faster than $\rho_v^+ \sim_{v \to +\infty} 1/v$ given by (6.3) resulting from the conditioning on a single variable. However, unfortunately, there is no qualitative change. Thus, the correlation coefficient conditioned on both variables does not yield new significant information and does not provide any special improvement with respect to the correlation coefficient conditioned on a single variable.

### 6.2.5 An Example of Empirical Implementation

Let us consider four national stock markets in Latin America, namely Argentina (MERVAL index), Brazil (IBOV index), Chile (IPSA index) and Mexico (MEXBOL index). We are particularly interested in the contagion effects which may have occurred across these markets. We will study this question for the market indexes expressed in US Dollar to emphasize the effect of the devaluations of local currencies and to account for monetary crises. Doing so, we follow the same methodology as in most contagion studies (see [178], for instance). Our sample contains the daily (log) returns of each stock in local currency and US dollar during the time interval from 15 January, 1992 to 15 June, 2002 and thus encompasses both the Mexican crisis as well as the more recent Argentinean crisis.

Before applying the theoretical results derived above, we need to test whether the distributions of the returns are not too fat-tailed so that the correlation coefficient exists. Recall that this is the case if and only if the tail of the distribution decays faster than a power law with tail index $\mu = 2$, and its estimator given by the Pearson's coefficient is well behaved if at least the fourth moment of the distribution is finite.

Figure 6.1 shows the complementary distribution of the positive and negative tails of the index returns of four Latin American countries in US dollars. The positive tail clearly decays faster than a power law with tail index $\mu = 2$. In fact, Hill's estimator provides a value ranging between 3 and 4 for the four indexes. The situation for the negative tail is slightly different, particularly for the Brazilian index. For the Argentina, the Chilean and the Mexican indexes,

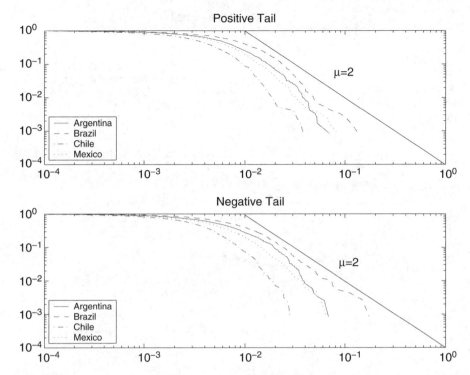

**Fig. 6.1.** The upper (respectively lower) panel graphs the complementary distribution of the positive (respectively the minus negative) returns in US dollar of the indices of four countries (Argentina, Brazil, Chile and Mexico). The straight line represents the slope of a power law with tail exponent $\mu = 2$

the negative tail behaves almost like the positive one, but for the Brazilian index, the negative tail exponent is hardly larger than two, as confirmed by Hill's estimator. This means that, in the Brazilian case, the estimates of the correlation coefficient will be particularly noisy and thus of weak statistical value.

We have checked that the fat-tailness of the indexes expressed in US dollar comes from the impact of the exchange rates. Thus, an alternative should be to consider the indexes in local currency, following the methodology of [314] and [315], but it would lead to focus on the linkages between markets only and to neglect the impact of the devaluations, which is precisely the main concern of studies on contagion.

Figures 6.2, 6.3 and 6.4 give the conditional correlation coefficient $\rho_v^{+,-}$ (plain thick line) for the pairs (Argentina/Brazil), (Brazil/Chile) and (Chile/Mexico) while the Figs. 6.5, 6.6 and 6.7 show the conditional correlation coefficient $\rho_v^s$ for the same pairs. For each figure, the thick dashed line gives the theoretical curve obtained under the bivariate Gaussian assumption whose analytical expressions can be found in Sect. 6.2.2. The unconditional corre-

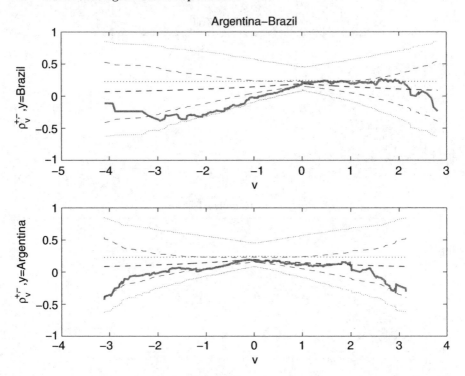

**Fig. 6.2.** In the upper panel, the thick plain curve depicts the correlation coefficient between the daily returns of the Argentinean and the Brazilian stock indices conditional on the Brazilian stock index daily returns larger than (smaller than) a given positive (negative) value $v$ (after normalization by the standard deviation). The thick dashed curve represents the theoretical conditional correlation coefficient $\rho_v^{+,-}$ calculated for a bivariate Gaussian model, while the two thin dashed curves define the area within which we cannot consider at the 95% confidence level that the estimated correlation coefficient is significantly different from its Gaussian theoretical value. The dotted curves provide the same information under the assumption of a bivariate Student's model with $\nu = 3$ degrees of freedom. The lower panel is the same as the upper panel but the conditioning is on the Argentinean stock index daily returns larger than (smaller than) a given positive (negative) value $v$ (after normalization by the standard deviation)

lation coefficient of the Gaussian model is set to the empirically estimated unconditional correlation coefficient. The two thin dashed lines represent the interval within which we cannot reject, at the 95% confidence level, the hypothesis according to which the estimated conditional correlation coefficient is equal to the theoretical one. This confidence interval has been estimated using the Fisher's statistics. Similarly, the thick dotted curve graphs the theoretical conditional correlation coefficient obtained under the bivariate Student's assumption with $\nu = 3$ degrees of freedom (whose expressions are given in

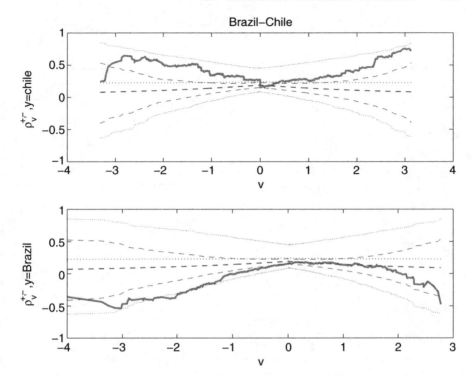

**Fig. 6.3.** Same as Fig. 6.2 for the (Brazil, Chile) pair. The upper (respectively lower) panel corresponds to a conditioning on the Chilean (respectively Brazilian) stock market index

Appendices 6.B.3 and 6.B.4) and the two thin dotted lines are its 95% confidence level. Here, the Fisher's statistics cannot be applied, since it requires at least that the fourth moment of the distribution exists. In fact, Meerschaert and Scheffler have shown that, for $\nu = 3$, the distribution of the sample correlation converges to a stable law with index $3/2$ [356]. This explains why the confidence interval for the Student's model with three degrees of freedom is much larger than the confidence interval for the Gaussian model. In the present case, we have used a bootstrap method to derive this confidence interval since the scale factor of the stable law is difficult to calculate.

In Figs. 6.2, 6.3 and 6.4, the changes in the conditional correlation coefficients $\rho_v^{+,-}$ are not significantly different, at the 95% confidence level, from those obtained with a bivariate Student's model with three degrees of freedom. In contrast, the Gaussian model is almost always rejected as expected, since marginal returns distributions are not Gaussian (as shown by Fig. 6.1). In fact, similar results hold (but are not depicted here) for the three others pairs (Argentina/Chile), (Argentina/Mexico) and (Brazil/Mexico). Since these results are compatible with a Student's model with constant correlation,

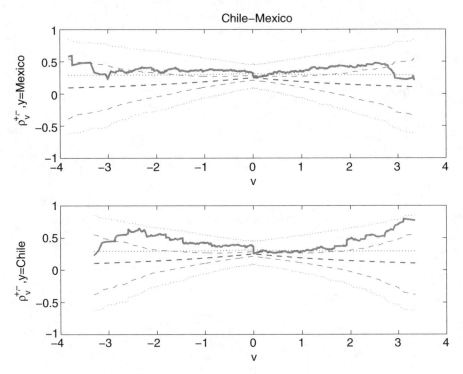

**Fig. 6.4.** Same as Fig. 6.2 for the (Chile, Mexico) pair. The upper (respectively lower) panel corresponds to a conditioning on the Mexican (respectively Chilean) stock market index

this suggests that no change in the correlations, and therefore no contagion mechanism, needs to be invoked to explain the data.

Let us now discuss the results obtained for the correlation coefficient conditioned on the volatility. Figures 6.5 and 6.7 show that the estimated correlation coefficients conditioned on volatility remain consistent with the Student's model with three degrees of freedom, while they still reject the Gaussian model. In contrast, Fig. 6.6 shows that the increase of the correlation cannot be explained by any of the Gaussian or Student models, when conditioning on the Mexican index volatility. Indeed, when the Mexican index volatility becomes larger than 2.5 times its standard deviation, none of these models can account for the increase of the correlation. The same discrepancy is observed for the pairs (Argentina/Chile), (Argentina/Mexico) and (Brazil/Mexico) which are not shown here. In each case, the Chilean and the Mexican markets have an impact on the Argentinean and the Brazilian markets which cannot be accounted for by neither the Gaussian model nor the Student model with constant correlation.

To conclude this empirical part, there is no significant increase in the real correlation between Argentina and Brazil on the one hand and between Chile

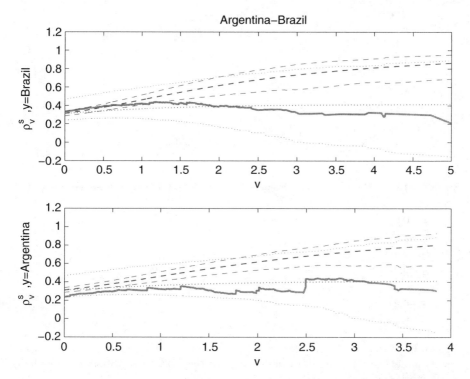

**Fig. 6.5.** In the upper panel, the thick plain curve gives the correlation coefficient between the daily returns of the Argentinean and the Brazilian stock indices conditioned on the daily volatility of the Brazilian stock index being larger than a given value $v$ (after normalization by the standard deviation). The thick dashed curve represents the theoretical conditional correlation coefficient $\rho_v^{+,-}$ calculated for a bivariate Gaussian model, while the two thin dashed curves delineate the area within which we cannot consider at the 95% confidence level that the estimated correlation coefficient is significantly different from its Gaussian theoretical value. The dotted curves provide the same information using a bivariate Student's model with $\nu = 3$ degrees of freedom. The lower panel is the same as the upper panel but the conditioning is on the Argentinean stock index

and Mexico on the other hand, when the volatility or the returns exhibit large moves. In contrast, in period of high volatility, the Chilean and Mexican market seem to have a genuine impact on the Argentinean and Brazilian markets. *A priori*, this should confirm the existence of a contagion across these markets. However, this conclusion is based only on two theoretical models. One should thus remain cautious before concluding positively on the existence of contagion on the sole basis of these results, in particular in view of the use of theoretical models which are all symmetric in their positive and negative tails. Such a symmetry is crucial for the derivation of the theoretical expressions of $\rho_v^s$. However, the empirical sample distributions are certainly not symmetric,

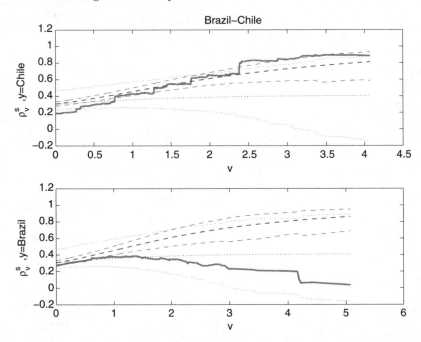

**Fig. 6.6.** Same as Fig. 6.5 for the (Brazil, Chile) pair. The upper (respectively lower) panel corresponds to a conditioning on the Chilean (respectively Brazilian) stock market index

as shown in Fig. 6.1. Using univariate and bivariate switching volatility models, Edwards and Susmel [142] have found strong volatility co-movements in Latin American but no clear evidence of contagion.

### 6.2.6 Summary

The previous sections have shown that the conditional correlation coefficients can exhibit all possible types of behavior, depending on their conditioning set and the underlying distributions of returns. More precisely, we have shown that the correlation coefficients, conditioned on large returns or volatility above a threshold $v$, can be either increasing or decreasing functions of the threshold, can go to any value between zero and one when the threshold goes to infinity and can produce contradictory results in the sense that accounting for a trend or not can lead to conclude on an absence of linear correlation or on a perfect linear correlation. Moreover, due to the large statistical fluctuations of the empirical estimates, one should be very careful when concluding on an increase or decrease of the genuine correlations.

Thus, from the general standpoint of the study of extreme dependences, but more particularly for the specific problem of the contagion across countries, the use of conditional correlation does not seem very informative and

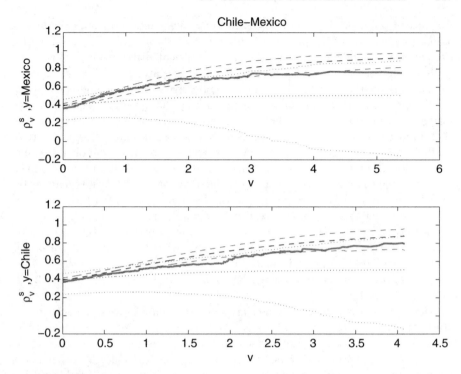

**Fig. 6.7.** Same as Fig. 6.5 for the (Chile, Mexico) pair. The upper (respectively lower) panel corresponds to a conditioning on the Mexican (respectively Chilean) stock market index

is sometimes misleading since it leads to spurious changes in the observed correlations: even when the unconditional correlation remains constant, conditional correlations yield artificial changes. Since one of the most commonly accepted and used definition of contagion is the detection of an increase of the conditional correlations during a period of turmoil, namely when the volatility increases, these results cast serious shadows on previous studies. In this respect, the conclusions of Calvo and Reinhart [87], about the occurrence of contagion across Latin American markets during the 1994 Mexican crisis, but more generally also the results of [271] or [299], on the effect of the October 1987 crash on the linkage of national markets, must be considered with some caution. It is quite desirable to find a more reliable tool for studying extreme dependences.

## 6.3 Conditional Concordance Measures

The (conditional) correlation coefficients, which have just been investigated, suffer from several theoretical as well as empirical deficiencies. From the theoretical point of view, they constitute just linear measures of dependence. Thus, as recalled in Chap. 4, they are fully satisfying only for the description of the dependence of variables with elliptical distributions. Moreover, we have seen that the correlation coefficient aggregates the information contained both in the marginal and in the collective behavior. The correlation coefficient is not invariant under an increasing change of variable, a transformation which is known to let unchanged the dependence structure. From the empirical standpoint, we have seen that, for some data, the correlation coefficient may not always exist, and even when it exits, it cannot always be accurately estimated, due to sometimes "wild" statistical fluctuations. Thus, it is desirable to find another measure of the dependence between two assets or more generally between two random variables, which, contrarily to the linear correlation coefficient, is always well-defined and only depends on the copula properties. This ensures that this measure is not affected by a change in the marginal distributions (provided that the mapping is increasing). It turns out that this desirable property is shared by all measures of *concordance*. Among these measures are the well-known Kendall's tau, Spearman's rho or Gini's beta (see Sect. 4.2).

However, these concordance measures are not well-adapted, as such, to the study of extreme dependence, because they are functions of the whole distribution, including the moderate and small returns. A simple idea to investigate the extreme concordance properties of two random variables is to calculate these quantities conditioned on values larger than a given threshold and let this threshold go to infinity.

In the sequel, we will only focus on the rank correlation which can be easily estimated empirically. It offers a natural generalization of the (linear) correlation coefficient. Indeed, Spearman's rho quantifies the degree of functional dependence, whatever the functional dependence between the two random variables may be. This represents a very interesting improvement. Perfect correlations (respectively anticorrelation) give a value 1 (respectively $-1$) both for the standard correlation coefficient and for the Spearman's rho. Otherwise, there is no general relation allowing us to deduce the Spearman's rho from the correlation coefficient and vice-versa.

### 6.3.1 Definition

Recall that Spearman's rho, denoted $\rho_s$ in the sequel, measures the difference between the probability of concordance and the probability of discordance for the two pairs of random variables $(X_1, Y_1)$ and $(X_2, Y_3)$, where the pairs $(X_1, Y_1)$, $(X_2, Y_2)$ and $(X_3, Y_3)$ are three independent realizations drawn from the same distribution:

$$\rho_s = 3 \left( \Pr[(X_1 - X_2)(Y_1 - Y_3) > 0] - \Pr[(X_1 - X_2)(Y_1 - Y_3) < 0] \right). \quad (6.21)$$

Thus, setting $U = F_X(X)$ and $V = F_Y(Y)$, we have seen that $\rho_s$ is nothing but the (linear) correlation coefficient of the uniform random variables $U$ and

$V$ (see Chap. 4):

$$\rho_s = \frac{\text{Cov}(U,\ V)}{\sqrt{\text{Var}(U)\text{Var}(V)}}\ , \tag{6.22}$$

which justifies its name as a *correlation coefficient of the rank*, and shows that it can easily be estimated.

An attractive feature of the Spearman's rho is to be independent of the margins, as we can see in equation (6.22). Thus, contrarily to the linear correlation coefficient, which aggregates the marginal properties of the variables with their collective behavior, the rank correlation coefficient takes into account only the dependence structure of the variables.

Using expression (6.22), a natural definition of the conditional rank correlation, conditioned on $V$ larger than a given threshold $\tilde{v}$, can be proposed:

$$\rho_s(\tilde{v}) = \frac{\text{Cov}(U, V \mid V \geq \tilde{v})}{\sqrt{\text{Var}(U \mid V \geq \tilde{v})\text{Var}(V \mid V \geq \tilde{v})}}\ , \tag{6.23}$$

whose expression in term of the copula $C(\cdot, \cdot)$ is given in Appendix 6.C.

Obviously, $\rho_s(v)$ is not a true concordance measure, as defined at the end of Sect. 4.2. An alternative definition of the conditional Spearman's rho [96] – and more generally of any conditional concordance measure – which preserves all the properties of concordance measures, can be obtained by considering the concordances measures of the conditional copula defined by (3.58). As an example, the conditional Kendall's tau would be defined by the Kendall's tau of the conditional copula. This idea has several advantages. In particular, when one focuses on the conditional Kendall's tau, asymptotic results can be straightforwardly derived for Archimedean copulas, in relation with result (3.61). Indeed, considering an Archimedean copula with a regularly varying generator $\phi$ (with tail index $\theta$), the conditional copula (3.58) converges to Clayton's copula with parameter $\theta$ as the threshold $u$ goes to zero. Therefore, Kendall's tau $\tau_u$ of the conditional copula converges to Kendall's tau of Clayton's copula, so that

$$\lim_{u \to 0} \tau_u = \frac{\theta}{\theta + 2}\ , \tag{6.24}$$

according to Table 4.1.

### 6.3.2 Example

Contrarily to the conditional correlation coefficient, it is difficult to obtain analytical expressions for the conditional Spearman's rho, for the Gaussian and Student distributions. Obviously, for many families of copulas known in closed form, equation (6.23) allows for an explicit calculation of $\rho_s(v)$.

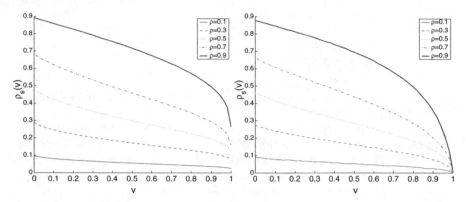

**Fig. 6.8.** Conditional Spearman's rho for a bivariate Gaussian copula (*left panel*) and a Student's copula with three degrees of freedom (*right panel*), with an unconditional linear correlation coefficient $\rho = 0.1, 0.3, 0.5, 0.7, 0.9$, as a function of the constraint level $v$

However, most copulas of interest in finance have no simple closed form, so that it is necessary to resort to numerical computations.

As an example, let us consider the bivariate Gaussian distribution (or copula) with unconditional correlation coefficient $\rho$. It is well-known that its unconditional Spearman's rho is given by

$$\rho_s = \frac{6}{\pi} \cdot \arcsin \frac{\rho}{2} \ . \tag{6.25}$$

The left panel of Fig. 6.8 shows the conditional Spearman's rho $\rho_s(v)$ defined by (6.23) obtained from a numerical integration. We observe the same bias as for the conditional correlation coefficient, namely the conditional rank correlation changes with $v$ even though the unconditional correlation is fixed to a constant value. Nonetheless, this conditional Spearman's rho seems more sensitive than the conditional correlation coefficient since one can observe in the left panel of Fig. 6.8 that, as $v$ goes to one, the conditional Spearman's rho $\rho_s(v)$ does not go to zero for all values of $\rho$ (at the precision of our bootstrap estimates), as previously observed with the conditional correlation coefficient (see (6.3)).

The right panel of Fig. 6.8 depicts the conditional Spearman's rho of Student's copula with three degrees of freedom. The biases are qualitatively the same as for the Gaussian copula, but $\rho_s(v)$ goes in this case to zero for all value of $\rho$ when $v$ goes to one. Thus, here again, several different behaviors can be observed depending on the underlying copula of the random variables. Moreover, these two examples show that the quantification of extreme dependence is a function of the tools used to quantify this dependence. Here, the conditional Spearman's $\rho$ goes to a nonvanishing constant for the Gaussian model, while the conditional (linear) correlation coefficient goes to zero, contrarily to the Student's distribution for which the situation is exactly the opposite.

### 6.3.3 Empirical Evidence

Figures 6.9, 6.10 and 6.11 give the conditional Spearman's rho respectively for the (Argentinean/Brazilian), the (Brazilian/Chilean), and the (Chilean/Mexican) stock markets. As previously, the plain thick line refers to the estimated correlation, while the dashed lines refer to the Gaussian copula and its 95% confidence levels and and dotted lines to Student's copula with three degrees of freedom and its 95% confidence levels.

Contrarily to the cases of the conditional (linear) correlation coefficient exhibited in Figs. 6.2, 6.3 and 6.4, the empirical conditional Spearman's $\rho$ does not always comply with the Student's model (neither with the Gaussian

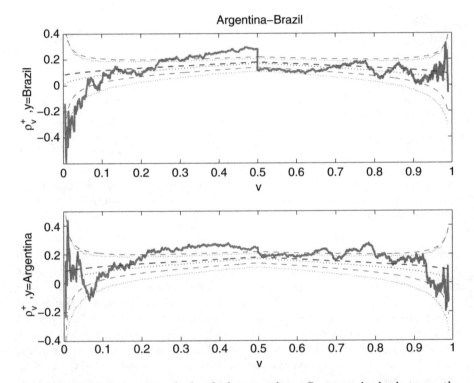

**Fig. 6.9.** In the upper panel, the thick curve shows Spearman's rho between the Argentinean stock index daily returns and the Brazilian stock index daily returns. Above the quantile $v = 0.5$, Spearman's rho is conditioned on the Brazilian index daily returns whose quantiles are larger than $v$, while below the quantile $v = 0.5$ it is conditioned on the Brazilian index daily returns whose quantiles are smaller than $v$. As in the above figures for the correlation coefficients, the dashed lines refer to the prediction of the Gaussian copula and its 95% confidence levels and the dotted lines to Student's copula with three degrees of freedom and its 95% confidence levels. The lower panel is the same as the upper panel but with the conditioning done on the Argentinean index daily returns

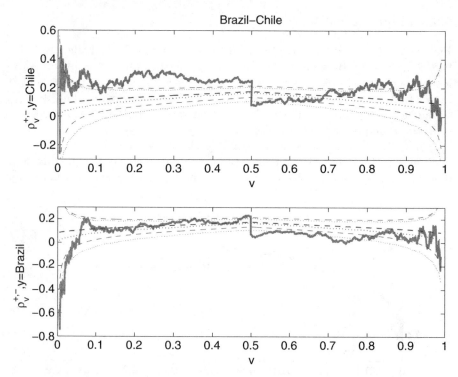

**Fig. 6.10.** Same as Fig. 6.9 for the (Brazil, Chile) pair. The upper (respectively lower) panel corresponds to a conditioning on the Chilean (respectively Brazilian) stock market index

one), and thus confirm the discrepancies observed in Figs. 6.5, 6.6 and 6.7. In all cases, for thresholds $v$ larger than the quantile 0.5 corresponding to the positive returns, the Student model with three degrees of freedom is almost always sufficient to explain the data. In contrast, for the negative returns and thus thresholds $v$ lower then the quantile 0.5, only the interaction between the Chilean and the Mexican markets is well described by the Student copula and does not need to invoke the contagion mechanism. For all other pairs, none of these models explain the data satisfyingly. Therefore, for these cases and from the perspective of these models, the contagion hypothesis seems to be needed.

There are however several caveats. First, even though we have considered the most natural financial models, there may be other models with constant dependence structure, that we have ignored, which could account for the observed evolutions of the conditional Spearman's $\rho$. If this is the case, then the contagion hypothesis would not be needed. Second, the main discrepancy between the empirical conditional Spearman's $\rho$ and the prediction of Student's model does not occur in the tails of the distribution, i.e for large and extreme movements, but in the bulk. Thus, during periods of turmoil, the

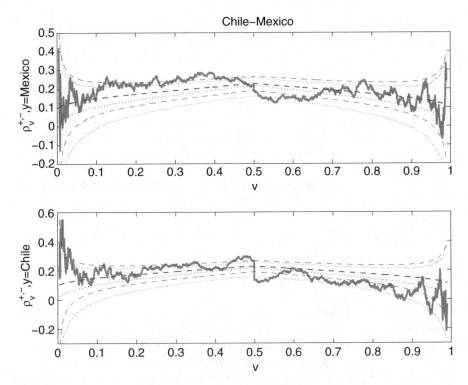

**Fig. 6.11.** Same as Fig. 6.9 for the (Chile, Mexico) pair. The upper (respectively lower) panel corresponds to a conditioning on the Mexican (respectively Chilean) stock market index

Student's model with three degrees of freedom seems to remain a good model of co-movements. Third, the contagion effect is never necessary for upwards moves. Indeed, we observe the same asymmetry or trend dependence as found in [315] for five major equity markets. This was apparent in Figs. 6.2, 6.3 and 6.4 for $\rho_v^{+,-}$, and is strongly confirmed on the conditional Spearman's $\rho$.

Interestingly, there is also an asymmetry or directivity in the mutual influence between markets. For instance, the Chilean and Mexican markets have an influence on the Argentinean and Brazilian markets, but the later do not have any impact on the Mexican and Chile markets. Chile and Mexico have no contagion effect on each other while Argentina and Brazil have.

These empirical results on the conditional Spearman's rho are different from and often opposite to the conclusion derived from the conditional correlation coefficients $\rho_v^{+,-}$. This puts in light the difficulty in obtaining reliable, unambiguous and sensitive estimations of conditional correlation measures. In particular, Pearson's coefficient usually employed to estimate the correlation coefficient between two variables is known to be not very efficient when the variables are fat-tailed and when the estimation is performed on a small sample. Indeed, with small samples, Pearson's coefficient is very sensitive to the largest value, which can lead to an important bias in the estimation. Moreover, even with large sample sizes, Meerschaert and Scheffler [356] have shown that the nature of convergence of the Pearson coefficient of two times series with tail index $\mu$ toward the theoretical correlation, as the sample size $T$ tends to infinity, is sensitive to the existence and strength of the theoretical correlation. If there is no theoretical correlation between the two times series, the sample correlation tends to zero with Gaussian fluctuations. If the theoretical correlation is nonzero, the difference between the sample correlation and the theoretical correlation times $T^{1-2/\mu}$ converges in distribution to a stable law with index $\mu/2$. These large statistical fluctuations are responsible for the lack of accuracy of the estimated conditional correlation coefficient encountered in the previous section. Thus, we think that the conditional Spearman's $\rho$ provides a good alternative both from a theoretical and an empirical viewpoint.

## 6.4 Extreme Co-movements

For the sake of completeness, and since it is directly related to the multivariate extreme value theory, we study the coefficient of tail dependence $\lambda$, which has been defined in Sect. 4.5. It would seem that the coefficient of tail dependence could provide a useful measure of the extreme dependence between two random variables for the analysis of contagion between markets. Two possibilities can occur. Either the whole data set does not exhibit tail dependence, and a contagion mechanism seems necessary to explain the occurrence of concomitant large movements during turmoil periods. Or, the data set exhibits tail dependence which by itself is enough to produce concomitant extremes (and contagion is not needed).

Unfortunately, the empirical estimation of the coefficient of tail dependence is a strenuous task. Indeed, a direct estimation of the conditional probability $\Pr\{X > F_X^{-1}(u) \mid Y > F_Y^{-1}(u)\}$, which should tend to $\lambda$ when $u \to 1$ is very difficult to implement in practice due to the combination of the curse of dimensionality and the drastic decrease of the number of realizations as $u$ become close to one. A better approach consists in using kernel methods, which generally provide smooth and accurate estimators [168, 284, 305]. However, these smooth estimators lead to copulas which are differentiable. This automatically gives vanishing tail dependence, as already mentioned in Chap. 5. Indeed, in order to obtain a nonvanishing coefficient of tail dependence, it is necessary for the corresponding copula to be nondifferentiable at the point $(1,1)$ (or at $(0,0)$). An alternative is then the fully parametric approach. One can choose to model dependence via a specific copula, and thus to determine the associated tail dependence [315, 334, 380]. The problem with such a method is that the choice of the parameterization of the copula amounts to choose *a priori* whether or not the data presents tail dependence.

In fact, there are three ways for estimating the tail dependence coefficient. The two first methods are specific to a class of copulas or of models, while the last one is very general, but less accurate. The first method is only reliable when the underlying copula is known to be Archimedean. In such a case, the limit theorem established by Juri and Wüthrich [260] (see Chap. 3.) allows one to estimate the tail dependence. The problem is that it is not obvious that the Archimedean copulas provide a good representation of the dependence structure for financial assets. For instance, the Archimedean copulas are generally inconsistent with a representation of assets by linear factor models. A second method – based upon results of Sect. 4.5.3 – offers good results by allowing to estimate the tail dependence in a semiparametric way, which solely relies on the estimation of marginal distributions, when the data can be explained by a factor model [332, 335].

When none of these situations occur, or when the factors are too difficult to extract, a third and fully nonparametric method exists, which is based upon the mathematical results of Ledford and Tawn [294, 295] and Coles *et al.* [106] and has recently been applied by Poon *et al.* [390]. The method consists in transforming the original random variables $X$ and $Y$ into Fréchet random variables denoted by $S$ and $T$ respectively. Then, considering the variable $Z = \min\{S, T\}$, its survival distribution is:

$$\Pr\{Z > z\} = \mathcal{L}(z) \cdot z^{1/\eta} \qquad \text{as} \quad z \to \infty , \tag{6.26}$$

where $\mathcal{L}$ denotes a slowly varying function. Now, assuming that

$$\lim_{z \to \infty} \mathcal{L}(z) = d \in (0, 1], \tag{6.27}$$

the coefficient of tail dependence $\lambda$ and the coefficient $\bar{\lambda}$, defined by (4.84), are simple functions of $d$ and $\eta$: $\bar{\lambda} = 2 \cdot \eta - 1$ with $\lambda = 0$ if $\eta < 1$, or $\bar{\lambda} = 1$ and $\lambda = d$ otherwise. The parameters $\eta$ and $d$ can be estimated by maximum likelihood, and deriving their asymptotic statistics allows one to test whether the hypothesis $\bar{\lambda} = 1$ can be rejected or not, and consequently, whether the data present tail dependence or not.

Let us implement this procedure on the four previously considered Latin American markets (Argentina, Brazil, Chile and Mexico). The results for the estimated values of the coefficient of tail dependence are given in Table 6.1 both for the positive and the negative tails. The tests show that one cannot reject the hypothesis of tail dependence between the four considered Latin American markets. Notice that the positive tail dependence is almost always slightly smaller than the negative one, which could be linked with the existence of trend asymmetry [315], but it turns out that these differences are not statistically significant. These results indicate that, according to this analysis of the extreme dependence coefficient, the propensity of extreme co-movements is almost the same for each pair of stock markets: even if the transmission mechanisms of a crisis are different from one country to another one, the propagation occurs with the same probability overall. Thus, the subsequent

**Table 6.1.** Coefficients of tail-dependence between pairs among four Latin American markets. The figure within parenthesis gives the standard deviation of the estimated value derived under the assumption of asymptotic normality of the estimators. Only the coefficients above the diagonal are indicated since they are symmetric

| Negative tail | Argentina | Brazil | Chile | Mexico |
|---|---|---|---|---|
| Argentina | – | 0.28 (0.04) | 0.25 (0.04) | 0.25 (0.05) |
| Brazil | | – | 0.19 (0.03) | 0.25 (0.05) |
| Chile | | | – | 0.24 (0.07) |
| Mexico | | | | – |

| Positive tail | Argentina | Brazil | Chile | Mexico |
|---|---|---|---|---|
| Argentina | – | 0.21 (0.06) | 0.20 (0.04) | 0.22 (0.04) |
| Brazil | | – | 0.28 (0.04) | 0.19 (0.04) |
| Chile | | | – | 0.19 (0.03) |
| Mexico | | | | – |

**Table 6.2.** Coefficients of tail dependence between pairs among four Latin American markets derived under the assumption of a Student copula with three degrees of freedom

| | Student hypothesis $\nu = 3$ | | | |
|---|---|---|---|---|
| | Argentina | Brazil | Chile | Mexico |
| Argentina | – | 0.24 | 0.25 | 0.27 |
| Brazil | | – | 0.24 | 0.27 |
| Chile | | | – | 0.28 |
| Mexico | | | | – |

risks are the same. Table 6.2 also gives the coefficients of tail dependence estimated under the Student's copula (or in fact any copula derived from an elliptical distribution – see Chap. 4) with three degrees of freedom, given by expression (4.91). One can observe a remarkable agreement between these values and the nonparametric estimates given in Table 6.1. This is consistent with the results given by the conditional Spearman's rho, for which we have remarked that the Student's copula seems to reasonably account for the extreme dependence.

## 6.5 Synthesis and Consequences

Table 6.3 summarizes the asymptotic dependences for large $v$ and $u$ of the signed conditional correlation coefficient $\rho_v^+$, the unsigned conditional correlation coefficient $\rho_v^s$ and the correlation coefficient $\rho_u$ conditioned on both variables for the bivariate Gaussian, the Student's model, the Gaussian factor model and the Student's factor model. These results provide a quantitative proof that conditioning on exceedance leads to conditional correlation coefficients that may be very different from the unconditional correlation. This provides a straightforward mechanism for fluctuations or changes of correlations, based on fluctuations of volatility or changes of trends. In other words, the many reported variations of correlation structure might be in large part attributed to changes in volatility (and statistical uncertainty).

The distinct dependences as a function of exceedance $v$ and $u$ of the conditional correlation coefficients offer novel tools for characterizing the statistical multivariate distributions of extreme events. Since their direct characterization is in general restricted by the curse of dimensionality and the scarcity of data, the conditional correlation coefficients provide reduced statistics which can be estimated with reasonable accuracy and reliability at least when the pdf of the data decays faster than any hyperbolic function with tail index equal to 2. In this respect, the empirical results suggest that a Student's copula, or more generally an elliptical copula, with a tail index of about three accounts for the main extreme dependence properties investigated here. This result is not really surprising since Chap. 5 has shown that Student's copula is a reasonable choice to account for the dependence structure between foreign exchange rates. In the present case, since the value of any domestic stock index has been converted into the US dollar, the influence of the dependence structure of foreign exchange rates can be considered as dominant in comparison with the dependence structure between each domestic stock index expressed in local currency. This dominance of the dependence structure of foreign exchange rates seems particularly true during turmoil periods.

Table 6.4 gives the asymptotic values of $\rho_v^+$, $\rho_v^s$ and $\rho_u$ for $v \to +\infty$ and $u \to \infty$ in order to compare them with the tail-dependence $\lambda$.

These two tables only scratch the surface of the rich sets of measures of tail and extreme dependences. We have already stressed that complete independence implies the absence of tail dependence: $\lambda = 0$, but that $\lambda = 0$ does not imply independence, at least in the intermediate range, since it is only an asymptotic property. Conversely, a nonzero tail dependence $\lambda$ implies the absence of asymptotic independence. Nonetheless, it does not imply necessarily that the conditional correlation coefficients $\rho_{v=\infty}^+$ and $\rho_{v=\infty}^s$ are nonzero, as one could have a priori expected.

Note that the examples of Table 6.4 are such that $\lambda = 0$ seems to go hand-in-hand with $\rho_{v\to\infty}^+ = 0$. However, the logical implication

$$(\lambda = 0) \implies (\rho_{v\to\infty}^+ = 0)$$

does not hold in general. A counter example is offered by the Student's factor model in the case where $\nu_Y > \nu_\epsilon$ (the tail of the distribution of the idiosyncratic noise is fatter than that of the distribution of the factor). In this case, $X$ and $Y$ have the same tail-dependence as $\epsilon$ and $Y$, which is zero by construction. But, $\rho_{v=\infty}^+$ and $\rho_{v=\infty}^s$ are both one because a large $Y$ almost

**Table 6.3.** Large $v$ and $u$ dependence of the conditional correlations $\rho_v^+$ (signed condition), $\rho_v^s$ (unsigned condition) and $\rho_u$ (on both variables) for the different models discussed in this chapter, described in the first column. The numbers in parentheses give the equation numbers from which the formulas are derived. The factor model is defined by (6.15), *i.e.*, $X = \beta Y + \epsilon$. $\rho$ is the unconditional correlation coefficient

| | $\rho_v^+$ | $\rho_v^s$ | $\rho_u$ |
|---|---|---|---|
| Bivariate Gaussian | $\dfrac{\rho}{\sqrt{1-\rho^2}} \cdot \dfrac{1}{v}$  (6.3) | $\mathrm{sgn}(\rho) \cdot \left(1 - \dfrac{1}{2}\dfrac{1-\rho^2}{\rho^2}\dfrac{1}{v^2}\right)$  (6.7) | $\rho\dfrac{1+\rho}{1-\rho} \cdot \dfrac{1}{u^2}$  (6.20) |
| Bivariate student's | $\dfrac{\rho}{\sqrt{\rho^2+(\nu-1)}}\sqrt{\dfrac{\nu-2}{\nu}}\,(1-\rho^2)$  (6.13) | $\dfrac{\rho}{\sqrt{\rho^2+\frac{1}{(\nu-1)}}\sqrt{\frac{\nu-2}{\nu}}\,(1-\rho^2)}$  (6.14) | – |
| Gaussian factor model | same as (6.3) | same as (6.7) | same as (6.20) |
| Student's factor model | $\mathrm{sgn}(\beta) \cdot \left(1 - \dfrac{K}{2v}\right)$  (6.18) | $\mathrm{sgn}(\beta) \cdot \left(1 - \dfrac{K}{2v^2}\right)$  (6.18) | – |

**Table 6.4.** Asymptotic values of $\rho_v^+$, $\rho_v^s$ and $\rho_u$ for $v \to +\infty$ and $u \to \infty$ and comparison with the tail-dependence $\lambda$ and $\bar\lambda$ for the four models indicated in the first column. The factor model is defined by (6.15), *i.e.*, $X = \alpha Y + \epsilon$. $\rho$ is the unconditional correlation coefficient. For Student's factor model, $Y$ and $\epsilon$ have centered Student's distributions with the same number $\nu$ of degrees of freedom and their scale factors are respectively equal to 1 and $\sigma$, so that $\rho = (1 + \frac{\sigma^2}{\beta^2})^{-1/2}$. For the Bivariate Student's distribution, we refer to Table 1 for the constant values of $\rho_{v=\infty}^+$ and $\rho_{v=\infty}^s$

| | $\rho_{v=\infty}^+$ | $\rho_{v=\infty}^s$ | $\rho_{u=\infty}$ | $\lambda$ | $\bar\lambda$ |
|---|---|---|---|---|---|
| Bivariate Gaussian | 0 | $\mathrm{sgn}(\rho)$ | 0 | 0 | $\rho$ |
| Bivariate student's | see Table 6.3 | see Table 6.3 | — | $2\cdot\bar T_{\nu+1}\left(\sqrt{\nu+1}\sqrt{\frac{1-\rho}{1+\rho}}\right)$ | 1 |
| Gaussian factor model | 0 | $\mathrm{sgn}(\rho)$ | 0 | 0 | $\rho$ |
| Student's factor model | $\mathrm{sgn}(\beta)$ | $\mathrm{sgn}(\beta)$ | — | $\dfrac{\rho^\nu}{\rho^\nu+(1-\rho^2)^{\nu/2}}\mathbf{1}_{\{\beta>0\}}$ | 1 |

always gives a large $X$ and the simultaneous occurrence of a large $Y$ and a large $\epsilon$ can be neglected. The reason for this absence of tail dependence (in the sense of $\lambda$) coming together with asymptotically strong conditional correlation coefficients stems from two facts:

- first, the conditional correlation coefficients put much less weight on the extreme tails that the tail-dependence parameter $\lambda$. In other words, $\rho_{v=\infty}^{+}$ and $\rho_{v=\infty}^{s}$ are sensitive to the marginals, $i.e.$, there are determined by the full bivariate distribution, while, as we said, $\lambda$ is a pure copula property independent of the marginals. Since $\rho_{v=\infty}^{+}$ and $\rho_{v=\infty}^{s}$ are measures of extreme dependence weighted by the specific shapes of the marginals, it is natural that they may behave differently.
- Secondly, the tail dependence $\lambda$ probes the extreme dependence property of the original copula of the random variables $X$ and $Y$. On the contrary, when conditioning on $Y$, one changes the copula of $X$ and $Y$, so that the extreme dependence properties investigated by the conditional correlations are not exactly those of the original copula. This last remark explains clearly what Boyer $et$ $al.$ [78] call a "bias" in the conditional correlations. Indeed, changing the dependence between two random variables obviously changes their correlations.

There are important consequences to these facts. Consider a situation in which one measure ($\lambda$) would conclude on asymptotic tail-independence while the other measures $\rho_{v=\infty}^{+}$ and $\rho_{v=\infty}^{s}$ would conclude the opposite. Therefore, before concluding on a change in the dependence structure with respect to a given parameter – the volatility or the trend, for instance – one should check that this change does not result from the tool used to probe the dependence. These results shed new light on recent controversial results about the occurrence or absence of contagion during the Latin American crises. As in every previous work, the analysis reported in this chapter finds no evidence of contagion between Chile and Mexico, but contrarily to [178], it is difficult to ignore the possibility of contagion toward Argentina and Brazil, in agreement with [87].

In fact, most of the discrepancies between these different studies probably stem from the fact that the conditional correlation coefficient does not provide an accurate tool for probing the potential changes of dependence. Indeed, even when the bias has been accounted for, the fat-tailness of the distributions of returns are such that the Pearson's coefficient is subjected to very strong statistical fluctuations which forbid an accurate estimation of the correlation. Moreover, when studying the dependence properties, it is interesting to free oneself from the marginal behavior of each random variable. This is why the conditional Spearman's rho seems a good tool: it only depends on the copula and is statistically well-behaved.

The conditional Spearman's rho has identified a change in the dependence structure during downward trends in Latin American markets, similar to that found by Longin and Solnik [315] in their study of the contagion across five major equity markets. It has also put in light the asymmetry in the contagion effects: Mexico and Chile can be potential sources of contagion toward Argentina and Brazil, while the reverse does not seem to hold. This phenomenon has been observed during the 1994 Mexican crisis and appears to remain true in the recent Argentinean crisis, for which only Brazil seems to exhibit the signature of a possible contagion.

The origin of the discovered asymmetry may lie in the difference between the more market-oriented countries and the more state-intervention oriented economies, giving rise to either currency floating regimes adapted to an important manufacturing sector which tend to deliver more competitive real exchange rates (Chile and Mexico) or to fixed rate pegs (Argentina until the 2001 crisis and Brazil until the early 1999 crisis) [187, 188, 189]. The asymmetry of the contagion is compatible with the view that fixed exchange rates tighten more strictly an economy and its stock market to external shocks (case of Argentina and Brazil) while a more flexible exchange rate seems to provide a cushion allowing a decoupling between the stock market and external influences.

Finally, the absence of contagion does not imply necessarily the absence of *contamination*. Indeed, the study of the coefficient of tail dependence has proven that with or without contagion mechanisms (*i.e.*, increase in the linkage between markets during crisis) the probability of extreme co-movements during the crisis (*i.e.*, the contamination) is almost the same for all pairs of markets. Thus, whatever the propagation mechanism may be – historically strong relationship or irrational fear and herd behavior – the observed effects are the same: the propagation of the crisis. From the practical perspective of risk management or regulatory policy, this last point is perhaps more important than the real knowledge of the occurrence or not of contagion.

# Appendix

## 6.A Correlation Coefficient for Gaussian Variables Conditioned on Both $X$ and $Y$ Larger Than $u$

Let us consider a pair of Normal random variables $(X, Y) \sim \mathcal{N}(0, \Sigma)$ where $\Sigma$ is their covariance matrix with unconditional correlation coefficient $\rho$. Without loss of generality, and for simplicity, we shall assume that $\Sigma$ has unconditional variances equal to 1. By definition, the conditional correlation coefficient $\rho_u$, conditioned on both $X$ and $Y$ larger than $u$, is

$$\rho_u = \frac{\text{Cov}[X, Y \mid X > u, Y > u]}{\sqrt{\text{Var}[X \mid X > u, Y > u]}\sqrt{\text{Var}[Y \mid X > u, Y > u]}} , \tag{6.A.1}$$

$$= \frac{m_{11} - m_{10} \cdot m_{01}}{\sqrt{m_{20} - m_{10}^2}\sqrt{m_{02} - m_{01}^2}} , \tag{6.A.2}$$

where $m_{ij}$ denotes $\text{E}[X^i \cdot Y^j \mid X > u, Y > u]$.

Using the proposition A.1 of [15] or the expressions in [252, p.113], we can assert that

$$m_{10}\ L(u,u;\rho) = (1+\rho)\ \varphi(u)\left[1 - \Phi\left(\sqrt{\frac{1-\rho}{1+\rho}}u\right)\right], \qquad (6.A.3)$$

$$m_{20}\ L(u,u;\rho) = (1+\rho^2)\ u\ \varphi(u)\left[1 - \Phi\left(\sqrt{\frac{1-\rho}{1+\rho}}u\right)\right]$$
$$+ \frac{\rho\ \sqrt{1-\rho^2}}{\sqrt{2\pi}}\ \varphi\left(\sqrt{\frac{2}{1+\rho}}u\right) + L(u,u;\rho), \qquad (6.A.4)$$

$$m_{11}\ L(u,u;\rho) = 2\rho\ u\ \varphi(u)\left[1 - \Phi\left(\sqrt{\frac{1-\rho}{1+\rho}}u\right)\right]$$
$$+ \frac{\sqrt{1-\rho^2}}{\sqrt{2\pi}}\ \varphi\left(\sqrt{\frac{2}{1+\rho}}u\right) + \rho\ L(u,u;\rho), \qquad (6.A.5)$$

where $L(\cdot,\cdot;\cdot)$ denotes the bivariate Gaussian survival (or complementary cumulative) distribution:

$$L(h,k;\rho) = \frac{1}{2\pi\sqrt{1-\rho^2}}\int_h^\infty dx \int_k^\infty dy\ \exp\left(-\frac{1}{2}\frac{x^2 - 2\rho xy + y^2}{1-\rho^2}\right), \qquad (6.A.6)$$

$\varphi(\cdot)$ is the Gaussian density:

$$\varphi(x) = \frac{1}{\sqrt{2\pi}}\ e^{-\frac{x^2}{2}}, \qquad (6.A.7)$$

and $\Phi(\cdot)$ is the cumulative Gaussian distribution:

$$\Phi(x) = \int_{-\infty}^x du\ \varphi(u). \qquad (6.A.8)$$

### 6.A.1 Asymptotic Behavior of $L(u,u;\rho)$

Let us focus on the asymptotic behavior of $L(u,u;\rho)$, where $L(h,k;\rho)$ is defined by (6.A.6), for large $u$. Performing the change of variables $x' = x - u$ and $y' = y - u$, we can write

$$L(u,u;\rho) = \frac{e^{-\frac{u^2}{1+\rho}}}{2\pi\sqrt{1-\rho^2}}\int_0^\infty dx' \int_0^\infty dy'\ \exp\left(-u\frac{x'+y'}{1+\rho}\right)$$
$$\times \exp\left(-\frac{1}{2}\frac{x'^2 - 2\rho x'y' + y'^2}{1-\rho^2}\right). \qquad (6.A.9)$$

Using the fact that

$$\exp\left(-\frac{1}{2}\frac{x'^2 - 2\rho x'y' + y'^2}{1-\rho^2}\right) = 1 - \frac{x'^2 - 2\rho x'y' + y'^2}{2(1-\rho^2)}$$
$$+ \frac{(x'^2 - 2\rho x'y' + y'^2)^2}{8(1-\rho^2)^2} - \frac{(x'^2 - 2\rho x'y' + y'^2)^3}{48(1-\rho^2)^3} + \cdots, \qquad (6.A.10)$$

and applying Theorem 3.1.1 in [247, p. 68] (Laplace's method), (6.A.9) and (6.A.10) yield

$$
L(u, u; \rho) = \frac{(1+\rho)^2}{2\pi\sqrt{1-\rho^2}} \cdot \frac{e^{-\frac{u^2}{1+\rho}}}{u^2} \left[ 1 - \frac{(2-\rho)(1+\rho)}{1-\rho} \cdot \frac{1}{u^2} \right.
$$
$$
+ \frac{(2\rho^2 - 6\rho + 7)(1+\rho)^2}{(1-\rho)^2} \cdot \frac{1}{u^4}
$$
$$
\left. -3 \frac{(12 - 13\rho + 8\rho^2 - 2\rho^3)(1+\rho)^3}{(1-\rho)^3} \cdot \frac{1}{u^6} + \mathcal{O}\left(\frac{1}{u^8}\right) \right], \tag{6.A.11}
$$

and

$$
1/L(u, u; \rho) = \frac{2\pi\, u^2\, \sqrt{1-\rho^2}}{(1+\rho)^2} \cdot e^{\frac{u^2}{1+\rho}} \left[ 1 + \frac{(2-\rho)(1+\rho)}{1-\rho} \cdot \frac{1}{u^2} \right.
$$
$$
- \frac{3 - 2\rho + \rho^2)(1+\rho)^2}{(1-\rho)^2} \cdot \frac{1}{u^4}
$$
$$
\left. + \frac{(16 - 13\rho + 10\rho^2 - 3\rho^3)(1+\rho)^3}{(1-\rho)^3} \cdot \frac{1}{u^6} + \mathcal{O}\left(\frac{1}{u^8}\right) \right]. \tag{6.A.12}
$$

## 6.A.2 Asymptotic Behavior of the First Moment $m_{10}$

The first moment $m_{10} = \mathrm{E}[X \mid X > u, Y > u]$ is given by (6.A.3). For large $u$,

$$
1 - \Phi\left(\sqrt{\frac{1-\rho}{1+\rho}}\, u\right) = \frac{1}{2}\, \mathrm{erfc}\left(\sqrt{\frac{1-\rho}{2(1+\rho)}}\, u\right) \tag{6.A.13}
$$
$$
= \sqrt{\frac{1+\rho}{1-\rho}}\, \frac{e^{-\frac{1-\rho}{2(1+\rho)}u^2}}{\sqrt{2\pi}\, u} \left[ 1 - \frac{1+\rho}{1-\rho} \cdot \frac{1}{u^2} + 3\left(\frac{1+\rho}{1-\rho}\right)^2 \cdot \frac{1}{u^4} \right.
$$
$$
\left. - 15\left(\frac{1+\rho}{1-\rho}\right)^3 \cdot \frac{1}{u^6} + \mathcal{O}\left(\frac{1}{u^8}\right) \right], \tag{6.A.14}
$$

so that multiplying by $(1+\rho)\,\phi(u)$, we obtain

$$
m_{10}\, L(u, u; \rho) = \frac{(1+\rho)^2}{\sqrt{1-\rho^2}}\, \frac{e^{-\frac{u^2}{1+\rho}}}{2\pi u} \left[ 1 - \frac{1+\rho}{1-\rho} \cdot \frac{1}{u^2} \right.
$$
$$
\left. + 3\left(\frac{1+\rho}{1-\rho}\right)^2 \cdot \frac{1}{u^4} - 15\left(\frac{1+\rho}{1-\rho}\right)^3 \cdot \frac{1}{u^6} + \mathcal{O}\left(\frac{1}{u^8}\right) \right]. \tag{6.A.15}
$$

Using the result given by equation (6.A.11), we can conclude that

$$
m_{10} = u + (1+\rho) \cdot \frac{1}{u} - \frac{(1+\rho)^2(2-\rho)}{(1-\rho)} \cdot \frac{1}{u^3}
$$
$$
+ \frac{(10 - 8\rho + 3\rho^2)(1+\rho)^3}{(1-\rho)^2} \cdot \frac{1}{u^5} + \mathcal{O}\left(\frac{1}{u^7}\right). \tag{6.A.16}
$$

In the sequel, we will also need the behavior of $m_{10}{}^2$:

$$m_{10}{}^2 = u^2 + 2\,(1+\rho) - \frac{(1+\rho)^2(3-\rho)}{(1-\rho)} \cdot \frac{1}{u^2}$$
$$+ 2\frac{(8-5\rho+2\rho^2)(1+\rho)^3}{(1-\rho)^2} \cdot \frac{1}{u^4} + \mathcal{O}\left(\frac{1}{u^6}\right). \tag{6.A.17}$$

## 6.A.3 Asymptotic Behavior of the Second Moment $m_{20}$

The second moment $m_{20} = \mathrm{E}[X^2 \mid X > u, Y > u]$ is given by expression (6.A.4). The first term in the right hand side of (6.A.4) yields

$$(1+\rho^2)\,u\,\varphi(u)\left[1 - \Phi\left(\sqrt{\frac{1-\rho}{1+\rho}}u\right)\right] = (1+\rho^2)\sqrt{\frac{1+\rho}{1-\rho}}\,\frac{e^{-\frac{u^2}{1+\rho}}}{2\pi}\,\times$$
$$\left[1 - \frac{1+\rho}{1-\rho} \cdot \frac{1}{u^2} + 3\left(\frac{1+\rho}{1-\rho}\right)^2 \cdot \frac{1}{u^4} - 15\left(\frac{1+\rho}{1-\rho}\right)^3 \cdot \frac{1}{u^6} + \mathcal{O}\left(\frac{1}{u^8}\right)\right] \tag{6.A.18}$$

while the second term gives

$$\frac{\rho\,\sqrt{1-\rho^2}}{\sqrt{2\pi}}\,\varphi\left(\sqrt{\frac{2}{1+\rho}}u\right) = \rho\,\sqrt{1-\rho^2}\,\frac{e^{-\frac{u^2}{1+\rho}}}{2\pi}. \tag{6.A.19}$$

Putting these two expressions together and factorizing the term $(1+\rho)/(1+\rho^2)$ gives

$$m_{20}\,L(u,u;\rho) = \frac{(1+\rho)^2}{\sqrt{1-\rho^2}}\,\frac{e^{-\frac{u^2}{1+\rho}}}{2\pi}\left[1 - \frac{1+\rho^2}{1-\rho} \cdot \frac{1}{u^2} + 3\frac{(1+\rho^2)(1+\rho)}{(1-\rho)^2} \cdot \frac{1}{u^4}\right.$$
$$\left. - 15\frac{(1+\rho^2)(1+\rho)^2}{(1-\rho)^3} \cdot \frac{1}{u^6} + \mathcal{O}\left(\frac{1}{u^8}\right)\right] + L(u,u;\rho), \tag{6.A.20}$$

which finally yields

$$m_{20} = u^2 + 2\,(1+\rho) - 2\frac{(1+\rho)^2}{1-\rho}\,\frac{1}{u^2}$$
$$+ 2\frac{(5+4\rho+\rho^3)(1+\rho)^2}{(1-\rho)^2}\,\frac{1}{u^4} + \mathcal{O}\left(\frac{1}{u^6}\right). \tag{6.A.21}$$

## 6.A.4 Asymptotic Behavior of the Cross Moment $m_{11}$

The cross moment $m_{11} = \mathrm{E}[X \cdot Y \mid X > u, Y > u]$ is given by expression (6.A.5). The first and second terms in the right-hand side of (6.A.5) respectively give

$$2\rho \, u \, \varphi(u)[1 - \Phi(u)] = 2\rho \, \sqrt{\frac{1+\rho}{1-\rho}} \, \frac{e^{-\frac{u^2}{1+\rho}}}{2\pi} \times$$

$$\left[1 - \frac{1+\rho}{1-\rho} \cdot \frac{1}{u^2} + 3\left(\frac{1+\rho}{1-\rho}\right)^2 \cdot \frac{1}{u^4} - 15\left(\frac{1+\rho}{1-\rho}\right)^3 \cdot \frac{1}{u^6} + \mathcal{O}\left(\frac{1}{u^8}\right)\right], \quad \text{(6.A.22)}$$

$$\frac{\sqrt{1-\rho^2}}{\sqrt{2\pi}} \, \phi\left(\sqrt{\frac{2}{1+\rho}}u\right) = \sqrt{1-\rho^2} \, \frac{e^{-\frac{u^2}{1+\rho}}}{2\pi} , \quad \text{(6.A.23)}$$

which, after factorization by $(1+\rho)/\rho$, yields

$$m_{11} \, L(u, u; \rho) = \frac{(1+\rho)^2}{\sqrt{1-\rho^2}} \, \frac{e^{-\frac{u^2}{1+\rho}}}{2\pi} \left[1 - 2\frac{\rho}{1-\rho} \cdot \frac{1}{u^2} + 6\frac{\rho(1+\rho)}{(1-\rho)^2} \cdot \frac{1}{u^4}\right.$$

$$\left. -30\frac{\rho(1+\rho)^2}{(1-\rho)^3} \cdot \frac{1}{u^6} + \mathcal{O}\left(\frac{1}{u^8}\right)\right] + \rho \, L(u, u; \rho), \quad \text{(6.A.24)}$$

and finally

$$m_{11} = u^2 + 2\,(1+\rho) - \frac{(1+\rho)^2(3-\rho)}{(1-\rho)} \cdot \frac{1}{u^2}$$

$$+ \frac{(16 - 9\rho + 3\rho^2)(1+\rho)^3}{(1-\rho)^2} \cdot \frac{1}{u^4} + \mathcal{O}\left(\frac{1}{u^6}\right). \quad \text{(6.A.25)}$$

## 6.A.5 Asymptotic Behavior of the Correlation Coefficient

The conditional correlation coefficient conditioned on both $X$ and $Y$ larger than $u$ is defined by (6.A.2). Using the symmetry between $X$ and $Y$, we have $m_{10} = m_{01}$ and $m_{20} = m_{02}$, which allows us to rewrite (6.A.2) as follows:

$$\rho_u = \frac{m_{11} - m_{10}^2}{m_{20} - m_{10}^2} . \quad \text{(6.A.26)}$$

Putting together the previous results, we have

$$m_{20} - m_{10}^2 = \frac{(1+\rho)^2}{u^2} - 2\frac{(4 - \rho + 3\rho^2 + 3\rho^3)(1+\rho)^2}{1-\rho} \frac{1}{u^4} + \mathcal{O}\left(\frac{1}{u^6}\right), \quad \text{(6.A.27)}$$

$$m_{11} - m_{10}^2 = \rho \, \frac{(1+\rho)^3}{1-\rho} \cdot \frac{1}{u^4} + \mathcal{O}\left(\frac{1}{u^6}\right), \quad \text{(6.A.28)}$$

which proves that

$$\rho_u = \rho \, \frac{1+\rho}{1-\rho} \cdot \frac{1}{u^2} + \mathcal{O}\left(\frac{1}{u^4}\right) \quad \text{and} \quad \rho \in [-1, 1). \quad \text{(6.A.29)}$$

## 6.B Conditional Correlation Coefficient for Student's Variables

### 6.B.1 Proposition

Let us consider a pair of Student's random variables $(X, Y)$ with $\nu > 2$ degrees of freedom and unconditional correlation coefficient $\rho$. Let $\mathcal{A}$ be a subset of $\mathbb{R}$ such that $\Pr\{Y \in \mathcal{A}\} > 0$. The correlation coefficient of $(X, Y)$, conditioned on $Y \in \mathcal{A}$ defined by

$$\rho_{\mathcal{A}} = \frac{\mathrm{Cov}(X, Y \mid Y \in \mathcal{A})}{\sqrt{\mathrm{Var}(X \mid Y \in \mathcal{A})} \, \sqrt{\mathrm{Var}(Y \mid Y \in \mathcal{A})}} \qquad (6.\mathrm{B}.30)$$

can be expressed as

$$\rho_{\mathcal{A}} = \frac{\rho}{\sqrt{\rho^2 + \frac{\mathrm{E}[\mathrm{E}(x^2 \mid Y) - \rho^2 Y^2 \mid Y \in \mathcal{A}]}{\mathrm{Var}(Y \mid Y \in \mathcal{A})}}} \;, \qquad (6.\mathrm{B}.31)$$

with

$$\mathrm{Var}(Y \mid Y \in \mathcal{A}) = \nu \left[ \frac{\nu - 1}{\nu - 2} \cdot \frac{\Pr\left\{\sqrt{\frac{\nu}{\nu-2}} Y \in \mathcal{A} \mid \nu - 2\right\}}{\Pr\{Y \in \mathcal{A} \mid \nu\}} - 1 \right]$$
$$- \left[ \frac{\int_{y \in \mathcal{A}} dy \; y \cdot t_y(y)}{\Pr\{Y \in \mathcal{A} \mid \nu\}} \right]^2 , \qquad (6.\mathrm{B}.32)$$

where $t_\nu(y)$ is given below by (6.B.36) and

$$\mathrm{E}[\mathrm{E}(X^2 \mid Y) - \rho^2 Y^2 \mid Y \in \mathcal{A}] = (1 - \rho^2) \frac{\nu}{\nu - 2} \cdot \frac{\Pr\left\{\sqrt{\frac{\nu}{\nu-2}} Y \in \mathcal{A} \mid \nu - 2\right\}}{\Pr\{Y \in \mathcal{A} \mid \nu\}}. \quad (6.\mathrm{B}.33)$$

### 6.B.2 Proof of the Proposition

Let the variables $X$ and $Y$ have a multivariate Student distribution with $\nu > 2$ degrees of freedom and a correlation coefficient $\rho$ :

$$P_{XY}(x, y) = \frac{\Gamma\left(\frac{\nu+2}{2}\right)}{\nu\pi \, \Gamma\left(\frac{\nu+1}{2}\right) \sqrt{1 - \rho^2}} \left( 1 + \frac{x^2 - 2\rho xy + y^2}{\nu \, (1 - \rho^2)} \right)^{-\frac{\nu+2}{2}} , \qquad (6.\mathrm{B}.34)$$

$$= \left( \frac{\nu + 1}{\nu + y^2} \right)^{1/2} \frac{1}{\sqrt{1 - \rho^2}} \, t_\nu(y) \cdot t_{\nu+1} \left[ \left( \frac{\nu + 1}{\nu + y^2} \right)^{1/2} \frac{x - \rho y}{\sqrt{1 - \rho^2}} \right], \quad (6.\mathrm{B}.35)$$

where $t_\nu(\cdot)$ denotes the univariate Student density with $\nu$ degrees of freedom

$$t_\nu(x) = \frac{\Gamma\left(\frac{\nu+1}{2}\right)}{\Gamma\left(\frac{\nu}{2}\right) (\nu\pi)^{1/2}} \cdot \frac{1}{\left(1 + \frac{x^2}{\nu}\right)^{\frac{\nu+1}{2}}} = \frac{C_\nu}{\left(1 + \frac{x^2}{\nu}\right)^{\frac{\nu+1}{2}}} \;. \qquad (6.\mathrm{B}.36)$$

Let us evaluate $\mathrm{Cov}(X, Y \mid Y \in \mathcal{A})$:

$$\mathrm{Cov}(X, Y \mid Y \in \mathcal{A}) = \mathrm{E}(X \cdot Y \mid Y \in \mathcal{A}) - \mathrm{E}(X \mid Y \in \mathcal{A}) \cdot \mathrm{E}(Y \mid Y \in \mathcal{A})$$
$$= \mathrm{E}(\mathrm{E}(X \mid Y) \cdot Y \mid Y \in \mathcal{A}) - \mathrm{E}(\mathrm{E}(X \mid Y) \mid Y \in \mathcal{A}) \cdot \mathrm{E}(Y \mid Y \in \mathcal{A}). \quad (6.B.37)$$

As it can be seen in equation (6.B.35), $\mathrm{E}(X \mid Y) = \rho Y$, which gives

$$\mathrm{Cov}(X, Y \mid Y \in \mathcal{A}) = \rho \cdot \mathrm{E}(Y^2 \mid Y \in \mathcal{A}) - \rho \cdot \mathrm{E}(Y \mid Y \in \mathcal{A})^2, \quad (6.B.38)$$
$$= \rho \cdot \mathrm{Var}(Y \mid Y \in \mathcal{A}). \quad (6.B.39)$$

Thus, we have

$$\rho_{\mathcal{A}} = \rho \sqrt{\frac{\mathrm{Var}(Y \mid Y \in \mathcal{A})}{\mathrm{Var}(X \mid Y \in \mathcal{A})}} . \quad (6.B.40)$$

Using the same method as for the calculation of $\mathrm{Cov}(X, Y \mid Y \in \mathcal{A})$, we find

$$\mathrm{Var}(X \mid Y \in \mathcal{A}) = \mathrm{E}[\mathrm{E}(X^2 \mid Y) \mid Y \in \mathcal{A}] - \mathrm{E}[\mathrm{E}(X \mid Y) \mid Y \in \mathcal{A}]^2,$$
$$= \mathrm{E}[\mathrm{E}(X^2 \mid Y) \mid Y \in \mathcal{A}] - \rho^2 \cdot \mathrm{E}[Y \mid Y \in \mathcal{A}]^2, \quad (6.B.41)$$
$$= \mathrm{E}[\mathrm{E}(X^2 \mid Y) - \rho^2 Y^2 \mid Y \in \mathcal{A}] - \rho^2 \cdot \mathrm{Var}[Y \mid Y \in A] ,$$

which yields (6.B.31).

To go one step further, we have to evaluate the three terms $\mathrm{E}(Y \mid Y \in \mathcal{A})$, $\mathrm{E}(Y^2 \mid Y \in \mathcal{A})$, and $\mathrm{E}[\mathrm{E}(X^2 \mid Y) \mid Y \in \mathcal{A}]$.

The first one is trivial to calculate :

$$\mathrm{E}(Y \mid Y \in \mathcal{A}) = \frac{\int_{y \in \mathcal{A}} dy \, y \cdot t_y(y)}{\Pr\{Y \in \mathcal{A} \mid \nu\}} . \quad (6.B.42)$$

The second one gives

$$\mathrm{E}(Y^2 \mid Y \in \mathcal{A}) = \frac{\int_{y \in \mathcal{A}} dy \, y^2 \cdot t_y(y)}{\Pr\{Y \in \mathcal{A} \mid \nu\}}, \quad (6.B.43)$$

$$= \nu \left[ \frac{\nu - 1}{\nu - 2} \cdot \frac{\Pr\left\{ \sqrt{\frac{\nu}{\nu-2}} Y \in \mathcal{A} \mid \nu - 2 \right\}}{\Pr\{Y \in \mathcal{A} \mid \nu\}} - 1 \right], \quad (6.B.44)$$

so that

$$\mathrm{Var}(Y \mid Y \in \mathcal{A}) = \nu \left[ \frac{\nu - 1}{\nu - 2} \cdot \frac{\Pr\left\{ \sqrt{\frac{\nu}{\nu-2}} Y \in \mathcal{A} \mid \nu - 2 \right\}}{\Pr\{Y \in \mathcal{A} \mid \nu\}} - 1 \right]$$

$$- \left[ \frac{\int_{y \in \mathcal{A}} dy \, y \cdot t_y(y)}{\Pr\{Y \in \mathcal{A} \mid \nu\}} \right]^2 . \quad (6.B.45)$$

To calculate the third term, we first need to evaluate $\mathrm{E}(X^2 \mid Y)$. Using equation (6.B.35) and the results given in [1], we find

$$E(X^2 \mid Y) = \int dx \left(\frac{\nu+1}{\nu+y^2}\right)^{1/2} \frac{x^2}{\sqrt{1-\rho^2}} \cdot t_{\nu+1}\left[\left(\frac{\nu+1}{\nu+y^2}\right)^{1/2} \frac{x-\rho y}{\sqrt{1-\rho^2}}\right],$$

$$= \frac{\nu+y^2}{\nu-1}(1-\rho^2) - \rho^2 y^2, \tag{6.B.46}$$

which yields

$$E[E(X^2 \mid Y) - \rho^2 Y^2 \mid Y \in \mathcal{A}] = \frac{\nu}{\nu-1}(1-\rho^2) + \frac{1-\rho^2}{\nu-1} E[Y^2 \mid Y \in \mathcal{A}], \tag{6.B.47}$$

and applying the result given in equation (6.B.44), we finally obtain

$$E[E(X^2 \mid Y) - \rho^2 Y^2 \mid Y \in \mathcal{A}] = (1-\rho^2)\frac{\nu}{\nu-2} \cdot \frac{\Pr\left\{\sqrt{\frac{\nu}{\nu-2}}Y \in \mathcal{A} \mid \nu-2\right\}}{\Pr\{Y \in \mathcal{A} \mid \nu\}}, \tag{6.B.48}$$

which concludes the proof.

### 6.B.3 Conditioning on $Y$ Larger Than $v$

The conditioning set is $\mathcal{A} = [v, +\infty)$, thus

$$\Pr\{Y \in \mathcal{A} \mid \nu\} = \bar{T}_\nu(v) = \nu^{\frac{\nu-1}{2}}\frac{C_\nu}{v^\nu} + \mathcal{O}\left(v^{-(\nu+2)}\right), \tag{6.B.49}$$

$$\Pr\left\{\sqrt{\frac{\nu}{\nu-p}}Y \in \mathcal{A} \mid \nu-p\right\} = \bar{T}_{\nu-p}\left(\sqrt{\frac{\nu-p}{\nu}}\,v\right)$$

$$= \frac{\nu^{\frac{\nu-p}{2}}}{(\nu-p)^{\frac{1}{2}}}\frac{C_{\nu-p}}{v^{\nu-p}} + \mathcal{O}\left(v^{-(\nu-p+2)}\right), \tag{6.B.50}$$

$$\int_{y\in\mathcal{A}} dy\, y \cdot t_y(y) = \sqrt{\frac{\nu}{\nu-2}}\, t_{\nu-2}\left(\sqrt{\frac{\nu-2}{\nu}}\,v\right)$$

$$= \frac{\nu^{\frac{\nu}{2}}}{\sqrt{\nu-2}}\frac{C_{\nu-2}}{v^{\nu-1}} + \mathcal{O}\left(v^{-(\nu-3)}\right), \tag{6.B.51}$$

where $t_\nu(\cdot)$ and $\bar{T}_\nu(\cdot)$ denote respectively the density and the Student survival distribution with $\nu$ degrees of freedom and $C_\nu$ is defined in (6.B.36).

Using equation (6.B.31), one can thus give the exact expression of $\rho_v^+$. Since it is very cumbersome, we will not write it explicitly. We will only give the asymptotic expression of $\rho_v^+$:

$$\mathrm{Var}(Y \mid Y \in \mathcal{A}) = \frac{\nu}{(\nu-2)(\nu-1)^2}v^2 + \mathcal{O}(1) \tag{6.B.52}$$

$$E[E(X^2 \mid Y) - \rho^2 Y^2 \mid Y \in \mathcal{A}] = \sqrt{\frac{\nu}{\nu-2}}\frac{1-\rho^2}{\nu-1}v^2 + \mathcal{O}(1). \tag{6.B.53}$$

Thus, for large $v$,

$$\rho_v^+ \longrightarrow \frac{\rho}{\sqrt{\rho^2 + (\nu - 1)\sqrt{\frac{\nu-2}{\nu}}\,(1 - \rho^2)}}. \tag{6.B.54}$$

### 6.B.4 Conditioning on $|Y|$ Larger Than $v$

The conditioning set is now $\mathcal{A} = (-\infty, -v] \cup [v, +\infty)$, with $v \in R_+$. Thus, the right-hand sides of equations (6.B.49) and (6.B.50) have to be multiplied by two while

$$\int_{y \in \mathcal{A}} dy \, y \cdot t_y(y) = 0 , \tag{6.B.55}$$

for symmetry reasons. So, equation (6.B.53) still holds while

$$\mathrm{Var}(Y \mid Y \in \mathcal{A}) = \frac{\nu}{(\nu - 2)} \, v^2 + \mathcal{O}(1) . \tag{6.B.56}$$

Thus, for large $v$,

$$\rho_v^s \longrightarrow \frac{\rho}{\sqrt{\rho^2 + \frac{1}{(\nu-1)}\sqrt{\frac{\nu-2}{\nu}}\,(1 - \rho^2)}} . \tag{6.B.57}$$

### 6.B.5 Conditioning on $Y > v$ Versus on $|Y| > v$

The results (6.B.54) and (6.B.57) are valid for $\nu > 2$, as one can expect since the second moment must exist for the correlation coefficient to be defined. Contrarily to the Gaussian case, the conditioning set is not really important. Indeed with both conditioning set, $\rho_v^+$ and $\rho_v^s$ go to constants different from zero and (plus or minus) one, when $v$ goes to infinity. This striking difference with the Gaussian case can be explained by the large fluctuations allowed by the Student's distribution, and can be related to the fact that the coefficient of tail dependence for this distribution does not vanish even though the variables are anticorrelated (see Sect. 4.5.3).

Contrarily to the Gaussian distribution which binds the fluctuations of the variables near the origin, the Student's distribution allows for "wild" fluctuations. These properties are thus responsible for the result that, contrarily to the Gaussian case for which the conditional correlation coefficient goes to zero when conditioned on large signed values and goes to one when conditioned on large unsigned values, the conditional correlation coefficient for Student's variables have a similar behavior in both cases. Intuitively, the large fluctuations of $X$ for large $v$ dominate and control the asymptotic dependence.

## 6.C Conditional Spearman's Rho

To obtain the conditional Spearman's rho defined in (6.23), we need a few intermediate calculations. We have

$$
\mathrm{E}[\cdot \mid V \geq \tilde{v}] = \frac{\int_{\tilde{v}}^1 \int_0^1 \cdot \, dC(u,v)}{\int_{\tilde{v}}^1 \int_0^1 dC(u,v)} = \frac{1}{1-\tilde{v}} \int_{\tilde{v}}^1 \int_0^1 \cdot \, dC(u,v) . \quad (6.C.58)
$$

Performing a simple integration by parts, we obtain

$$
\mathrm{E}[U \mid V \geq \tilde{v}] \quad = 1 + \frac{1}{1-\tilde{v}} \left[ \int_0^1 du \, C(u,\tilde{v}) - \frac{1}{2} \right], \quad (6.C.59)
$$

$$
\mathrm{E}[V \mid V \geq \tilde{v}] \quad = \frac{1+\tilde{v}}{2}, \quad (6.C.60)
$$

$$
\mathrm{E}[U^2 \mid V \geq \tilde{v}] \quad = 1 + \frac{2}{1-\tilde{v}} \left[ \int_0^1 du \, u \, C(u,\tilde{v}) - \frac{1}{3} \right], \quad (6.C.61)
$$

$$
\mathrm{E}[V^2 \mid V \geq \tilde{v}] \quad = \frac{\tilde{v}^2 + \tilde{v} + 1}{3}, \quad (6.C.62)
$$

$$
\mathrm{E}[U \cdot V \mid V \geq \tilde{v}] = \frac{1+\tilde{v}}{2} + \frac{1}{1-\tilde{v}} \left[ \int_{\tilde{v}}^1 dv \int_0^1 du \, C(u,v) \right.
$$
$$
\left. + \tilde{v} \int_0^1 du \, C(u,\tilde{v}) - \frac{1}{2} \right], \quad (6.C.63)
$$

which yields

$$
\mathrm{Cov}(U,V \mid V \geq \tilde{v}) = \frac{1}{1-\tilde{v}} \int_{\tilde{v}}^1 dv \int_0^1 du \, C(u,v) - \frac{1}{2} \int_0^1 du \, C(u,\tilde{v}) - \frac{1}{4} ,
$$

$$
\mathrm{Var}(U \mid V \geq \tilde{v}) \quad = \frac{1-4\tilde{v}}{12\,(1-\tilde{v})^2} + \frac{2}{1-\tilde{v}} \int_0^1 du \, u \, C(u,\tilde{v})
$$
$$
+ \frac{2\tilde{v}-1}{(1-\tilde{v})^2} \int_0^1 du \, C(u,\tilde{v}) - \frac{1}{(1-\tilde{v})^2} \left( \int_0^1 du \, C(u,\tilde{v}) \right)^2 ,
$$

$$
\mathrm{Var}(V \mid V \geq \tilde{v}) \quad = \frac{(1-\tilde{v})^2}{12} ,
$$

so that

$$
\rho_s(\tilde{v}) = \frac{\frac{12}{1-\tilde{v}} \int_{\tilde{v}}^1 dv \int_0^1 du \, C(u,v) - 6 \int_0^1 du \, C(u,\tilde{v}) - 3}{\sqrt{D}}, \quad (6.C.64)
$$

with

$$
D = 1 - 4\tilde{v} + 24\,(1-\tilde{v}) \int_0^1 du \, u \, C(u,\tilde{v})
$$
$$
+ 12\,(2\tilde{v}-1) \int_0^1 du \, C(u,\tilde{v}) - 12 \left( \int_0^1 du \, C(u,\tilde{v}) \right)^2 .
$$

# 7

# Summary and Outlook

## 7.1 Synthesis

A common theme underlying the chapters of this book is that many important applications of risk management rely on the assessment of the positive or negative outcomes of uncertain positions. The probability theory and statistics, together with the valuation of losses incurred for a given exposition to various risk factors, take a predominant place in this process. However, they are not, by far, the sole ingredients needed in an efficient risk management system. Quoting Andrew Lo [311], one can assert that

> [Although most] current risk-management practices are based on probabilities of extreme dollar losses (*e.g.*, measures like Value-at-Risk), [...] these measures capture only part of the story. Any complete risk management system must address two other important factors: prices and preferences. Together with probabilities, these comprise the three P's of Total Risk Management. [Understanding] how the three P's interact [allows] to determine sensible risk profiles for corporations and for individuals, guidelines for how much risk to bear and how much to hedge. By synthesizing existing research in economics, psychology, and decision sciences, and through an ambitious research agenda to extend this synthesis into other disciplines, a complete and systematic approach to rational decision making in an uncertain world is within reach.

Among the three P's, Probability constitutes today, in our opinion, the most solid pillar of risk management, because it has reached the highest level of maturation. Compared with Price and Preference, the Probability theory is clearly the most developed in terms of its mathematical formulation, providing important and accurate quantitative results.

Asset valuation – and therefore Price assessment – is also very developed quantitatively, but it remains, for a large part, subordinated to the quality of the estimation of the probabilities. Indeed, a cornerstone of modern finance theory holds that the (fair) value of a given investment vehicle is nothing but the mathematical expectation – under a suitable probability measure – of the future discounted cash-flows generated by this investment vehicle. The assessment of future cash-flows for complex investment vehicles is nothing but an exercise of pure financial analysis but, without a correct probability measure, this exercise has little value. This indubitably shows that Prices and Probabilities are inextricably entangled and that an accurate price assessment requires an accurate determination of the probabilities.

Preferences are also of crucial importance – in fact the most important of the three P's, according to Lo – since under this term is embodied the entire human decision making process. But here, in contrast to the two other P's, our knowledge is still in its infancy. The pioneering theoretical work by Von Neuman and Morgenstern [482] has laid the foundations of a rational decision theory. However, this theory has been undermined over the years by several paradoxes and deficiencies [4, 5], when tested against real human preferences. Most of the recent theories, notably those directly inspired by psychological studies [204, 263, 352, and references therein], attempt to cure the original rational decision theory from its inconsistencies. But, one should recognize that, while significant qualitative progress has been obtained, there is not yet a satisfying fully operational theory of decision making.

For all these reasons, Probability still plays a dominant role in current risk management practice. And we firmly believe that this supremacy will extend well into the future, in view of the still large remaining potential for improvement. Of course, the modern science of human psychology and decision making is in constant progression and accounts better and better for the many anomalies observed on financial markets. However, its fusion with finance, which has given birth to the field of "behavioral finance," will provide useful practical tools only with the development of accurate *quantitative* predictions. Until then, behavioral finance will continue to be mostly the playground of academic research. We thus believe that, in the next few years, the most important improvements in applied risk management will occur through more elaborate modeling of financial markets and more generally of the economic environment.

In spite of the key role of *Price* and *Preference*, this book has mainly focused on the role of *Probability* in the risk assessment and management processes. The different probabilistic concepts presented in the core chapters of this book should provide a better understanding and modeling of the various sources of uncertainty and therefore of risk factors that investors are facing. Our presentation has been organized around the key idea that the risk of a set of positions can be decomposed into two major components:

(i)   the marginal risks associated with the variations of wealth of each risky position,

(ii)  the cross-dependence between the change in the wealth of each position.

This decomposition has been justified in Chap. 3 by the introduction of the notion of *copula*, which is the pivot of the book.

With respect to the marginal risks, Chap. 2 has highlighted the weaknesses of traditional methods used to assess large downside risks, which are generally based upon tools derived from the extreme value theory. As an alternative, we have advocated using comprehensive parametric distributions which provide a good compromise between weak model errors – inherent to any parametric risk measurement – and accurate risk estimates. Accounting for model error is of vital importance for risk management purposes. This aspect is often forgotten, but it really plays a prominent role in the risk assessment process. Consider for instance the VaR estimates obtained under the Gaussian hypothesis. We have seen that, for large confidence levels, they have only little value and, yet, this class of models is still promoted by regulating institutions such as the Bank for International Settlements [42].

Concerning the cross-dependence between assets, we have emphasized that copulas are the most fundamental concept and tool. They should therefore constitute a cornerstone of modern risk management practices. Among many properties, this results in particular from the unique and optimal separation between individual risks and collective dependence that they provide. As a consequence, copulas have been shown to exhibit an unsurpassed flexibility and versatility for the elaboration of scenarios. This comes, however, at the cost of the preliminary calibration of the best copula, a problem which has not yet received a fully satisfying answer, as reviewed in Chap. 5. Nevertheless, the few families of copulas, which have been surveyed in Chap. 3, appear to be reasonable candidates for modeling the dependence structures of arbitrary baskets of financial assets and therefore allow for a relatively easy and useful generation of case studies.

Chapters 5 and 6 have unearthed an important and somewhat surprising difference in the dependence between currencies and between stocks. Overall, the analyses which have been presented find a much weaker dependence between stocks than between currencies. This is reflected quantitatively by the fact that the dependence between currencies can be described by Student's copulas with a low number of degrees of freedom (typically 4 to 6). In contrast, the dependence between stock returns require a larger number of degrees of freedom (10 or more). This observation raises important questions for internationally diversified portfolios. Indeed, if the dependence between the returns of foreign exchange rates is much stronger than the dependence between stock returns as reported in Chap. 5, the benefits of international diversification can vanish. It is true that various national stock markets exhibit anticorrelations between some stocks and at some epochs, which *a priori* would justify holding stocks from different national markets. However, if the corresponding currencies are positively associated with strong tail dependences, as in the Latin American markets (see Chap. 6), the diversification effect mostly disappears once the gains or losses of the stocks are translated into the same monetary unit. In particular, once expressed in the domestic currency of the holder of an internationally allocated portfolio, the diversification effect seems to disappear at times of turmoil, that is, exactly when the investor needs it the most.

Chapter 6 has examined in depth this question of possible changes of dependence during financial crises. Two possible explanations have been considered: (i) contamination operating via the sensitivity of dependence measures to the changing volatility level, or (ii) contagion reflecting a genuine change of dependence. The later case leads to inefficient allocations when neglected. However, the most important conclusion for risk management is not so much the distinction between contamination or contagion but the presence of a strong tail dependence that may exist between national markets, because it destroys the benefits of diversification across countries when one market goes in crisis.

## 7.2 Outlook and Future Directions

Our exposition has mainly focused on the concept of cross-sectional dependence between several random variables, but there are many aspects of the question, such as time dependencies, which have been only barely touched or which have been actually neglected. It may be useful to discuss them so as to provide a better appreciation of the limits of the methods proposed in this book and consequently of their domains of application. It is also useful to delineate possible future exciting directions for future improvements of the risk management practice.

### 7.2.1 Robust and Adaptive Estimation of Dependences

A major concern, especially for practitioners, is whether this whole mathematical edifice, its algorithmic implementations and its rigorous statistical tests are relevant and useful for the really important risks, such as global market moves and crashes. It is indeed a common experience that the dependence estimated and predicted by standard models change dramatically at certain times, not only during crashes, but also when the market exhibits a collective downward plunge. A quite common observation is that investment strategies, which have some moderate $\beta$ (coefficient of regression to the market) for normal times, can see their $\beta$ jumps to a much larger value (close to 1 or larger depending on the leverage of the investment) at certain times when the market collectively dives. However, investments which are thought to be hedged against negative global market trends may actually lose as much or more than the global market, at certain times when a large majority of stocks plunge simultaneously.

This question has been touched upon in Chap. 4 when discussing the possible strategies for preventing the large downward moves of a portfolio, based upon its tail dependence with the market. We will also revisit this problem in the context of the occurrence of "outliers" and of time-varying dependence. But it contains two other components: (1) Is the estimation of the dependence meaningful and really robust? (2) If not, what can be done?

The first question deals with the development of robust estimation techniques, defined as methods which are insensitive to small departures from the idealized assumptions which have been used to optimize the algorithm. Such techniques include M-estimates (which follow from maximum likelihood considerations), L-estimates (which are linear combinations of order statistics), and R-estimates (based on statistical rank tests) [238, 479, 484].

The second question requires novel approaches. A possible one is inspired from Herbert A. Simon, the famous economist and cognitive scientist studying how people make real-world decisions, who observed that they seldom optimize. "Rather people seek strategies that will work well enough, that include hedges against various potential outcomes and that are adaptive. Tomorrow will bring information unavailable today; therefore, people plan on revising their plans" summarize Popper *et al.* [391]. In this spirit, people have developed an approach to look not for optimal strategies but for robust ones, defined as strategies which perform well when compared with the alternatives across a wide range of plausible futures. "It need not be the optimal strategy in any future; it will, however, yield satisfactory outcomes in both easy-to-envision futures and hard-to-anticipate contingencies. This approach replicates the way people often reason about complicated and uncertain decisions in everyday life" says Popper *et al.* The process of decision-making under conditions of deep uncertainty requires first to consider ensembles of scenarios, then to seek robust and adaptive strategies, and finally to combine machine and human capabilities interactively. Outstanding questions involve the compromise between near-term objectives and long-term sustainability, and the characterization of irreducible risks and of "surprises" [301].

In the same vein, the approach in terms of *universal portfolios* initiated by Cover [113] a decade ago has opened the way to many studies [63, 227, 261]. Assuming for instance that one invests in a constantly rebalanced strategy, the question amounts to determining the best weights (the fraction of wealth invested in each stock) of this strategy for an investment horizon $T$. Since the optimal weights can only be assessed *ex-ante* – once the time $T$ has been reached – it seems impossible to design *ex-ante* a strategy whose performance will compare with the performance of the best strategy with hindsight. However, it turns out that the universal approach promoted by Cover circumvents this problem. It consists in investing uniformly in all constantly rebalanced portfolio strategies. This results in a strategy that is nearly optimal in the sense that, for any sequence of stock market outcomes, this particular investment strategy has a performance comparable to the best constantly rebalanced portfolio in the long run. More generally, any universal strategy is such that its average logarithmic performance over horizon $T$ approaches the best *expost* average logarithmic performance over the same horizon $T$, in the limit of long horizon $T$, irrespective of the market price sequence from now to time $T$.

## 7.2.2 Outliers, Kings, Black Swans and Their Dependence

In its conclusion, Chap. 2 notes the existence of "outliers" (also called "kings" or "black swans"), in the distribution of financial risks measured at variable time scales such as with drawdowns. These outliers are identified only with metrics adapted to take into account transient increases of the time dependence in the time series of returns of individual financial assets [249] (see also Chap. 3 of [450]). These outliers seem to belong to a statistical population which is different from the bulk of the distribution and require some additional amplification mechanisms active only at special times.

Chapter 5 shows that two exceptional events in the period from January 1989 to December 1998 stand out in statistical tests determining the relevance of the Gaussian copula to describe the dependence between the German Mark and the Swiss Franc. The first of the two events is the coup against Gorbachev in Moscow on 19 August, 1991 for which the German mark (respectively the Swiss Franc) lost 3.37% (respectively 0.74%) against the US dollar. The second event occurred on 10 September, 1997, and corresponds to an appreciation of the German Mark of 0.60% against the US dollar while the Swiss Franc lost 0.79% which represents a moderate move for each currency, but a large joint move.

The presence of such outliers both in marginal distributions and in concomitant moves, together with the strong impact of crises and of crashes, suggests the need for novel measures of dependence between drawdowns and other time-varying metrics across different assets. This program is part of the more general need for a joint multi-time-scale and multi-asset approach to dependence. Examples of efforts in this direction include multidimensional GARCH models [23, 45, 46, 154, 296, 400, 477] and the multivariate multifractal random walk [366]. It also epitomizes the need for new multi-period risk measures, which would account for this class of events. Several avenues of research have recently been opened by attempting to generalize the notions of Value-at-Risk and of coherent measures of risk within a multi-period framework [21, 405, 483].

## 7.2.3 Endogeneity Versus Exogeneity

The presence of outliers such as those mentioned in the previous section poses the problem of exogeneity versus endogeneity. An event identified as anomalous could perhaps be cataloged as resulting from exogenous influences.[1] The same issue has been investigated in Chap. 6 when testing for contagion versus contamination in the Latin American crises. Contamination refers to an endogenous dependence described by an approximately constant copula. In contrast, contagion is by definition the concept that the dependence has

---

[1] However, outliers may also have an endogenous origin, as described for financial crashes [250, 449, 450].

changed, either transiently or with lasting effects, due to some influence or mechanism which is exogenous (in a sense that needs to be defined precisely) to the previous regime.

The concept of exogeneity[2] is fundamental in empirical econometric modeling and statistical estimation (see for instance [152, 156]). Here, we refer to the question of exogeneity versus endogeneity in the broader context of self-organized criticality[3] [32, 231, 246, 451], inspired in particular from the physical and natural sciences. According to self-organized criticality, extreme events are seen to be endogenous, in contrast with previous prevailing views (see for instance the discussion in [33, 448]). But, how can one assert with 100% confidence that a given extreme event is really due to an endogenous self-organization of the system, rather than to the response to an external shock? Most natural and social systems are indeed continuously subjected to external stimulations, noises, shocks, solicitations, and forcing, which can widely vary in amplitude. It is thus not clear *a priori* if a given large event is due to a strong exogenous shock, to the internal dynamics of the system, or maybe to a combination of both. Addressing this question is fundamental for understanding the relative importance of self-organization versus external forcing in complex systems and underpins much of the problem of dependence between variables.

The question, whether distinguishing properties characterize endogenous versus exogenous shocks, permeates many systems, for instance, biological extinctions such as the Cretaceous/Tertiary KT boundary (meteorite versus extreme volcanic activity versus self-organized critical extinction cascades), commercial successes (progressive reputation cascade versus the result of a well-orchestrated advertisement), immune system deficiencies (external viral/bacterial infections versus internal cascades of regulatory breakdowns), the aviation industry recession (9/11 versus structural endogenous problems), discoveries (serendipity versus the outcome of slow endogenous maturation processes), cognition and brain learning processes (role of external inputs versus internal self-organization and reinforcements) and recovery after wars (internally generated – i.e., civil wars – versus imported from the outside) and so on. In economics, endogeneity versus exogeneity has been hotly debated for decades. A prominent example is the theory of Schumpeter [432] on the importance of technological discontinuities in economic history. Schumpeter argued that "evolution is lopsided, discontinuous, disharmonious by nature... studded with violent outbursts and catastrophes... more like a series of explosions than a gentle, though incessant, transformation". Endogeneity versus exogeneity is also paramount in economic growth theory [415].

Several evidences of quantitative signatures distinguishing exogenous from endogenous shocks have recently been described. Concerning the way the

---

[2] In a nutshell, conditioning on an exogenous variable does not decrease the amount of information in parameter estimation [152].

[3] Self-organized criticality is part of the theory of complex systems.

continuous stream of news gets incorporated into market prices for instance, it has recently been shown how one can distinguish the effects of events like the 11 September, 2001 attack or the coup against Gorbachev on 19 August, 1991 from events like financial crashes such as October, 1987 as well as smaller volatility bursts. Based on a stochastic volatility model with long range dependence (the so-called "multifractal random walk", whose main properties are given in Appendix 2.A), Sornette *et al.* [456] have predicted different response functions of the volatility to large external shocks compared with what we term endogenous shocks, *i.e.*, which result from the cooperative accumulation of many small news. This theory, which has been successfully tested against empirical data with no adjustable parameters, suggests a general classification into two classes of events (endogenous and exogenous) with specific signatures and characteristic precursors for the endogenous class. It also proposes a simple origin for endogenous shocks as the accumulations, in certain circumstances, of tiny bad news that add *coherently* due to their persistence.

Another example supporting the existence of specific signatures distinguishing endogenous and exogenous events has been provided by a recent investigation concerning the origin of the success of best sellers [128, 455]. The question is whether the latest best seller is simply the product of a clever marketing campaign or if it has truly permeated society? In other words, can one determine whether a book's popularity will wane as quickly as it appeared or will it become a classic for future generations? The study in [455, 128] describes a simple and generic method that distinguishes exogenous shocks (*e.g.*, very large news impact) from endogenous shocks (*e.g.*, book that becomes a best seller by word of mouth) within the network of online buyers. An endogenous shock appears slowly but results in a long-lived growth and decay of sales due to small but very extensive interactions in the network of buyers. In contrast, while an exogenous shock appears suddenly and propels a book to best seller status, these sales typically decline rapidly as a power law with exponent larger than for endogenous shocks. These results suggest that the network of human acquaintances is close to "critical," with information neither propagating nor disappearing but spreading marginally between people. These results have interesting potential for marketing agencies, which could measure and maximize the impact of their publicity on the network of potential buyers, for instance.

These two examples show that the concepts of endogeneity and exogeneity should have many applications including the modeling and prediction of financial crashes [250, 458], Initial Public Offerings (IPO) [245], the movie industry [119] and many other domains related to marketing [452], for which the mechanism of information cascade derives from the fact that agents can observe box office revenues and communicate word of mouth about the quality of the movies they have seen. The formulation of a comprehensive theory of (time) dependence allowing to characterize endoneity and exogeneity and to distinguish between them is thus of great importance in future developments.

### 7.2.4 Nonstationarity and Regime Switching in Dependence

The general problem of the application of mathematical statistics to nonstationary data (including nonstationary time series) is very important, but alas, not much can be done. There are only a few approaches which may be used and only in specific conditions:

- Use of algorithms and methods which are robust with respect to possible nonstationarity in data, such as normalization procedures or the use of quantile samples instead of initial samples.
- Model nonstationarity by some low-frequency random processes, such as, e.g., a narrow-band random process $X(t) = A(t) \cos(\omega t + \phi(t))$ where $\omega \ll 1$ and $A(t)$ and phase $\phi(t)$ are slowly varying amplitude and phase. In this case, the Hilbert transform can be very useful to characterize $\phi(t)$ nonparametrically [79, 379].
- The estimation of the parameters of a low-frequency process based on a "short" realization is often hopeless. In this case, the only quantity which can be evaluated is the uncertainty (or scatter) of the results due to the nonstationarity.

Regime switching popularized by Hamilton [221] for autoregressive time series models is a special case of nonstationary, which can be handled with specific methods. Regime switching has been extensively used in business cycle analysis in order to describe the economic fluctuations in a rigorous statistical framework. The key idea is that the parameters of a model may switch between two (or more) regimes, where the switching is governed by a time-dependent state variable $S_t$ which takes typically two values 0 or 1. When $S_t = 0$, the parameters of the model are different from those when $S_t = 1$. Clearly, if $S_t$ were an observed variable, the parameters could simply be estimated using dummy variable methods. Regime-switching methods rely on the observation by Hamilton that, even when the state is unobservable, the parameters of the model in each state can be estimated provided that restrictions are placed on the probability process governing $S_t$. The simplest such restriction is to assume that $S_t$ obeys the dynamics of a first-order Markov chain, which means that any persistence in the state is completely embodied in the value of the state in the previous period. Many generalizations are under development.

In recent studies, many workers have extended this idea to model regime switches in the dependence structure of financial assets. We refer to [382, 498] for a perspective of recent efforts and references therein. Developing these ideas in the framework of copulas is a promising avenue for future research.

Regime switching is also appealing from a microeconomic viewpoint as it may reflect the changing conventions used by investors. Conventions can be formed by the belief of agents on the existence of correlations between information and returns for instance. Following this belief, agents try to estimate this correlation from past time series and act on it, thus creating it [494]. Another mechanism for conventions is based on imitation and moods [460]. Both mechanisms predict the existence of random abrupt changes. For the future, it would be interesting to combine both mechanisms as they are arguably present together in real markets, in order to clarify their relative importance and interplay. Another important field of research is to combine these microeconomic models with the tools developed to detect regime switching.

### 7.2.5 Time-Varying Lagged Dependence

Determining the arrow of causality between two time series $X(t)$ and $Y(t)$ has a long history, especially in economics, econometrics and finance and it is often asked which economic variable might influence other economic phenomena [93, 199]. This question is raised in particular for the relationships between respectively inflation and GDP, inflation and growth rate, interest rate and stock market returns, exchange rate and stock prices, bond yields and stock prices, returns and volatility [95], advertising and consumption and so on. One simple naive measure is the lagged cross-correlation function

$$C_{X,Y}(\tau) = \frac{\text{Cov}\left[X(t)Y(t+\tau)\right]}{\sqrt{\text{Var}[X]\text{Var}[Y]}} \ .$$

Then, a maximum of $C_{X,Y}(\tau)$ at some nonzero positive time lag $\tau$ implies that the knowledge of $X$ at time $t$ gives some information on the future realization of $Y$ at the later time $t + \tau$. However, such correlations do not imply necessarily causality in a strict sense as a correlation may be mediated by a common source influencing the two time series at different times. The concept of Granger causality bypasses this problem by taking a pragmatic approach based on predictability: if the knowledge of $X(t)$ and of its past values improves the prediction of $Y(t + \tau)$ for some $\tau > 0$, then it is said that $X$ Granger causes $Y$ [22, 199] (see [98] for a recent extension to nonlinear time series). Such a definition does not address the fundamental philosophical and epistemological question of the real causality links between $X$ and $Y$ but has been found useful in practice.

However, most economic and financial time series are not strictly stationary and the lagged correlation/dependence and/or causality between two time series may be changing as a function time, for instance reflecting regime switches and/or changing agent expectations. It is thus important to define tests of causality or of lagged dependence which are sufficiently reactive to such regime switches, allowing to follow almost in real time the evolving structure of the causality. Cross-correlation methods and Granger causality tests require rather substantial amount of data in order to obtain reliable conclusions. In addition, cross-correlation techniques are fundamentally linear measures of dependence and may miss important nonlinear dependence properties. Granger causality tests are most often formulated using linear parametric autoregressive models. It may thus be that many of the paradoxes in macroeconomics concerning the causal relationship between such variables as inflation, inflation change, GDP growth rate and unemployment rate could result from an inadequate description of the time-varying lag cross-sectional dependence.

Recently, a new method, called "Optimal thermal causal path", has been introduced [459]. It is both nonparametric and sufficiently general so as to detect *a priori* arbitrary nonlinear dependence structures. Moreover, it is specifically conceived so as to adapt to the time evolution of the causality structure. The "Optimal thermal causal path" can be viewed as an extension of the "time distance" measure which amounts to compare trend lines upon horizontal differences of two time series [212].

The development of generalized dependence measures using such time-adaptive lag structure seems to be another promising domain of future developments.

### 7.2.6 Toward a Dynamical Microfoundation of Dependences

The need for the rather sophisticated statistical methods described in this book, as well as the developments suggested in this concluding chapter, reflect in our opinion the absence of a fundamental genuine economic understanding. To make a comparison with Natural Sciences, the need of such statistical methods has been less important, probably because most of the fundamental equations are known (at least at the macroscopic level) and the challenge lies more in understanding the emergence of complex solutions from seemingly simple mathematical formulations. In physics, for instance, the issues of dependences raised in this book are better and more simply attacked from a study of the fundamental dynamical equations. In contrast, we lack a deep underpinning for understanding the mechanisms at the origin of the dynamical behavior of financial markets. It is thus possible that the emerging field of behavioral finance, with its sister fields of neuroeconomics and evolutionary psychology, and their exploration of the impact on decision making of imperfect bounded subjective probability perceptions [36, 206, 437, 439, 474], may provide a fundamental shift in our understanding and therefore in the formulation of dependence between assets. This will have major impacts on risk assessment and its optimization.

# References

1. Abramovitz, E. and I.A. Stegun (1972) *Handbook of Mathematical Functions.* Dover Publications, New York. 267
2. Acerbi, C. (2002) Spectral measures of risk: A coherent representation of subjective risk aversion. *Journal of Banking & Finance* **26**, 1505–1518. 11
3. Alexander, G.J. and A.M. Baptista (2002) Economic implications of using a mean-VaR model for portfolio selection: A comparison with mean-variance analysis. *Journal of Economic Dynamics & Control* **26**, 1159–1193. 15
4. Allais, M. (1953) Le comportement de l'homme rationel devant le risque, critique des postulats de l'école américaine. *Econometrica* **21**, 503–546. 4, 272
5. Allais, M. (1990) Allais Paradox. In *The New Palgrave, Utility and Probability.* Macmillan, 3–9. 4, 272
6. Andersen, J.V. and D. Sornette (2001) Have your cake and eat it too: Increasing returns while lowering large risks! *Journal of Risk Finance* **2**, 70–82. 58, 233
7. Andersen, J.V. and D. Sornette (2005) *A Mechanism for Pockets of Predictability in Complex Adaptive Systems*, Europhys. Lett., **70**(5), 697–703. 23
8. Andersen, T.G., T. Bollerslev, F.X. Diebold and P. Labys (2001) The distribution of realized exchange rate volatility. *Journal of the American Statistical Association* **96**, 42–55. 219
9. Anderson, P.W. (1972) More is different (Broken symmetry and the nature of the hierarchical structure of science). *Science* **177**, 393–396. 22
10. Anderson, P.W., K.J. Arrow and D. Pines (1988) *The Economy as an Evolving Complex System.* Santa Fe Institute Studies in the Sciences of Complexity **5**. Westview Press, Addison-Wesley, Redwood City CA, 1988. 14
11. Anderson, T.W. and D.A. Darling (1952) Asymptotic theory of certain "goodness of fit" criteria. *Annals of Mathematical Statistics* **23**, 193–212. 61
12. Andersson, M., B. Eklund and J. Lyhagen (1999) A simple linear time series model with misleading nonlinear properties. *Economics Letters* **65**, 281–285. 43
13. Andreanov, A., F. Barbieri and O.C. Martin (2004) Large deviations in spin glass ground state energies. *European Physical Journal B* **41**, 365–375. 44
14. Ang, A. and G. Bekaert (2002) International asset allocation with regime shifts. *Review of Financial Studies* **15**, 1137–1187. 76, 231
15. Ang, A. and J. Chen (2002) Asymmetric correlations of equity portfolios. *Journal of Financial Economics* **63**, 443–494. 231, 240, 262

16. Arneodo, A., E. Bacry and J.F. Muzy (1998) Random cascades on wavelet dyadic trees. *Journal of Mathematical Physics* **39**, 4142–4164. 82
17. Arneodo, A., J.F. Muzy and D. Sornette (1998) Direct causal cascade in the stock market. *European Physical Journal B* **2**, 277–282. 82
18. Arthur, W.B., S.N. Durlauf and D.A. Lane (1997) *The Economy As an Evolving Complex System II.* Santa Fe Institute Studies in the Sciences of Complexity **27** Westview Press, Addison-Wesley, Redwood City CA. 14
19. Artzner, P., F. Delbaen, J.M. Eber and D. Heath (1997) Thinking coherently. *Risk* **10** (November), 68–71. 2, 4, 5
20. Artzner, P., F. Delbaen, J.M. Eber and D. Heath (1999) Coherent measures of risk. *Mathematical Finance* **9**, 203–228. 2, 4, 5
21. Artzner, P., F. Delbaen, J.M. Eber, D. Heath and H. Ku (2004) Coherent multiperiod risk adjusted values and Bellman's principle. *Working Paper*. 276
22. Ashley, R., C.W.J. Granger and R. Schmalensee (1980) Advertising and aggregate consumption: An analysis of causality. *Econometrica* **48**, 1149–1167. 280
23. Audrino, F. and G. Barone-Adesi (2004) Average conditional correlation and tree structures for multivariate GARCH models. *Working Paper*. Available at `http://papers.ssrn.com/paper.taf?abstract_id=553821` 276
24. Axelrod, R. (1997) *The Complexity of Cooperation.* Princeton University Press, Princeton, NJ. 22
25. Axtell, R. (2001) Zipf distribution of U.S. firm sizes. *Science* **293**, 1818–1820. 41
26. Bachelier, L. (1900) Théorie de la spéculation. *Annales Scientifiques de l'Ecole Normale Supérieure* **17**, 21–86. VIII, 37, 80
27. Bacry, E., J. Delour and J.-F. Muzy (2001) Multifractal random walk. *Physical Review E* **64**, 026103. 39, 40, 84, 85, 86
28. Bacry, E. and J.-F. Muzy (2003) Log-infinitely divisible multifractal processes. *Communications in Mathematical Physics* **236**, 449–475. 41, 84
29. Baig, T. and I. Goldfajn (1999) Financial market contagion in the Asian crisis. *IMF Staff Papers* **46**(2), 167–195. Available at `http://www.imf.org/external/Pubs/FT/staffp/1999/06-99/baig.htm` 211, 238
30. Baillie, R.T. (1996) Long memory processes and fractional integration in econometrics, *Journal of Econometrics* **73**, 5–59. 87
31. Baillie, R.T., T. Bollerslev and H.O. Mikelsen (1996) Fractionally integrated generalized autoregressive conditional heteroskedasticity. *Journal of Econometrics* **74**, 3–30. 37, 80
32. Bak, P. (1996) *How Nature Works: The Science of Self-organized Criticality.* Copernicus, New York. 277
33. Bak, P. and M. Paczuski (1995) Complexity, contingency and criticality. *Proceedings of the National Academy of Science USA* **92**, 6689–6696. 277
34. Bali, T.G. (2003) An extreme value approach to estimating volatility and Value-at-Risk. *Journal of Business* **76**, 83–108. 79
35. Barbe, P., C. Genest, K. Ghoudi and B. Rémillard (1996) On Kendall's process. *Journal of Multivariate Analysis* **58**, 197–229. 204
36. Barberis, N. and R. Thaler (2003) A survey of behavioral finance. In *Handbook of the Economics of Finance*, **1**(B), G.M. Constantinides, M. Harris and R.M. Stulz, eds. Elsevier, Amsterdam, 1053–1123. 4, 281
37. Barndorff-Nielsen, O.E. (1997) Normal inverse Gaussian distributions and the modeling of stock returns. *Scandinavian Journal of Statistics* **24**, 1–13. 43

38. Barndorff-Nielsen, O.E. and N. Shephard (2002) Econometric analysis of realised volatility and its use in estimating stochastic volatility models. *Journal of the Royal Statistical Society, Series B* **64**, 253–280. 219

39. Barral, J. and B.B. Mandelbrot (2002) Multifractal products of cylindrical pulses. *Probability Theory & Related Fields* **124**, 409–430. 37, 41

40. Barro, R.J., E.F. Fama, D.R. Fischel, A.H. Meltzer, R. Roll and L.G. Telser (1989) *Black Monday and the Future of Financial Markets*, R.W. Kamphuis Jr., R.C. Kormendi and J.W.H. Watson, eds. Mid American Institute for Public Policy Research, Inc. and Dow Jones-Irwin, Inc. 26

41. Basle Committee on Banking Supervision (1996) *Amendment to the Capital Accord to Incorporate Market Risks*. Available at http://www.bis.org/publ/bcbs24.pdf VIII

42. Basle Committee on Banking Supervision (2001) *The New Basel Capital Accord*. Available at http://www.bis.org/publ/bcbsca.htm VIII, 138, 273

43. Basrak, B., R.A. Davis and T. Mikosch (1999) The sample ACF of a simple bilinear process. *Stochastic Processes & their Application* **83**, 1–14. 232

44. Bates, D.S. (1991) The crash of 87: Was it expected? The evidence from options markets. *Journal of Finance* **46**, 1009–1044. 132

45. Baur, D. (2005) A flexible dynamic correlation model. In *Advances in Econometrics: Econometric Analysis of Economic and Financial Time Series* **20**(A), T.B. Fomby and D. Terrell eds. Elsevier, Kidlington, Oxford 276

46. Bauwens, L., S. Laurent and J. Rombouts (2003) Multivariate GARCH models: A survey. *Working Paper*. Available at http://ssrn.com/abstract=411062 276

47. Beck, U. (1992) *Risk Society: Toward a New Modernity*. Sage Publications, Thousand Oaks, CA. VII

48. Berman, S.M. (1962) A law of large numbers for the maximum in a stationary Gaussian sequence. *Annals of Mathematical Statistics* **33**, 93–97. 44

49. Berman, S.M. (1992) *Sojourns and Extremes of Stochastic Processes*. Wadsworth. Chapman & Hall/CRC. 44

50. Bernardo, A. and O. Ledoit (2001) Gain, loss and asset pricing. *Journal of Political Economy* **108**, 144–172. 13

51. Bernstein, P.L. (1998) *Against the Gods: The Remarkable Story of Risk*, New edition. Wiley, New York. 13

52. Bernstein, P.L. (1993) *Capital Ideas: The Improbable Origins of Modern Wall Street*, Reprint edition. Free Press. 13, 14

53. Bhansali, V. and M.B. Wise (2001) Forecasting portfolio risk in normal and stressed markets. *The Journal of Risk* **4**(1), 91–106. 232

54. Biagini F., P. Guasoni and M. Prattelli (2000) Mean-variance hedging for stochastic volatility models. *Mathematical Finance* **10**, 109–123. 136

55. Biham, O., O. Malcai, M. Levy and S. Solomon (1998) Generic emergence of power law distributions and Levy–Stable intermittent fluctuations in discrete logistic systems. *Physical Review E* **58**, 1352–1358. 39, 57

56. Bikos, A. (2000) Bivariate FX PDFs: A sterling ERI application. *Working Paper*, Bank of England. 132

57. Bingham, N.H., C.M. Goldie and J.L. Teugel (1987) *Regular variation*. Cambridge University Press, Cambridge. 39, 171

58. Bjerve, S. and K. Doksum (1993) Correlation curves: Measures of association as functions of covariate values. *Annals of Statistics* **21**, 890–902. 151

59. Black, F., M.C. Jensen and M.S. Scholes (1972) The capital asset pricing model: Some empirical tests. In *Studies in the Theory of Capital Markets*, M.C. Jensen, ed. Praeger, New York, 79–121. 14

60. Black, F. and M. Scholes (1973) The pricing of options and corporate liabilities. *Journal of Political Economy* **81**, 637–653. VIII, 38

61. Blanchard, O.J. and M.W. Watson (1982) Bubbles, rational expectations and speculative markets. In *Crisis in Economic and Financial Structure: Bubbles, Bursts, and Shocks*, P. Wachtel, ed. Lexington Books, Lexington. 39

62. Blattberg, R. and Gonnedes, N. (1974) A comparison of stable and Student distribution as statistical models for stock prices. *Journal of Business* **47**, 244–280. 42

63. Blum, A. and A. Kalai (1999) Universal portfolios with and without transaction costs. *Machine Learning* **35**, 193–205. 275

64. Blum, P., A. Dias and P. Embrechts (2002) The ART of dependence modeling: The latest advances in correlation analysis. In *Alternative Risk Strategies*. Risk Books, Morton Lane, London, 339–356. 99

65. Bollerslev, T. (1986) Generalized autoregressive conditional heteroscedasticity. *Journal of Econometrics* **31**, 307–327. 43

66. Bollerslev, T., R.F. Engle and D.B. Nelson (1994) ARCH models. In *Handbook of Econometrics*, **4**, R.F. Engle and D.L. McFadden, eds. North-Holland, Amsterdam, 2959–3038. 43

67. Bonabeau, E., M. Dorigo and G. Théraulaz (1999) *Swarm Intelligence: From Natural to Artificial Systems*. Oxford University Press, Oxford. 22

68. Bonabeau, E., J. Kennedy and R.C. Eberhart (2001) *Swarm Intelligence*. Academic Press, New York. 22

69. Bonabeau, E. and C. Meyer (2001) Swarm intelligence: A whole new way to think about business. *Havard Business Review* (May), 106–114. 22

70. Bonabeau, E. and G. Théraulaz (2000) Swarm smarts. *Scientific American* (March), 72–79. 22

71. Bookstaber, R. (1997) Global risk management: Are we missing the point? *Journal of Portfolio Management* **23**, 102–107. 228, 231

72. Bouchaud, J.-P. and M. Potters (2003) *Theory of Financial Risks: From Statistical Physics to Risk Management*, 2nd edition. Cambridge University Press, Cambridge, New York. 38, 42, 59, 176

73. Bouchaud, J.-P., D. Sornette, C. Walter and J.-P. Aguilar (1998) Taming large events: Optimal portfolio theory for strongly fluctuating assets. *International Journal of Theoretical & Applied Finance* **1**, 25–41. 2

74. Bouyé, E., V. Durrleman, A. Nikeghbali, G. Riboule and T. Roncalli (2000) Copulas for finance: A reading guide and some applications. *Technical Document*, Groupe de Recherche Opérationelle, Crédit Lyonnais. 103

75. Bovier, A. and D.M. Mason (2001) Extreme value behavior in the Hopfield model. *Annals of Applied Probability* **11**, 91–120. 44

76. Bowyer, K. and P.J. Phillips (1998) *Empirical Evaluation Techniques in Computer Vision*. IEEE Computer Society, Los Alamos, CA. 3

77. Box, G.E.P. and M.E. Muller (1958) A note on the generation of random normal deviates. *Annals of Mathematics & Statistics* **29**, 610–611. 121

78. Boyer, B.H., M.S. Gibson and M. Lauretan (1997) Pitfalls in tests for changes in correlations. *Board of the Governors of the Federal Reserve System*, International Finance Discussion Paper 597. 228, 231, 232, 233, 234, 260

79. Bracewell, R. (1999) The Hilbert transform. In *The Fourier Transform and Its Applications*, 3rd edition. McGraw-Hill, New York, 267–272. 279

80. Bradley, B. and M. Taqqu (2004) Framework for analyzing spatial contagion between financial markets. *Finance Letters* **2**(6), 8–15. 232

81. Bradley, B. and M. Taqqu (2005) Empirical evidence on spatial contagion between financial markets. *Finance Letters* **3**(1), 64–76. 232

82. Bradley, B. and M. Taqqu (2005) How to estimate spatial contagion between financial markets. *Finance Letters* **3**(1), 77–86. 232

83. Breymann, W., A. Dias and P. Embrechts (2003) Dependence structures for multivariate high-frequency data in finance. *Quantitative Finance* **3**, 1–14. 215

84. Brock, W.A., W.D. Dechert, J.A. Scheinkman and B. Le Baron (1996) A test for independence based on the correlation dimension. *Econometric Reviews* **15**, 197–235. 38

85. Brockwell, P.J. and R.A. Davis (1996) *Introduction to Time Series and Forecasting*. Springer Series in Statistics, Springer, New York. 3, 80

86. Buhmann, M.D. (2003) *Radial Basis Functions: Theory and Implementations*. Cambridge University Press, Cambridge, New York. 3

87. Calvo, S. and C.M. Reinhart (1995) Capital flows to Latin America: Is there evidence of contagion effects? In *Private Capital Flows to Emerging Market After the Mexican Crisis*, G.A. Calvo, M. Goldstein and E. Haochreiter, eds. Institute for International Economics, Washington, DC. 247, 260

88. Campbell, J.Y., A.W. Lo and A.C. MacKinlay (1997) *The Econometrics of Financial Markets*. Princeton University Press, Princeton, NJ. 38

89. Carpentier, D. and P. Le Doussal (2001) Glass transition of a particle in a random potential, front selection in nonlinear renormalization group, and entropic phenomena in Liouville and Sinh–Gordon models. *Physical Review E* **63**, 026110. 44

90. Carr, P., H. Geman, D.B. Madan and M. Yor (2002) The fine structure of assets returns: An empirical investigation. *Journal of Business* **75**, 305–332. 43

91. Challet, D. and M. Marsili (2003) Criticality and market efficiency in a simple realistic model of the stock market. *Physical Review E* **68**, 036132. 57

92. Challet, D., M. Marsili and Y.-C. Zhang (2004) *The Minority Game*. Oxford University Press, Oxford. 23

93. Chamberlain, G. (1982) The general equivalence of Granger and Sims causality. *Econometrica* **50**, 569–582. 280

94. Champenowne, D.G. (1953) A model of income distribution. *Economic Journal* **63**, 318–351. 39

95. Chan, K.C., L.T.W. Cheng and P.P. Lung (2001) Implied volatility and equity returns: Impact of market microstructure and cause–effect relation. *Working Paper*. 280

96. Charpentier, A. (2004) *Multivariate risks and copulas*. Ph.D. thesis, University of Paris IX. 249

97. Chen, K. and S.-H. Lo (1997) On a mapping approach to investigating the bootstrap accuracy. *Probability Theory & Related Fields* **107**, 197–217. 207

98. Chen, Y., G. Rangarajan, J. Feng and M. Ding (2004) Analyzing multiple nonlinear time series with extended Granger causality. *Physics Letters A* **324**, 26–35. 280

99. Cherubini, U. and Luciano, E. (2002) Bivariate option pricing with copulas. *Applied Mathematical Finance* **9**, 69–86. 100, 124, 131

100. Cherubini, U.E. Luciano and W. Vecchiato (2004) *Copula Methods for Finance*. Wiley, New York. 124

101. Cizeau, P., M. Potters and J.P. Bouchaud (2001) Correlation structure of extreme stock returns. *Quantitative Finance* **1**, 217–222. 232

102. Claessen, S., R.W. Dornbush and Y.C. Park (2001) Contagion: Why crises spread and how this can be stopped. In *International Financial Contagion*, S. Cleassens and K.J. Forbes, eds. Kluwer Academic Press, Dordrecht, Boston. 231

103. Clayton, D.G. (1978) A model for association in bivariate life tables and its application in epidemiological studies of familial tendency in chronic disease incidence. *Biometrika* **65**, 141–151. 113

104. Cochran, J.H. (2001) *Asset Pricing*. Princeton University Press, Princeton, NJ.

105. Cohen, E., R.F. Riesenfeld and G. Elber (2001) *Geometric Modeling with Splines: An Introduction*. AK Peters, Natick, MA. 3

106. Coles, S., J. Heffernan and J. Tawn (1999) Dependence measures for extreme value analyses. *Extremes* **2**, 339–365. 169, 255

107. Coles, S. and J.A. Tawn (1991) Modeling extreme multivariate events. *Journal of the Royal Statistical Society, Series B* **53**, 377–392. 116

108. Cont, R., Potters, M. and J.-P. Bouchaud (1997) Scaling in stock market data: Stable laws and beyond. In *Scale Invariance and Beyond*, B. Dubrulle, F. Graner and D. Sornette, eds. Springer, Berlin. 42

109. Cont, R. and P. Tankov (2003) *Financial Modeling with Jump Processes*. Chapman & Hall, London. 35

110. Cossette, H., P. Gaillardetz, E. Marceau and J. Rioux (2002) On two dependent individual risk models. *Insurance: Mathematics & Economics* **30**, 153–166. 100

111. Cox, D.R. (1972) Regression models in life tables (with discussion). *Journal of the Royal Statistical Society, Series B* **34**, 187–220. 113

112. Coutant, S., V. Durrleman, G. Rapuch and T. Roncalli (2001) Copulas, multivariate risk-neutral distributions and implied dependence functions. *Technical Document*, Groupe de Recherche Opérationelle, Crédit Lyonnais. 100, 135

113. Cover, T.M. (1991) Universal portfolios. *Mathematical Finance* **1**, 1–29. 275

114. Credit-Suisse-Financial-Products (1997) CreditRisk⁺: A credit risk management framework. *Technical Document*. Available at http://www.csfb.com/creditrisk. 137

115. Cramer, H. (1946) *Mathematical Methods of Statistics*. Princeton University Press, Princeton, NJ. 157

116. Cromwell, J.B., W.C. Labys and M. Terraza (1994) *Univariate Tests for Time Series Models*. Sage, Thousand Oaks, CA, 20–22. 38

117. Danielsson, J., P. Embrechts, C. Goodhart, C. Keating, F. Muennich, O. Renault and H.-S. Shin (2001) *An academic response to Basel II*. FMG and ESRC, 130 (London). VIII

118. Dawkins, R. (1989) *The Selfish Gene*, 2nd edition. Oxford University Press, Oxford. 21

119. De Vany, A. & Lee, C. (2001) Quality signals in information cascades and the dynamics of the distribution of motion picture box office revenues. *Journal of Economic Dynamics & Control* **25**, 593–614. 278

120. DeGregori, T.R. (2002) *The zero risk fiction*. April 12, American Council on Science and Health. Available at http://www.acsh.org VII

121. Deheuvels, P. (1979) La fonction de dépendence empirique et ses propriétés – Un test non paramétrique d'indépendance. *Académie Royale de Belgique – Bulletin de la Classe des Sciences, $5^{eme}$ Série* **65**, 274–292. 190, 203

122. Deheuvels, P. (1981) A non parametric test for independence. *Publications de l'Institut de Statistique de l'Université de Paris* **26**, 29–50. 190, 203

123. Delbaen, F. (2002) Coherent risk measures on general probability spaces. In *Advances in Finance and Stochastics, Essays in Honour of Dieter Sondermann*, K. Sandmann and P.J. Schonbucher, eds. Springer, New York. 5

124. Diether, K.B., C.J. Malloy and A. Scherbina (2002) Difference of opinion and the cross-section of stock returns. *Journal of Finance* **57**, 2113–2141. 20

125. Delbaen, F., P. Monat, W. Schachermayer, M. Schweizer and C. Stricker (1997) Weighted norm inequalities and hedging in incomplete markets. *Finance & Stochastics* **1**, 181–227. 136

126. Denuit, M., C. Genest and E. Marceau (1999) Stochastic bounds on sums of dependent risks. *Insurance: Mathematics & Economics* **25**, 85–104. 118, 126

127. Denuit, M. and O. Scaillet (2004) Non parametric tests for positive quadrant dependence *Journal of Financial Econometrics* **2**, 422–450. 167

128. Deschatres, F. and D. Sornette (2005) The dynamics of book sales: endogenous versus exogenous shocks in complex networks. *Physical Review E* **72**, 016112. 278

129. Devroye, L. (1986) *Non-uniform Random Variate Generation*. Springer-Verlag, New York. 123

130. Dhaene, J. and M.J. Goovaerts (1996) Dependency of risks and stop-loss order. *ASTIN Bulletin* **26**, 201–212. 149, 165

131. Dhaene, J., R.J.A., Laeven, S. Vanduffel, G. Darkiewicz and M.J. Goovaerts (2004) Can a coherent risk measure be too subadditive? *Working Paper.* 5

132. Diebolt, J. and D. Guegan (1991) Le modèle de série chronologique autorégressive $\beta$-ARCH. *Compte-rendus de l'Académie des Sciences* **312**, 625–630. 37, 57

133. Dodge, Y. (1985) *Analysis of Experiments with Missing Data*. Wiley, New York. 3

134. Doksum, K., S. Blyth, E. Bradlow, X. Meng and H. Zhao (1994) Correlation curves as local measures of variance explained by regression. *Journal of the American Statistical Association* **89**, 571–582. 151

135. Dorigo, M., G. Di Caro, and M. Sampels (2002) *Ant Algorithms*. Springer, Heidelberg. 22

136. Dragulescu, A.A. and V.M. Yakovenko (2002) Probability distribution of returns in the Heston model with stochastic volatility. *Quantitative Finance* **2**, 443–453. 58

137. Drozdz, S., J. Kwapien, F. Gruemmer, F. Ruf and J. Speth (2003) Are the contemporary financial fluctuations sooner converging to normal? *Physica Polonica B* **34**, 4293–4306. 78

138. Durrleman, V., A. Nikeghbali and T. Roncalli (2000) Copulas approximation and new families. *Technical Document*, Groupe de Recherche Opérationelle, Crédit Lyonnais. 192

139. Dyn, N., D. Leviatan, D. Levin and A. Pinkus (2001) *Multivariate Approximation and Applications*. Cambridge University Press, Cambridge, New York. 3

140. Eberlein, E., Keller, U. and Prause, K. (1998) New insights into smile, mispricing and value at risk: The hyperbolic model. *Journal of Business* **71**, 371–405. 43, 58

141. Eberlein, E. and F. Özkan (2003) Time consistency of Lévy models. *Quantitative Finance* **3**, 40–50. 35

142. Edwards, S. and R. Susmel (2001) Volatility dependence and contagion in emerging equity markets. *Working Paper*. Available at http://papers.ssrn.com/sol3/papers.cfm?abstract_id=285631 246

143. Efron, B. and R.J. Tibshirani (1986) Bootstrap method for standard errors, confidence intervals and other measures of statistical accuracy. *Statistical Science* **1**, 54–77. 207

144. Efron, B. and R.J. Tibshirani (1993) *An Introduction to the Bootstrap*. Chapman & Hall, CRC. 3

145. Embrechts, P., A. Hoeing and A. Juri (2003) Using copulae to bound the Value-at-Risk for functions of dependent risk. *Finance & Stochastics* **7**, 145–167. 100, 118, 119, 124

146. Embrechts, P., C.P. Klüppelberg and T. Mikosh (1997) *Modelling Extremal Events*. Springer-Verlag, Berlin. 46, 58

147. Embrechts, P., F. Lindskog and A. McNeil (2003) Modelling dependence with copulas and applications to risk management. In *Handbook of Heavy Tailed Distributions in Finance*, S. Rachev, ed. Elsevier, Amsterdam, 329–384. 167, 169, 232

148. Embrechts, P., A.J. McNeil and D. Straumann (1999) Correlation: Pitfalls and alternatives. *Risk* **12**(May), 69–71. 99

149. Embrechts, P., A.J. McNeil and D. Straumann (2002) Correlation and dependence in risk management: Properties and pitfalls. In *Risk Management: Value at Risk and Beyond*, M.A.H. Dempster, ed. Cambridge University Press, Cambridge, 176–223. 99, 150, 169, 172, 220, 232

150. Engle, R.F. (1982) Autoregressive conditional heteroskedasticity with estimation of the variance of UK inflation. *Econometrica* **50**, 987–1008. 35

151. Engle, R.F. (1984) Wald, likelihood ratio, and lagrange multiplier tests in econometrics. In *Handbook of Econometrics*, II, Z. Griliches and M.D. Intriligator, eds. North-Holland, Amsterdam. 38

152. Engle, R.F., D.F. Hendry and J.-F. Richard (1983) Exogeneity. *Econometrica* **51**, 277–304. 277

153. Engle, R.F. and A.J. Patton (2001) What good is a volatility model? *Quantitative Finance* **1**, 237–245. 66

154. Engle, R.F. and K. Sheppard (2001) Theoretical and empirical properties of dynamic conditional correlation multivariate GARCH. *NBER Working Papers number 8554*. Available at http://www.nber.org/papers/w8554.pdf 276

155. Erdös, P., A. Rényi and V.T. Sós (1966) On a problem of graph theory. *Studia Scientiarum Mathematicarum Hungarica* **1**, 215–235. 26

156. Ericsson, N. and J.S. Irons (1994) *Testing Exogeneity: Advanced Texts in Econometrics*. Oxford University Press, Oxford. 277

157. Fama, E.F. (1965) The behavior of stock market prices. *Journal of Business* **38**, 34–105. 42

158. Fama, E.F. (1970) Efficient capital markets: A review of theory and empirical work. *Journal of Finance* **25**, 383–417. 20

159. Fama, E.F. (1991) Efficient capital markets II. *Journal of Finance* **46**, 1575–1617. 20

160. Fama, E.F. and K.R. French (1992) The cross-section of expected stock returns. *Journal of Finance* **47**, 427–465. 3, 14, 18, 19, 24

161. Fama E.F. and K.R. French (1996) Multifactor explanations of asset pricing anomalies. *Journal of Finance* **51**, 55–84. 3, 24, 37

162. Fang, H.B. and T. Lai (1997) Co-kurtosis and capital asset pricing. *Financial Review* **32**, 293–307. 15, 58

163. Fang, H.B., K.T. Fang and S. Kotz (2002) The meta-elliptical distributions with given marginals. *Journal of Multivariate Analysis* **82**, 1–16. 108, 109

164. Farmer, J.D. (2002) Market force, ecology and evolution. *Industrial & Corporate Change* **11**(5), 895–953. 22

165. Farmer, J.D. and Lillo, F. (2004) On the origin of power-law tails in price fluctuations. *Quantitative Finance* **4**, 11–15. 41

166. Farmer, J.D., P. Patelli and I. I. Zovko (2005) The predictive power of zero intelligence models in financial markets. *Proceedings of the National Academy of Sciences* **102**(6), 2254–2259. 21

167. Feller, W. (1971) *An Introduction to Probability Theory and its Applications*, II. Wiley, New York. 58

168. Fermanian, J.D. and O. Scaillet (2003) Nonparametric estimation of copulas for time series. *Journal of Risk* **5**(4), 25–54. 192, 193, 194, 254

169. Fermanian, J.D. and O. Scaillet (2005) Some statistical pitfalls in copula modeling for financial applications. In *Capital Formation, Governance and Banking*, E. Klein, ed. Nova Publishers, Hauppauge, NY. 204

170. Figlewski, S. and X. Wang (2001) Is the "Leverage Effect" a Leverage Effect? *Unpublished working paper*, New York University. 85

171. Fishman, G.S. (1996) *Monte Carlo*. Springer-Verlag, New York. 120

172. Flood, P. and P.M. Garber (1994) *Speculative Bubbles, Speculative Attacks, and Policy Switching*. MIT Press, Cambridge, MA. 3

173. Flury, B. (1997) *A First Course in Multivariate Statistics*. Springer, New York. 3

174. Föllmer, H. and A. Schied (2002) Convex measures of risk and trading constraints. *Finance & Stochastics* **6**, 429–447. 4, 7

175. Föllmer, H. and A. Schied (2003) Robust preferences and convex measures of risk. In *Advances in Finance and Stochastics, Essays in Honour of Dieter Sondermann*, K. Sandmann and P.J. Schonbucher, eds. Springer-Verlag, New York. 4, 7

176. Föllmer, H. and M. Schweizer (1991) Hedging of contingent claims under incomplete information. In *Applied Stochastic Analysis, Stochastic Monographs*, **5**, M.H.A. Davis and R.J. Elliot, eds. Gordon and Breach, New York. 136

177. Föllmer, H. and D. Sondermann (1986) Hedging of non-redundant contingent claims. In *Contributions to Mathematical Economics*, W. Hildenbrand and A. Mascolell, eds. North-Holland, Amsterdam. 136

178. Forbes, K.J. and R. Rigobon (2002) No contagion, only interdependence: Measuring stock market co-movements. *Journal of Finance* **57**, 2223–2261. 232, 238, 240, 260

179. Frank, M.J., R.B. Nelsen and B. Schweizer (1987) Best-possible bounds for the distribution for a sum – A problem of Kolmogorov. *Probability Theory & Related Fields* **74**, 199–211. 118, 124, 127

180. Franses, P.H. and D. van Dijk (2000) *Nonlinear Time Series Models in Empirical Finance*. Cambridge University Press, Cambridge, New York. 3

181. Füredi, Z. and J. Komlós (1981) The eigenvalues of random symmetric matrices *Combinatorica* **1**, 233–241.

182. Frees, W.E., J. Carriere and E.A. Valdez (1996) Annuity valuation with dependent mortality. *Journal of Risk & Insurance* **63**, 229–261. 113

183. Frees, W.E. and E.A. Valdez (1998) Understanding relationship using copulas. *North American Actuarial Journal* **2**, 1–25. 100, 103, 111, 124, 201

184. Frey, R. and A. McNeil (2001) Modelling dependent defaults. *ETH E-Collection*. Available at `http://e-collection.ethbib.ethz.ch/show?type=bericht&nr=273` 100, 124

185. Frey, R. and A. McNeil (2002) VaR and expected shortfall in portfolios of dependent credit risks: Conceptual and practical insights. *Journal of Banking & Finance* **26**, 1317–1334. 181

186. Frey, R., A. McNeil and M. Nyfeler (2001) Credit risk and copulas. *Risk* **14**, 111–114. 100, 138, 181

187. Frieden, J.A. (1992) *Debt, Development, and Democracy: Modern Political Economy and Latin America, 1965–1985*. Princeton University Press, Princeton, NJ. 261

188. Frieden, J.A., P. Ghezzi and E. Stein (2001) Politics and exchange rates: A cross-country approach. In *The Currency Game: Exchange Rate Politics in Latin America*, J.A. Frieden, P. Ghezzi, E. Stein, eds. Inter-American Development Bank, Washington, DC. 261

189. Frieden, J.A. and E. Stein (2000) The political economy of exchange rate policy in Latin America: An analytical overview. *Working Paper*, Harvard University. 261

190. Frisch, U. (1995) *Turbulence*. Cambridge University Press, Cambridge. 82

191. Frisch, U. and D. Sornette (1997) Extreme deviations and applications. *Journal de Physique I, France* **7**, 1155–1171. 57

192. Frittelli, M. (2000) The minimal entropy martingale measure and the valuation problem in incomplete markets. *Mathematical Finance* **10**, 39–52. 136

193. Gabaix, X. (1999) Zipf's law for cities: An explanation. *Quarterly Journal of Economics* **114**, 739–767. 39

194. Gabaix, X., P. Gopikrishnan, V. Plerou and H.E. Stanley (2003) A theory of power-law distributions in financial market fluctuations. *Nature* **423**, 267–270. 41

195. Gabaix, X., P. Gopikrishnan, V. Plerou and H.E. Stanley (2003) A theory of large fluctuations in stock market activity. *MIT Department of Economics Working Paper 03-30*. Available at `http://ssrn.com/abstract=442940` 41

196. Geman, H. (2002) Pure jump Lévy process for asset price modelling. *Journal of Banking & Finance* **26**, 1297–1316. 35

197. Genest, C., K. Ghoudi and L.P. Rivest (1995) A semiparametric estimation procedure of dependence parameters in multivariate families of distributions. *Biometrika* **82**, 543–552. 197, 198, 200, 222, 223

198. Genest, C. and J. MacKay (1986) The joy of copulas: Bivariate distribution with uniform marginals. *American Statistician* **40**, 28–283. 123, 155

199. Geweke, J. (1984) Inference and causality in economic time series models. In *Handbook of Economics*, Vol. II, Z. Griliches and M.D. Intriligator, eds. Elsevier Science Publisher BV, Amsterdam, 1101–1144. 280

200. Ghoudi, K. and B. Remillard (1998) Empirical processes based on pseudo-observations. In *Asymptotic Methods in Probability and Statistics: A Volume in Honour of Miklos Csorgo*, B. Szyskowicz, ed. Elsevier, Amsterdam. 206

201. Ghoudi, K. and B. Remillard (2004) Empirical processes based on pseudo-observations II: The multivariate case. *Fields Institute Communications Series* **44**, 381–406. 206

202. Gilboa, I. (1987) Expected utility with purely subjective non-additive probabilities. *Journal of Mathematical Economics* **16**, 65–88. 6

203. Gilboa, I. and D. Schmeidler (1989) Maxmin expected utility with non-unique prior. *Journal of Mathematical Economics* **18**, 141–153. 6

204. Glimcher, P.W. and A. Rustichini (2004) Neuroeconomics: The consilience of brain and decision. *Science* **306**, 447–452. 272

205. Godambe, V.P. (1960) An optimum property of regular maximum likelihood estimation. *Annals of Mathematical Statistics* **31**, 1208–1211. 202

206. Goldberg, J. and R. von Nitzsch (2001) *Behavioral Finance*, translated by A. Morris. Wiley, New York. 4, 281

207. Goldie, C.M. (1991) Implicit renewal theory and tails of solutions of random equations. *Annals of Applied Probability* **1**, 126–166. 39

208. Gouriéroux, C. and J. Jasiak (1998) Truncated maximum likelihood, goodness of fit tests and tail analysis. *Working Paper*, CREST. 42

209. Gouriéroux, C. and A. Monfort (1994) Testing non-nested hypothesis. In *Handbook of Econometrics*, **4**, R.F. Engle and D. McFadden, eds. North-Holland, Amsterdam, 2585–2637. 74, 216, 224

210. Gouriéroux, C. and A. Montfort (1997) *Time Series and Dynamic Models*, translated and edited by G.M. Gallo. Cambridge University Press, Cambridge, New York. 3

211. Gopikrishnan, P., V. Plerou, L.A.N. Amaral, M. Meyer and H.E. Stanley (1999) Scaling of the distributions of fluctuations of financial market indices. *Physical Review E* **60**, 5305–5316. 51

212. Granger, C.W.J. and Y. Jeon (1997) Measuring lag structure in forecasting models – The introduction of Time Distance. *UCSD Economics Discussion Paper 97-24*. Available at http://ssrn.com/abstract=56515 281

213. Granger, C.W.J. and R. Joyeux (1980) An introduction to long memory time series models and fractional differencing. *Journal of Time Series Analysis* **1**, 15–29. 80

214. Granger, C.W.J., E. Maasoumi and J. Racine (2004) A dependence metric for possibly nonlinear processes. *Journal of Time Series Analysis* **25**, 649–669. 162, 164, 203

215. Granger, C.W.J. and T. Teräsvirta (1993) *Modelling Nonlinear Economic Relationships (Advanced Texts in Econometrics)*. Oxford University Press, Oxford. 3

216. Granger, C.W.J. and T. Teräsvirta (1999) A simple nonlinear model with misleading properties. *Economics Letters* **62**, 741–782. 43

217. Guillaume, D.M., M.M. Dacorogna, R.R. Davé, U.A. Müller, R.B. Olsen and O.V. Pictet (1997) From the bird's eye to the microscope: A survey of new stylized facts of the intra-day foreign exchange markets. *Finance & Stochastics* **1**, 95–130. 42, 44

218. Haas, C.N. (1999) On modeling correlated random variables in risk assessment. *Risk Analysis* **19**, 1205–1214. 120

219. Hall, P.G. (1979) On the rate of convergence of normal extremes. *Journal of Applied Probabilities* **16**, 433–439. 48

220. Hall, W.J. and J.A. Wellnel (1979) The rate of convergence in law of the maximum of an exponential sample. *Statistica Neerlandica* **33**, 151–154. 48, 49

294    References

221. Hamilton, J.D. (1989) A new approach to the economic analysis of non-stationary time series and the business cycle. *Econometrica* **57**, 357–384. 279
222. Hamilton, J.D. (1994) *Time Series Analysis*. Princeton University Press, Princeton, NJ. 3, 80
223. Harvey, C.R. and A. Siddique (2000) Conditional skewness in asset pricing tests. *Journal of Finance* **55**, 1263–1295. 14, 15
224. Hauksson, H.A., M.M. Dacorogna, T. Domenig, U.A. Müller and G. Samorodnitsky (2001) Multivariate extremes, aggregation and risk estimation. *Quantitative Finance* **1**, 79–95. 232
225. Havrda, J. and F. Charvat (1967) Quantification method of classification processes: Concept of structural $\alpha$-entropy. *Kybernetica Cislo I. Rocnik* **3**, 30–34. 163
226. Heath, D., E. Platen and M. Schweizer (2001) A comparison of two quadratic approaches to hedging in incomplete markets. *Mathematical Finance* **11**, 385–413. 136
227. Helmbold, D.P., R.E. Schapire, Y. Singer and M.K. Warmuth (1998) On-line portfolio selection using multiplicative updates. *Mathematical Finance* **8**, 325–347, 275
228. Hennessy, D. and H.E. Lapan (2002) The use of Archimedean copulas to model portfolio allocations. *Mathematical Finance* **12**, 143–154. 124
229. Herffernan, J.E. (2000) A directory of tail dependence. *Extremes* **3**, 279–290. 172
230. Henderson, V., D. Hobson, S. Howison and T. Kluge (2005) A comparison of option prices under different pricing measures in a stochastic volatility model with correlation. *Review of Derivatives Research* **8**, 5–25. 136
231. Hergarten, S. (2002) *Self-organized Criticality in Earth Systems*. Springer-Verlag, Heidelberg. 277
232. Heston, S.L. (1993) A closed-form solution for options with stochastic volatility with applications to bond and currency options. *Review of Financial Studies* **6**, 327–343. 37, 144
233. Hill, B.M. (1975) A simple general approach to inference about the tail of a distribution. *Annals of Statistics* **3**, 1163–1174. 64
234. Hobson, D. (2004) Stochastic volatility models, correlation and the q-optimal measure. *Mathematical Finance* **14**, 537–556. 136
235. Hotelling, H. (1936) Relations between two sets of variates. *Biometrika* **28**, 321–377. 152
236. Houggard, P. (1984) Life table methods for heterogeneous populations: Distributions describing for heterogeneity. *Biometrika* **71**, 75–83. 113
237. Houggard, P., B. Harvald and N.V. Holm (1992) Measuring the similarities between the lifetimes of adult Danish twins born between 1881–1930. *Journal of the American Statistical Association* **87**, 17–24. 113
238. Hubert, P.J. (2003) *Robust Statistics*. Wiley-Intersecience, New York. 275
239. Hult, H. and F. Lindskog (2001) Multivariate extremes, aggregation and dependence in elliptical distributions. *Advances in Applied Probability* **34**, 587–609. 174
240. Hwang, S. and M. Salmon (2002) An analysis of performance measures using copulae. In *Performance Measurement in Finance: Firms, Funds and Managers*, J. Knight and S. Satchell, eds. Butterworth-Heinemann, London. 124

241. Hwang, S. and S. Satchell (1999) Modelling emerging market risk premia using higher moments. *International Journal of Finance & Economics* **4**, 271–296. 13, 15, 58

242. Ingersoll, J.E. (1987) *The Theory of Financial Decision Making*. Rowman & Littlefield, Totowa, NJ.

243. Iman, R.L. and W.J. Conover (1982) A distribution-free approach to inducing rank correlation among input variables. *Communications in Statistics Simulation & Computation* **11**, 311–334. 120

244. Jackel, P. (2002) *Monte Carlo Methods in Finance*. Wiley, New York. 120

245. Jenkinson, T. and Ljungqvist, A. (2001) *Going Public: The Theory and Evidence on How Companies Raise Equity Finance*, 2nd edition. Oxford University Press, Oxford. 278

246. Jensen, H.J. (1998) *Self-organized Criticality: Emergent Complex Behavior in Physical and Biological Systems*, Cambridge Lecture Notes in Physics. Cambridge University Press, Cambridge. 277

247. Jensen, J.L. (1995) *Saddlepoint Approximations*. Oxford University Press, Oxford. 263

248. Joe, H. (1997) *Multivariate Models and Dependence Concepts*. Chapman & Hall, London. 103, 117, 137, 202

249. Johansen, A. and D. Sornette (2001) Large stock market price Drawdowns are outliers. *Journal of Risk* **4**, 69–110, Hauppauge, NY. 3, 23, 36, 79, 276

250. Johansen, A. and D. Sornette (2004) Endogenous versus exogenous crashes in financial markets. In *Contemporary Issues in International Finance*. Nova Science Publishers. 3, 36, 276, 278

251. Johnson, N.F., P. Jefferies and P.M. Hui (2003) *Financial Market Complexity*. Oxford University Press, Oxford. 23

252. Johnson, N.L. and S. Kotz (1972) *Distributions in Statistics: Continuous Multivariate Distributions*. Wiley, New York. 107, 121, 240, 262

253. Johnson, N.L., S. Kotz and N. Balakrishnan (1997) *Discrete Multivariate Distributions*. Wiley, New York. 3

254. Johnson, R.A. and D.W. Wichern (2002) *Applied Multivariate Statistical Analysis*, 5th edition. Prentice Hall, Upper Saddle River, NJ. 3

255. Johnson, T.C. (2004) Forecast dispersion and the cross section of expected returns. *Journal of Finance* **59**, 1957–1978. 20

256. Jondeau, E. and M. Rockinger (2003) Testing for differences in the tails of stock-market returns. *Journal of Empirical Finance* **10**, 559–581. 53, 78

257. Jorion, P. (1997) *Value-at-Risk: The New Benchmark for Controlling Derivatives Risk*. Irwin Publishing, Chicago, IL. 2

258. Jouini, M.N. and R.T. Clemen (1996) Copula models for aggregating expert opinions. *Operation Research* **43**, 444–457. 124

259. Jurczenko, E. and B. Maillet (2005) The 4-CAPM: in between Asset Pricing and Asset Allocation. In *Multi-Moment Capital Pricing Models and Related Topics*, C. Adcock, B. Maillet and E. Jurzenko, eds. Springer. Forthcoming 58

260. Juri, A. and M.V. Wüthrich (2002) Copula convergence theorem for tail events. *Insurance: Mathematics & Economics* **30**, 405–420. 115, 232, 255

261. Kalai, A. and S. Vempala (2000) Efficient algorithms for universal portfolios. In *Proceedings of the 41st Annual IEEE Symposium on Foundations of Computer Science*, 486–491. 275

262. Kaminsky, G.L. and S.L. Schmukler (1999) What triggers market jitters? A chronicle of the Asian crisis. *Journal of International Money & Finance* **18**, 537–560. 211

263. Kahneman, D. (2002) Maps of bounded rationality: A perspective on intuitive judgment an choice. *Nobel Prize Lecture.* Available at `http://nobelprize.org/economics/laureates/2002/kahnemann-lecture.pdf` 272

264. Karatzas, I. and S.E. Shreve (1991) *Brownian Motion and Stochastic Calculus.* Springer-Verlag, New York. 219

265. Karlen, D. (1998) Using projection and correlation to approximate probability distributions. *Computer in Physics* **12**, 380–384. 108, 219

266. Kass, R.E. and A.E. Raftery (1995) Bayes factors. *Journal of the American Statistical Association* **90**, 773–795. 74

267. Kearns, P. and A. Pagan (1997) Estimating the density tail index for financial time series. *Review of Economics & Statistics* **79**, 171–175. 43, 48, 51, 52, 65, 76

268. Kesten, H. (1973) Random difference equations and renewal theory for products of random matrices. *Acta Mathematica* **131**, 207–248. 39

269. Kim, C.-J. and C.R. Nelson (1999) *State-Space Models with Regime Switching: Classical and Gibbs-Sampling Approaches with Applications.* MIT Press, Cambridge, MA. 3

270. Kimeldorf, G. and A. Sampson (1978) Monotone dependence. *Annals of Statistics* **6**, 895–903. 101

271. King, M. and S. Wadhwani (1990) Transmission of volatility between stock markets. *Review of Financial Studies* **3**, 5–330. 231, 247

272. Klugman, S.A. and R. Parsa (1999) Fitting bivariate loss distributions with copulas. *Insurance: Mathematics & Economics* **24**, 139–148. 124, 201, 203

273. KMV Corporation (1997) Modelling default risk. *Technical Document.* Available at `http://www.kmv.com` 137

274. Knopoff, L. and D. Sornette (1995) Earthquake death tolls. *Journal de Physique I, France* **5**, 1681–1688. 126

275. Kon, S. (1984) Models of stock returns: A comparison. *Journal of Finance* **39**, 147–165. 42

276. Kotz, S. (2000) *Continuous Multivariate Distributions*, 2nd edition. Wiley, New York. 3

277. Kotz, S. and S. Nadarajah (2000) *Extreme Value Distribution: Theory and Applications.* Imperial College Press, London. 44

278. Krauss, A. and R. Litzenberger (1976) Skewness preference and the valuation of risk assets. *Journal of Finance* **31**, 1085–1099. 15

279. Krivelevich, M. and V.H. Vu (2002) On the concentration of eigenvalues of random symmetric matrices. *Israel Journal of Mathematics* **131**, 259–268. 26

280. Krugman, P. (1996) *Self-organizing Economy.* Blackwell publishers, Cambridge, MA, and Oxford. 14

281. Kruskal, W.H. (1958) Ordinal measures of association. *Journal of the American Statistical Association* **53**, 814–861. 160

282. Krzanowski, W.J. (2000) Principles of Multivariate Analysis: A User's Perspective, revised edition. Clarendon Press, Oxford; Oxford University Press, New York. 3

283. Krzysztofowicz, R. and K.S. Kelly (1996) A meta-Gaussian distribution with specified marginals. *Technical Document*, University of Virginia. 108

284. Kulpa, T. (1999) On approximations of copulas. *International Journal of Mathematics & Mathematical Sciences* **22**, 259–269. 254

285. Kusuoka, S. (2001) On law invariant coherent risk measures. In *Advances in Mathematical Economics*, **3**, S. Kusuoka, T. Maruyama, R. Anderson, C. Castaing, F.H. Clarke, G. Debreu, E. Dierker, D. Duffie, L.C. Evans, T. Fujimoto, J.-M. Grandmont, N. Hirano, L. Hurwicz, eds. Springer, Tokyo. 83–95. 6

286. Laherrère, J. and D. Sornette (1999) Stretched exponential distributions in nature and economy: Fat tails with characteristic scales. *European Physical Journal B* **2**, 525–539. 36, 43

287. Lakonishok, J. and A.C. Shapiro (1986) Systematic risk, total risk and size as determinants of stock market returns. *Journal of Banking & Finance* **10**, 115–132. 14

288. Laloux, L., P. Cizeau, M. Potters and J.P. Bouchaud (2000) Random matrix theory and financial correlations. *International Journal of Theoretical & Applied Finance* **3**, 391–397. 24, 27

289. Lamper, D., S.D. Howison and N.F. Johnson (2001) Predictability of large future changes in a competitive evolving population. *Physical Review Letters* **88**, 017902. 23

290. Lancaster, H.O. (1963) Correlation and complete dependence of random variables. *Annals of Mathematical Statistics* **34**, 1315–1321. 101

291. Laurent, J.P. and J. Gregory (2004) In the core of correlation. *Risk* **17**(10), 87–91. 181

292. Laurent, J.P. and H. Pham (1999) Dynamic programming and mean-variance hedging. *Finance & Stochastics* **3**, 83–101. 136

293. Leadbetter, M.R. (1974) On extreme values in stationary processes. *Wahrscheinlichkeits* **28**, 289–303. 44, 46

294. Ledford, A.W. and J.A. Tawn (1996) Statistics for near independence in multivariate extreme values. *Biometrika* **83**, 169–187. 255

295. Ledford, A.W. and J.A. Tawn (1998) Concomitant tail behavior for extremes. *Advances in Applied Probability* **30**, 197–215. 255

296. Ledoit, O., P. Santa-Clara and M. Wolf (2003) Flexible multivariate GARCH modeling with an application to international stock markets. *Review of Economics & Statistics* **85**, 735–747. 276

297. Ledoit, O. and M. Wolf (2004) Honey, I shrunk the sample covariance matrix – Problems in mean-variance optimization. *Journal of Portfolio Management* **30**(4), 110–119. 3

298. Ledoit, O. and M. Wolf (2004) A well-conditioned estimator for large-dimensional covariance matrices. *Journal of Multivariate Analysis* **88**, 365–411. 3

299. Lee, S.B. and K.J. Kim (1993) Does the October 1987 crash strengthen the co-movements among national stock markets? *Review of Financial Economics* **3**, 89–102. 247

300. Lehmann, E. (1966) Some concepts of dependence. *Annals of Mathematical Statistics* **37**, 1137–1153. 164, 165

301. Lempert, R.J., S.W. Popper, and S.C. Bankes (2003) *Shaping the next one hundred years: New methods for quantitative, long-term policy analysis*. RAND Pardee Center Publications MR-1626-CR. 275

302. Li, D.X. (1999) The valuation of basket credit derivatives. *CreditMetrics Monitor* (April), 34–50. 124

303. Li, D.X. (2000) On default correlation: A copula approach. *Journal of Fixed Income* **9**, 43–54. 124, 137

304. Li, D.X., Mikusinski, P., H. Sherwood and M.D. Taylor (1997) On approximation of copulas. In *Distributions with Given Marginals and Moments Problems*, V. Benes and J. Stephan, eds. Kluwer Academic Publisher, Dordrecht, Boston. 191

305. Li, D.X., Mikusinski, P. and Taylor, M.D. (1998) Strong approximation of copulas. *Journal of Mathematical Analysis & Applications* **225**, 608–623. 191, 254

306. Lim, K.G. (1989) A new test for the three-moment capital asset pricing model. *Journal of Financial & Quantitative Analysis* **24**, 205–216. 14, 15

307. Lindskog, F., A.J. McNeil and U. Schmock (2003) Kendall's tau for elliptical distributions. In *Credit Risk – Measurement, Evaluation and Management*, G. Bol, G. Nakhaeizadeh, S. Rachev, T. Ridder and K.-H. Vollmer, eds. Physica-Verlag, Heidelberg. 111, 157

308. Lintner, J. (1975) The valuation of risk assets and the selection of risky investments in stock portfolios and capital budgets. *Review of Economics & Statistics* **13**, 13–37. 14

309. Litterman, R. and K. Winkelmann (1998) *Estimating Covariance Matrices.* Risk Management Series, Goldman Sachs. 2

310. Little, R.J.A. and D.B. Rubin (1987) *Statistical Analysis with Missing Data.* Wiley, New York. 3

311. Lo, A.W. (1999) The three P's of total risk management. *Financial Analysts Journal* **55** (January/February), 13–26. 271

312. Longin, F.M. (1996) The asymptotic distribution of extreme stock market returns. *Journal of Business* **96**, 383–408. 42, 43, 44, 52, 61

313. Longin, F.M. (2000) From value at risk to stress testing: The extreme value approach. *Journal of Banking & Finance* **24**, 1097–1130. 43, 46, 79

314. Longin, F.M. and B. Solnik (1995) Is the correlation in international equity returns constant: 1960–1990? *Journal of International Money & Finance* **14**, 3–26. 231, 241

315. Longin F.M. and B. Solnik (2001) Extreme correlation of international equity markets. *Journal of Finance* **56**, 649–676. 231, 232, 236, 241, 253, 254, 255, 260

316. Loretan, M. (2000) Evaluating changes in correlations during periods of high market volatility. *Global Investor* **135**, 65–68. 231, 232

317. Loretan, M. and W.B. English (2000) Evaluating "correlation breakdowns" during periods of market volatility. *BIS Quarterly Review* (June), 29–36. 231, 232

318. Loynes, R.M. (1965) Extreme values in uniformly mixing stationary stochastic processes. *Annals of Mathematical Statistics* **36**, 993–999. 44

319. Lux, T. (1996) The stable Paretian hypothesis and the frequency of large returns: An examination of major German stocks. *Applied Financial Economics* **6**, 463–475. 78

320. Lux, T. (2000) On moment condition failure in German stock returns: An application of recent advances in extreme value statistics. *Empirical Economics* **25**, 641–652. 42

321. Lux, T. (2003) The multifractal model of asset returns: Its estimation via GMM and its use for volatility forecasting. *Working Paper*, University of Kiel. 40

322. Lux, T. (2001) The limiting extreme behavior of speculative returns: An analysis of intra-daily data from the Francfurt stock exchange. *Journal of Economic Behavior & Organization* **46**, 327–342. 42, 44

323. Lux, T. and D. Sornette (2002) On rational bubbles and fat tails. *Journal of Money Credit & Banking* **34**, 589–610. 39

324. Lyubushin Jr., A.A. (2002) Robust wavelet-aggregated signal for geophysical monitoring problems. *Izvestiya, Physics of the Solid Earth* **38**, 1–17. 152

325. Majumdar, S.N. and P.L. Krapivsky (2002) Extreme value statistics and traveling fronts: Application to computer science. *Physical Review E* **65**, 036127. 44

326. Majumdar, S.N. and P.L. Krapivsky (2003) Extreme value statistics and traveling fronts: Various applications. *Physica A* **318**, 161–170. 44

327. Makarov, G.D. (1981) Estimates for the distribution function of a sum of two random variables when the marginal distributions are fixed. *Theory of Probability & its Applications* **26**, 803–806. 118

328. Malamud, B.D., G. Morein and D.L. Turcotte (1998) Forest fires – An example of self-organized critical behavior. *Science* **281**, 1840–1842. 126

329. Malevergne, Y., V.F. Pisarenko and D. Sornette (2003) On the power of generalized extreme value (GEV) and generalized pareto distribution (GPD) estimators for empirical distributions of log-returns. *Applied Financial Economics*. Forthcoming. 38, 48, 49, 50, 51, 52, 55, 56

330. Malevergne, Y., V.F. Pisarenko and D. Sornette (2005) Empirical distribution of log-returns: Between the stretched-exponential and the power law? *Quantitative Finance* **5**. Forthcoming. 40, 54, 63, 64, 66, 68, 70, 72, 74, 75

331. Malevergne, Y. and D. Sornette (2001) Multi-dimensional rational bubbles and fat tails. *Quantitative Finance* **1**, 533–541. 39

332. Malevergne, Y. and D. Sornette (2002) Minimizing extremes. *Risk* **15** (11), 129–132. X, 174, 175, 178, 179, 255

333. Malevergne, Y. and D. Sornette (2005) Multi-moment methods for portfolio management: Generalized capital asset pricing model in homogeneous and heterogeneous markets. In *Multi-Moment Capital Pricing Models and Related Topics*, C. Adcock, B. Maillet and E. Jurzenko, eds. Springer. Forthcoming. X, 8, 16, 17, 31, 58

334. Malevergne, Y. and Sornette, D. (2003) Testing the Gaussian copula hypothesis for financial assets dependences. *Quantitative Finance* **3**, 231–250. 204, 207, 208, 211, 212, 214, 215, 216, 232, 254

335. Malevergne, Y. and D. Sornette (2004) How to account for extreme co-movements between individual stocks and the market. *Journal of Risk* **6**(3), 71–116. 174, 175, 255

336. Malevergne, Y. and D. Sornette (2004) VaR-efficient portfolios for a class of super- and sub-exponentially decaying assets return distributions. *Quantitative Finance* **4**, 17–36. X, 124, 128, 130, 141

337. Malevergne, Y. and D. Sornette (2004) Collective origin of the coexistence of apparent RMT noise and factors in large sample correlation matrices. *Physica A* **331**, 660–668. 24, 27, 28

338. Malevergne, Y. and D. Sornette (2005) Higher-moment portfolio theory (Capitalizing on behavioral anomalies of stock markets). *Journal of Portfolio Management* **31**(4), 49–55. 124

339. Mandelbrot, B.B. (1963) The variation of certain speculative prices. *Journal of Business* **36**, 392–417. 42

340. Mandelbrot, B.B. (1997) *Fractals and Scaling in Finance: Discontinuity, Concentration, Risk.* Springer, New York. 82

341. Mandelbrot, B.B., A. Fisher and L. Calvet (1997) A multifractal model of asset returns. *Coles Fundation Discussion Paper #1164.* 37, 41, 82, 84

342. Mansilla, R. (2001) Algorithmic complexity of real financial markets. *Physica A* **301**, 483–492. 232

343. Mantegna, R.N. (1999) Hierarchical structure in financial markets. *European Physical Journal B* **11**, 193–197. 28

344. Mantegna, R.N. and H.E. Stanley (1994) Stochastic process with ultra slow convergence to a Gaussian: The truncated Lévy flight. *Physical Review Letters* **73**, 2946–2949. 75

345. Mantegna, R.N. and H.E. Stanley (1995) Scaling behavior of an economic index. *Nature* **376**, 46–55. 42

346. Mantegna, R.N. and H.E. Stanley (2000) *An Introduction to Econophysics, Correlations and Complexity in Finance.* Cambridge University Press, Cambridge. 42

347. Markovitz, H. (1959) *Portfolio Selection: Efficient Diversification of Investments.* Wiley, New York. VIII, 1, 2, 13, 31, 38

348. Marshall, A. and I. Olkin (1988) Families of multivariate distributions. *Journal of the American Statistical Association* **83**, 834–841. 113, 123

349. Marsili, M. (2002) Dissecting financial markets: Sectors and states. *Quantitative Finance* **2**, 297–302. 28

350. Mashal, R. and A.J. Zeevi (2002) Beyond correlation: Extreme co-movements between financial assets. *Working Paper*, Columbia Business School. 110, 200, 212, 215, 216, 224, 226

351. Matia, K., L.A.N. Amaral, S.P. Goodwin and H.E. Stanley (2002) Different scaling behaviors of commodity spot and future prices. *Physical Review E* **66**, 045103. 51

352. McClure, S.M., D.I. Laibson, G. Loewenstein and J.D. Cohen (2004) Separate neural systems value immediate and delayed monetary rewards. *Science* **306**, 503–507. 272

353. McDonald, J.B. and W.K. Newey (1988) Partially adaptive estimation of regression models via the generalized *t*-distribution. *Econometric Theory* **4**, 428–457. 54

354. McLeish, D.L. and C.G. Small (1988) *The Theory and Application of Statistical Inference Functions.* Springer-Verlag, Berlin. 202

355. McNeil, A. and R. Frey (2000) Estimation of tail-related risk measures for heteroscedastic financial time series: An extreme value approach. *Journal of Empirical Finance* **7**, 271–300. 79

356. Meerschaert, M.M. and H. Scheffler (2001) Sample cross-correlations for moving averages with regularly varying tails. *Journal of Time Series Analysis* **22**, 481–492. 58, 148, 243, 254

357. Mehta, M.L. (1991) *Random Matrices*, 2nd edition. Academic Press, Boston. 26

358. Merton, R.C. (1974) On the pricing of corporate debt: The risk structure of interest rates. *Journal of Finance* **29**, 449–470. 20, 137

359. Merton, R.C. (1990) *Continuous-Time Finance.* Blackwell, Cambridge. 14

360. Mézard, M., G. Parisi and M.A. Virasoro (1987) *Spin Glass Theory and Beyond,* World Scientific Lecture Notes in Physics, Vol. 9. World Scientific, Singapore. 164

361. Mills, T.C. (1993) *The Econometric Modelling of Financial Time Series.* Cambridge University Press, Cambridge, New York. 3

362. Mittnik S., S.T. Rachev and M.S. Paolella (1998) Stable Paretian modeling in finance: Some empirical and theoretical aspects. In *A Practical Guide to Heavy Tails*, R.J. Adler, R.E. Feldman and M.S. Taqqu, eds. Birkhauser, Boston, 79–110. 42

363. Morrison, D.F. (1990) *Multivariate Statistical Methods*, 3rd edition. McGraw-Hill, New York. 3

364. Mossin, J. (1966) Equilibrium in a capital market. *Econometrica* **34**, 768–783. 14

365. J.-F. Muzy, A. Kozhemyak and E. Bacry (2005) Extreme values and fat tails of multifractal fluctuations. *Working Paper.* 40, 41

366. Muzy, J.F., D. Sornette, J. Delour and A. Arnéodo, (2001) Multifractal returns and hierarchical portfolio theory. *Quantitative Finance* **1**, 131–148. 40, 217, 276

367. Müller, U.A., M.M. Dacorogna, O.V. Pictet (1998) Heavy tails in high-frequency financial data. In *A Practical Guide to Heavy Tails*, R.J. Adler, R.E. Feldman and M.S. Taqqu, eds. Birkhauser, Boston, 55–78. 42, 44

368. Nagahara, Y. and G. Kitagawa (1999) A non-Gaussian stochastic volatility model. *Journal of Computational Finance* **2**, 33–47. 42

369. Naveau, P. (2003) Almost sure relative stability of the maximum of a stationary sequence. *Advances in Applied Probability* **35**, 721–736. 44

370. Nelsen, R.B. (1998) *An Introduction to Copulas*, Lectures Notes in statistic, **139**. Springer Verlag, New York. 103, 112, 115, 118, 155, 191

371. Neudecker, H., R. Heijmans, D.S.G. Pollock, A. Satorra (2000) *Innovations in Multivariate Statistical Analysis.* Kluwer Academic Press, Dordrecht, Boston. 3

372. Newey, W.K. and D. McFadden (1994) Large sample estimation and hypothesis testing. In *Handbook of Econometrics*, **4**, R.F. Engle and D. McFadden, eds. North-Holland, Amsterdam. 202, 216, 224

373. Noh, J.D. (2000) Model for correlations in stock market. *Physical Review E* **61**, 5981–5982. 27

374. O'Brien, G.L. (1987) Extreme values for stationary and Markov sequences. *Annals of Probability* **15**, 281–291. 44

375. Oakes, D. (1982) A model for association in bivariate survival data. *Journal of the Royal Statistical Society, Series B* **44**, 414–422. 196

376. Okuyama, K., M. Takayasu and H. Takayasu (1999) Zipf's law in income distribution of companies. *Physica A* **269**, 125–131. 41

377. Osborne, M.F.M. (1959) Brownian motion in the stock market. *Operations Research* **7**, 145–173. Reprinted in *The Random Character of Stock Market Prices*, P. Cootner ed. MIT Press, Cambridge, MA (1964), 100–128. 38, 80

378. Panja, D. (2004) Fluctuating fronts as correlated extreme value problems: An example of Gaussian statistics. *Physical Review E* **70**, 036101. 44

379. Papoulis, A. (1962) Hilbert transforms. In *The Fourier Integral and Its Applications.* McGraw-Hill, New York, 198–201. 279

380. Patton, J.A (2005) Estimation of multivariate models for time series of possibly different lengths. *Journal of Applied Econometrics.* Forthcoming. 100, 203, 217, 254

381. Patton, J.A. (2005) Modelling asymmetric exchange rate dependence. *International Economic Review.* Forthcoming. 203

382. Pelletier, D. (2005) Regime switching for dynamic correlations. *Journal of Econometrics*. Forthcoming. 279

383. Pfingsten, A., P. Wagner and C. Wolferink (2004) An empirical investigation of the rank correlation between different risk measures. *Journal of Risk* **6**(4), 55–74. 4

384. Pham, H., T. Rheinländer and M. Schweizer (1998) Mean-variance hedging for continuous processes: New proofs and examples. *Finance & Stochastics* **2**, 173–198. 136

385. Picoli Jr., S., R.S. Mendes and L.C. Malacarne (2003) q-exponential, Weibull, and q-Weibull distributions: An empirical analysis. *Physica A* **324**, 678–688. 43, 54

386. Pisarenko, V.F. (1998) Non-linear growth of cumulative flood losses with time. *Hydrological Processes* **12**, 461–470. 126

387. Plerou, V., P. Gopikrishnan, X. Gabaix and H.E. Stanley (2004) On the origin of power-law fluctuations in stock prices. *Quantitative Finance* **4**, 7–11. 41

388. Pochart, B. and J.-P. Bouchaud (2002) The skewed multifractal random walk with applications to option smiles. *Quantitative Finance* **2**, 303–314. 85

389. Polimenis, V. (2002) The distributional CAPM: Connecting risk premia to return distributions. *Working Paper*. 15

390. Poon, S.H., M. Rockinger and J. Tawn (2004) Extreme-value dependence in financial markets: Diagnostics, models and financial implications. *Review of Financial Studies* **17**, 581–610. 177, 229, 232, 255

391. Popper, S.W., R.J. Lempert and S.C. Bankes (2005) Shaping the future. *Scientific American* (April) 28. 275

392. Potters, M.J.-P. Bouchaud, L. Laloux and P. Cizeau (1999) Random matrix theory. *Risk* **12**(3), 69. 24, 27, 28

393. Prause, K. (1998) *The generalized hyperbolic model*, Ph.D. dissertation, University of Freiburg. 43, 58

394. President's Working Group on Financial Markets (1999) *Hedge Funds, Leverage, and the Lessons of Long-Term Capital Management*, Report of the US Department of Treasury. 24

395. Quintos, C.E. (2004) Dating breaks in the extremal behavior of financial time series. *Working Paper*. Available at `www.ssb.rochester.edu/fac/quintos/` 231, 232

396. Quintos, C.E., Z.H. Fan and P.C.B. Phillips (2001) Structural change tests in tail behaviour and the Asian crisis. *Review of Economic Studies* **68**, 633–663. 231

397. Ramchand, L. and R. Susmel (1998) Volatility and cross correlation across major stock markets. *Journal of Empirical Finance* **5**, 397–416. 76, 231

398. Rao, C.R. (1965) *Linear Statistical Inference and Its Applications*. Wiley, New York. 3, 152, 154

399. Raychaudhuri, S., Cranston, M., Przybyla, C., Shapir, Y. (2001) Maximal height scaling of kinetically growing surfaces. *Physical Review Letters* **87**, 136101. 44

400. Reese, C. and B. Rosenow (2003) Predicting multivariate volatility. Preprint at `http://arXiv.org/abs/cond-mat/0304082` 276

401. Reinganum, M.R. (1981) A new empirical perspective on the CAPM. *Journal of Financial & Quantitative Analysis* **16**, 439–462. 14

402. Reisman, H. (2002) Some comments on the APT. *Quantitative Finance* **2**, 378–386. 19

403. Rheinländer, T. (2005) An entropy approach to the Stein and Stein model with correlation. *Finance & Stochastics* **9**, 399–413. 136

404. Richardson, M. and T. Smith (1993) A test for multivariate normality in stocks. *Journal of Business* **66**, 295–321. 99

405. Riedel, F. (2004) Dynamic coherent risk measures. *Stochastic Processes & their Applications* **112**, 185–200. 276

406. RiskMetrics Group (1997) CreditMetrics. *Technical Document*. Available at http://www.riskmetrics.com/research 137

407. Rockafellar, R.T., S. Uryasev and M. Zabarankin (2005) Generalized Deviations in Risk Analysis. *Finance & Stochastics* **9**. Forthcoming. 4, 7, 8, 9

408. Rockafellar, R.T., S. Uryasev and M. Zabarankin (2005) Master Funds in Portfolio Analysis with General Deviation Measures. *Journal of Banking & Finance* **29**. Forthcoming. 4, 18

409. Rockinger, M. (2005) Modeling the Dynamics of Conditional Dependency between Financial Series. In *Multi-Moment Capital Pricing Models and Related Topics*, C. Adcock, B. Maillet and E. Jurzenko, eds. Springer. Forthcoming. 100

410. Rockinger, M. and E. Jondeau (2002) Entropy densities with an application to autoregressive conditional skewness and kurtosis. *Journal of Econometrics* **106**, 119–142. 219

411. Rodkin, M.V. and V.F. Pisarenko (2001) Earthquake losses and casualties: A statistical analysis. In Problems in Dynamics and Seismicity of the Earth: Coll. Sci. Proc. Moscow. *Computational Seismology* **31**, 242–272 (in Russian). 126

412. Rodriguez, J.C. (2003) Measuring financial contagion: A copula approach. *Working Paper*, Eurandom. 100

413. Roll, R. (1988) The international crash of October 1987. *Financial Analysts Journal* **4**(5), 19–35. 26

414. Roll, R. (1994) What every CFO should know about scientific progress in financial economics: What is known and what remains to be resolved. *Financial Management* **23**(2), 69–75. 1, 18, 19, 20, 24

415. Romer, D. (1996) *Advanced Macroeconomics*. McGraw-Hill, New York. 277

416. Rootzen, H., M.R. Leadbetter and L. de Haan (1998) On the distribution of tail array sums for strongly mixing stationary sequences. *Annals of Applied Probability* **8**, 868–885. 43

417. Rosenberg, J.V. (2003) Nonparametric pricing of multivariate contingent claims. *Journal of Derivatives*, Spring. 124, 131

418. Ross, S. (1976) The arbitrage theory of capital asset pricing. *Journal of Economic Theory* **17**, 254–286. 19

419. Rothschild, M. and J.E. Stiglitz (1970) Increasing risk I: A definition. *Journal of Economic Theory* **2**, 225–243. 4

420. Rothschild, M. and J.E. Stiglitz (1971) Increasing risk II: Its economic consequences. *Journal of Economic Theory* **3**, 66–84. 4

421. Rubinstein, M. (1973) The fundamental theorem of parameter-preference security valuation. *Journal of Financial & Quantitative Analysis* **8**, 61–69. 15, 16, 58

422. Rüschendorf, L. (1974) Asymptotic distributions of multivariate rank order statistics. *Annals of Mathematical Statistics* **4**, 912–923. 198, 222

423. Ruymgaart, F.H. (1974) Asymptotic normality of nonparametric tests for independence. *Annals of Mathematical Statistics* **2**, 892–910. 198, 222

424. Ruymgaart, F.H., G.R. Shorack and W.R. van Zwet (1972) Asymptotic normality of nonparametric tests for independence. *Annals of Statistics* **44**, 1122–1135. 198, 222

425. Samuelson, P.A. (1965) Rational theory of warrant pricing. *Industrial Management Review* **6**(Spring), 13–31. 38, 80

426. Sarfraz, M. (2003) *Advances in Geometric Modeling.* Wiley, Hoboken. 3

427. Sargent, T.J. (1987) *Dynamic Macroeconomic Theory.* Harvard University Press, Cambridge, MA. 19

428. Scott, D.W. (1992) *Multivariate Density Estimation: Theory, Practice, and Visualization.* Wiley, New York. 3

429. Sharpe, W. (1964) Capital asset prices: A theory of market equilibrium under condition of risk. *Journal of Finance* **19**, 425–442. 14, 16, 38

430. Schloegl, L. and D. O'Kane (2004) Pricing and risk-managing CDO tranches. *Quantitative Credit Research*, Lehman Brothers. 181

431. Schmeidler, D. (1986) Integral representation without additivity. *Proceedings of the American Mathematical Society* **97**, 255–261. 7

432. Schumpeter, J.A. (1939) *Business Cycles: A Theoretical, Historical and Statistical Analysis of the Capitalist Process.* McGraw-Hill, New York. 277

433. Schürmann, J. (1996) *Pattern Classification: A Unified View of Statistical and Neural Approaches.* Wiley, New York. 3

434. Schweizer, M. (1995) On the minimal martingale measure and the Föllmer-Schweizer decomposition. *Stochastic Analysis & Applications* **13**, 573–599. 136

435. Schweizer, M. (1999) A minimality property of the minimal martingale measure. *Statistics & Probability Letters* **42**, 27–31. 136

436. Serva, M., U.L. Fulco, M.L. Lyra and G.M. Viswanathan (2002) Kinematics of stock prices. Preprint at http://arXiv.org/abs/cond-mat/0209103. 78

437. Shefrin, H. (2000) *Beyond Greed and Fear: Understanding Behavioral Finance and the Psychology of Investing.* Harvard Business School Press, Boston, MA. 4, 281

438. Shiller, R.J. (2000) *Irrational Exuberance.* Princeton University Press, Princeton, NJ. 23

439. Shleifer, A. (2000) *Inefficient Markets: An Introduction to Behavioral Finance,* Clarendon Lectures in Economics. Oxford University Press, Oxford. 4, 281

440. Silvapulle, P. and C.W.J. Granger (2001) Large returns, conditional correlation and portfolio diversification: A Value-at-Risk approach. *Quantitative Finance* **1**, 542–551. 231

441. Simon, H.A. (1957) *Models of Man: Social and Rational; Mathematical Essays on Rational Human BEHAVIOR in a Social Setting.* Wiley, New York. 39

442. Skaug, H. and D. Tjostheim (1996) Testing for serial independence using measures of distance between densities. In *Athen Conference on Applied Probability and Time Series*, P. Robinson and M. Rosenblatt, eds. Springer, New York. 163

443. Sklar, A. (1959) Fonction de répartition à n dimensions et leurs marges. *Publication de l'Institut de Statistique de l'Université de Paris* **8**, 229–231. 103, 104

444. Smith, R.L. (1985) Maximum likelihood estimation in a class of non-regular cases. *Biometrika* **72**, 67–90. 44, 48

445. Sobehart, J. and R. Farengo (2003) A dynamical model of market under- and overreaction. *Journal of Risk* **5**(4). 57

446. Sornette, D. (1998) Linear stochastic dynamics with nonlinear fractal properties. *Physica A* **250**, 295–314. 39
447. Sornette, D. (1998) Large deviations and portfolio optimization. *Physica A* **256**, 251–283. 2
448. Sornette, D. (2002) Predictability of catastrophic events: Material rupture, earthquakes, turbulence, financial crashes and human birth. *Proceedings of the National Academy of Science USA* **99** (Suppl), 2522–2529. 277
449. Sornette, D. (2003) Critical market crashes. *Physics Reports* **378**, 1–98. 276
450. Sornette, D. (2003) *Why Stock Markets Crash, Critical Events in Complex Financial Systems*. Princeton University Press, Princeton, NJ. VII, 23, 79, 211, 276
451. Sornette, D. (2004) *Critical Phenomena in Natural Sciences, Chaos, Fractals, Self-organization and Disorder: Concepts and Tools*, 2nd enlarged edition. Springer Series in Synergetics, Heidelberg. VIII, 22, 42, 46, 58, 78, 277
452. Sornette, D. (2005) Endogenous versus exogenous origins of crises. In *Extreme Events in Nature and Society*, S. Albeverio, V. Jentsch and H. Kantz, eds. Springer, Heidelberg. 278
453. Sornette, D., J.V. Andersen and P. Simonetti (2000) Portfolio theory for "fat tails". *International Journal of Theoretical & Applied Finance* **3**, 523–535. 2, 58, 78, 108, 219, 233
454. Sornette, D. and R. Cont (1997) Convergent multiplicative processes repelled from zero: Power laws and truncated power laws. *Journal de Physique I France* **7**, 431–444. 39
455. Sornette, D., F. Deschatres, T. Gilbert and Y. Ageon (2004) Endogenous versus exogenous shocks in complex networks: An empirical test. *Physical Review Letters* **93**(22), 228701. 278
456. Sornette, D., Y. Malevergne and J.F. Muzy (2003) What causes craches? *Risk* **16** (2), 67–71. http://arXiv.org/abs/cond-mat/0204626. 87, 210, 278
457. Sornette, D., Simonetti, P. and Andersen, J.V. (2000) $\phi^q$-field theory for portfolio optimization: "Fat-tails" and non-linear correlations. *Physics Reports* **335**, 19–92. 34
458. Sornette D. and W.-X. Zhou (2005) Predictability of large future changes in complex systems. *International Journal of Forecasting*. Forthcoming. Available at http://arXiv.org/abs/cond-mat/0304601. 278
459. Sornette, D. and W.-X. Zhou (2004) Non-parametric determination of real-time lag structure between two time series: The "optimal thermal causal path" method. *Working Paper*. Available at http://arXiv.org/abs/cond-mat/0408166 281
460. Sornette, D. and W.-X. Zhou (2005) Importance of positive feedbacks and over-confidence in a self-fulfilling Ising model of financial Markets. *Working Paper*. Available at http://arxiv.org/abs/cond-mat/0503607. 279
461. Srivastava, M.S. (2002) *Methods of Multivariate Statistics*. Wiley-Interscience, New York. 3
462. Starica, C. (1999) Multivariate extremes for models with constant conditional correlations. *Journal of Empirical Finance* **6**, 515–553. 232
463. Starica, C. and O. Pictet (1999) The tales the tails of GARCH(1,1) process tell. *Working Paper*, University of Pennsylvania. 66
464. Stollnitz, E.J., T.D. DeRose and D.H. Salesin (1996) *Wavelets for Computer Graphics: Theory and Applications*. Morgan Kaufmann Publishers, San Francisco, CA. 3

465. Stuart, A. and K. Ord (1994) *Kendall's Advances Theory of Statistics*. Wiley, New York. 10, 58

466. Stulz, R.M. (1982) Options on the minimum or the maximum of two risky assets: Analysis and applications. *Journal of Financial Economics* **10**, 161–185. 135

467. Szergö, G. (1999) A critique to Basel regulation, or how to enhance (im)moral hazards. In *Proceedings of the International Conference on Risk Management and Regulation in Banking*. Bank of Israel, Kluwer Academic Press, Dordrecht, Boston. VIII

468. Tabachnick, B.G. and L.S. Fidell. (2000) *Using Multivariate Statistics*, 4th edition. Pearson Allyn & Bacon, Boston and New York. 3

469. Taleb, N.N. (2004) *Fooled by Randomness: The Hidden Role of Chance in Life and in the Markets*, 2nd edition. Texere, New York. 21

470. Taleb, N. (2004) Learning to expect the unexpected. *The New York Times*, April 8. 36

471. Tashe, D. (2002) Expected shortfall and beyond. *Journal of Banking & Finance* **26**, 1519–1533. 6

472. Tasche, D. and L. Tibiletti (2001) Approximations for the Value-at-Risk approach to risk-return analysis. *The ICFAI Journal of Financial Risk Management* **1**(4), 44–61. 128

473. Taylor, S.J. (1994) Modeling stochastic volatility: A review and comparative study. *Mathematical Finance* **2**, 183–204. 37

474. Thaler, R.H. (1993) *Advances in Behavioral Finance*. Russell Sage Foundation, New York. 4, 281

475. Toulouse, G. (1977) Theory of the frustration effect in spin glasses. *Communications in Physics* **2**, 115–119. 164

476. Tsallis, C. (1988) Possible generalization of Boltzmann–Gibbs statistics. *Journal of Statistical Physics* **52**, 479–487. For an updated bibliography on this subject, see http://tsallis.cat.cbpf.br/biblio.htm 163

477. Tsui, A.K. and Q. Yu (1999) Constant conditional correlation in a bivariate GARCH model: Evidence from the stock markets of China. *Mathematics & Computers in Simulation* **48**, 503–509. 231, 276

478. Valdez, E.A. (2001) Bivariate analysis of survivorship and persistency. *Insurance: Mathematics & Economics* **29**, 357–373. 100

479. van der Vaart, A.W., R. Gill, B.D. Ripley, S. Ross, B. Silverman and M. Stein (2000) *Asymptotic Statistics*. Cambridge University Press, Cambridge. 275

480. Vaupel, J.W., K.G. Manton and E. Stallard (1979) The impact of heterogeneity in individual frailty on the dynamics of mortality. *Demography* **16**, 439–454. 113

481. Vannimenus, J. and G. Toulouse (1977) Theory of the frustration effect: II. Ising spins on a square lattice. *Journal of Physics C: Solid State Physics* **10**, 537–541. 164

482. Von Neuman, J. and O. Morgenstern (1944) *Theory of Games and Economic Behavior*. Princeton University Press, Princeton, NJ. 4, 272

483. Wang, T. (2002) A class of dynamic risk measures. *Working Paper*, Faculty of Commerce and Business Administration, U.B.C. 276

484. Wilcox, R.R. (2004) *Introduction to Robust Estimation and Hypothesis Testing*, 2nd edition. Academic Press, New York. 275

485. Wilks, S.S. (1938) The large sample distribution of the likelihood ratio for testing composite hypotheses. *Annals of Mathematical Statistics* **9**, 60–62. 71

486. Williamson, R.C. and T. Downs (1990) Probabilistic arithmetic: Numerical methods for calculating convolutions and dependency bounds. *Journal of Approximate Reasoning* **4**, 89–158. 118, 127

487. Willekens, E. (1988) The structure of the class of subexponential distributions. *Probability Theory & Related Fields* **77**, 567–581. 57

488. Wilson, E.O. (1971) *The Insect Societies* Belknap Press of the Harvard University Press, Cambridge, MA. 22

489. Wilson, K.G. (1979) Problems in physics with many scales of length. *Scientific American* **241** (August), 158–179. 22

490. Wilson, K.G. and J. Kogut (1974) The renormalization group and the $\epsilon$ expansion. *Physics Report* **12**, 75–200. 22

491. White, H. (1982) Maximum likelihood estimation of misspecified models. *Econometrica* **50**, 1–25. 224

492. White, H. (1994) *Estimation, Inference and Specification Analysis*, Econometric Society Monograph, **22**. Cambridge University Press, Cambridge, New York. 202

493. Wu, L. (2006) Dampened power law: Reconciling the tail behavior of financial security returns. *Journal of Business* **79**(6). 75

494. Wyart, M. and J.-P. Bouchaud (2003) Self-referential behavior, overreaction, and conventions in financial markets. *Working Paper*. Available at http://arxiv.org/abs/cond-mat/0303584. 279

495. Yaari, M.E. (1987) The dual theory of choice under risk. *Econometrica* **55**, 95–115. 7

496. Zajdenweber, D. (1996) Extreme values in business interruption insurance. *Journal of Risk & Insurance* **63**, 95–110. 126

497. Zajdenweber, D. (2000) *Economie des Extrêmes*. Flammarion, Paris. 126

498. Zhang, J. and R.A. Stine (2001) Autocovariance structure of Markov regime switching models and model selection. *Journal of Time Series Analysis* **22**, 107–124. 279

# Index

Akaike information criterion 215
ALAE 201
Anderson-Darling distance 61
Arbitrage 35, 132, 133
Arbitrage pricing theory X, 19
ARCH *see* GARCH
Archimedean copula 111, 204, 255
  Clayton 112, 114
  Frank 112
  Gumbel 112
  Kendall's tau 155
  orthant dependence 166
  tail dependence 170
Asian crisis 211, 230
Associativity 114
Asymptotic independence 169

Bank for International Settlements
  VIII, 35
Bernstein polynomial 191
Bhattacharya-Matusita-Hellinger
  dependence metric 163, 203
Black swan *see* Outlier
Black-Merton-Scholes' option pricing
  model VIII, 20, 38
Book-to-market 18
Bootstrap 62, 192, 207
British Pound 208, 210

Canonical coefficient of $N$-correlation
  153, 154
Capital asset pricing model IX, 14, 38
Central limit theorem VIII

Clayton's copula 112, 114, 200, 217,
  249
  Kendall's tau 156
  simulation 123
  tail dependence 171
Coefficient of tail dependence *see* Tail
  dependence
Coherent measures of risk 4, 276
Comonotonicity 101, 102, 107, 149,
  155, 160
Complete market 133, 136
Complete monotonicity 111
Concordance measure 154–162, 165
Conditional correlation coefficient 233
Consistent measures of risk 7
Contagion XI, 231, 260
Contingent claim *see* Option
Convex measure of risk 7
Copula X, 34, 35, 103, 273
  Archimedean *see* Archimedean
    copula
  dual 104, 118
  elliptical *see* Elliptical copula
  extreme value *see* Extreme value
    copula
  Fréchet-Hoeffding bounds 106
  survival 104, 114, 132, 140, 166, 215
Correlation coefficient 2, 24, 99, 105,
  147–154, 165, 173, 174, 189, 219,
  220
  Hoeffding identity 149
Countermonotonicity 107, 149, 155,
  160, 164

Coup against Gorbachev   209, 276
Covariance matrix   3, 24, 33
CreditMetrics   138
CreditRisk$^+$   137
CSFB/Tremont index   167
Currency
   British Pound   208, 210
   Euro   195, 203, 217
   German Mark   197–199, 208–210,
      214, 215, 276
   Japanese Yen   197, 199, 208, 215, 217
   Malaysian Ringit   208, 210, 214
   Swiss Franc   208–210, 214, 276
   Thai Baht   208, 210
   US Dollar   208–210, 240, 276

Default risk   100, 137, 180
Dependence   2, 101
   mutual complete   101
   positive orthant   119, 164
Dependence measure   147, 161
Dependence metric   162
Dependence structure   see copula
Derivative   see Option
Digital option   131
Distribution function
   Exponential   58, 90, 175
   Fréchet   45, 255
   Gamma   58, 90, 175
   Gaussian   2, 37, 148, 169, 175, 233
   Generalized Pareto   39, 44–47, 116
   GEV   45, 47, 116
   Gumbel   45, 48
   Lévy stable law   2, 39, 42, 148, 243
   Log-normal   37, 60, 78, 150
   Log-Weibull   60, 69, 91
   Meta-elliptical   109
   Meta-Gaussian   108
   Modified-Weibull   128
   Pareto   39, 57, 64, 88
   Pearson type-VII   42
   Shifted-Pareto   126
   Stretched exponential   43, 50, 57
   Student $t$   42, 108, 157, 175, 233
   Weibull   50, 57, 67, 88
Diversification   19, 180
Dow Jones Industrial Average Index
      44, 53, 62, 78
Drawdown   23, 36, 79, 276

Dual copula   104, 118

Efficient market hypothesis   20
Elliptical copula   107, 196, 200
   Gaussian copula   108
   Kendall's tau   157
   simulation   120
   Student's copula   109
   tail dependence   172
Empirical copula   190
Endogeneity   276
Euro   195, 203, 217
European Monetary System   210
Evolutionary stable equilibrium   20
Exogeneity   277
Expectation-bounded measures of risk
      7
Expected utility   4, 21, 165
Expected-Shortfall   47, 79
Exponential distribution   58, 90, 175
Extremal index   46
Extreme value copula   116, 117
Extreme value theory   43, 45, 254, 273

Factor model   3, 19, 24, 29, 111, 138,
      174, 233, 238, 255
Fat tail   see Heavy tail
Federal Reserve Board   208
Firm size   37
Foreign exchange rate   215
Fréchet distribution   45, 255
Fréchet-Hoeffding bounds   106, 115,
      117, 119
Fractality   80
Fractional Brownian motion   81
Frailty model   113
Frank's copula   112
   Kendall's tau   156
   simulation   123
   tail dependence   171
Friendship theorem   26

Gain–loss ratio   13
Gamma distribution   58, 90, 175
GARCH   35, 37, 43, 108, 205, 217, 219,
      231
Gaussian copula   108, 128, 130, 131,
      135, 137, 204, 212, 217

Gaussian distribution    2, 37, 148, 169, 175, 233
General deviation measures    8
Generalized Extreme Value distribution    45, 47, 116
Generalized Pareto distribution    39, 44–47, 116
German Mark    197–199, 208–210, 214, 215, 276
Gini's gamma    161, 165
Girsanov theorem    144
Gnedenko theorem    45, 76
Goodness of fit    61, 164, 189
GPBH theorem    115
Great Depression    53
Gumbel distribution    45, 48
Gumbel's copula    112, 117
    Kendall's tau    156
    simulation    123
    tail dependence    171

Heavy tail    2, 12, 15, 36, 38, 42, 57, 157
Heteroscedasticity    see Volatility clustering
High frequency data    35, 37, 44
Hill estimator    43, 48, 64, 240
Hoeffding identity    149

Ibov index    240
Incomplete gamma function    57
Inflation    VII
Information matrix
    Fisher    198, 201, 222
    Godambe    202
Internet bubble    VII, VIII
Invariance theorem    105
Ipsa index    240

Japanese Yen    197, 199, 208, 215, 217

Kendall's tau    154, 165, 196, 200, 249
    Archimedean copula    155
    elliptical copula    157
Kernel estimator    192
King    see Outlier
KMV    138
Kolmogorov distance    61
Kullback-Leibler divergence    61, 163

Lévy process    35

Lévy stable law    2, 39, 42, 148, 243
Lambert function    172
Laplace transform    113
Latin American crisis    228, 260
    Argentinean crisis    231, 233, 240, 261
    Mexican crisis    230, 233, 240, 247, 261
Linear dependence    see Correlation coefficient
Local correlation coefficient    151
Log infinitely divisible process    41
Log-normal distribution    37, 60, 78, 150
Log-Weibull distribution    60, 69, 91
LTCM    24
Lunch effect    53

Malaysian Ringit    208, 210, 214
Market crash    VII, 23, 36, 38
    April 2000    230
    October 1987    VII, VIII, 26, 230, 247
Market index
    CSFB/Tremont    167
    Dow Jones Industrial Average    44, 53, 62, 78
    Ibov    240
    Ipsa    240
    Merval    240
    Mexbol    240
    Nasdaq Composite    53, 77, 230
    Standard & Poor's 500    39, 60, 129, 167, 177, 216
Market liquidity    6, 41
Market trend    231, 234, 253, 255
Markowitz' portfolio selection    see Mean-variance portfolio theory
Maximum domain of attraction    45, 46
Mean-variance portfolio theory    VIII, 33, 38, 58
Merton model of credit risk    20, 137
Merval index    240
Meta-elliptical distribution    109
Meta-Gaussian distribution    108
Mexbol index    240
Micro-structure    189, 215
Minimum option    135
Minority game    22
Mixture model    137
Modified-Weibull distribution    128, 131
Monte Carlo    120, 192

Multifractal Random Walk    39, 84, 210, 219, 278
Mutual complete dependence    101

Nasdaq Composite index    53, 77, 230
New economy    230
Normal law    see Gaussian distribution

Occam's razor    212
Option    33, 100, 131, 192
  digital    131
  minimum    135
  rainbow    135
Outlier    23, 30, 36, 80, 206, 209, 210, 276

Pareto distribution    39, 57, 64, 88
Pearson estimator    148, 157
Pearson type-VII distribution    42
Pickands estimator    43, 47–49
Portfolio    3, 100, 179, 189, 200, 216
  analysis    205
  insurance    14, 23
  management    VIII, 180, 212, 217, 231
  risk    3, 33, 124, 127, 128, 177, 192
  theory    3, 33
Positive orthant dependence    119, 164
  Archimedean copula    166
Pseudo likelihood    197, 215, 222
Pseudo-sample    197, 206

Quantile    35

Rainbow option    135
Regular variation    39, 171
Risk    VII, VIII, 1
  analysis    13
  assessment    78, 124, 212, 231
  aversion    4, 16, 17
  management    VIII, 35, 79, 100, 205, 271
  measure    10
    coherent measures of risk    4, 276
    consistent measures of risk    7
    expectation-bounded measures of risk    7
    general deviation measures    8
    spectral measures of risk    6
  premium    37
Russian crisis    230

Securization    100

Self-organized criticality    277
Self-similarity    80
Semi-invariant    10
Shannon entropy    163
Sharpe's market equilibrium model
    see Capital asset pricing model
Shifted-Pareto distribution    126
Sklar's theorem    104, 107, 120, 190
Spearman's rho    159, 165, 196, 248
Spectral measures of risk    6
Standard & Poor's 500    39, 60, 129, 167, 177, 216
Stone-Weierstrass theorem    192
Stress testing    10, 43, 210, 214
Student $t$ distribution    42, 108, 157, 175, 233
Student's copula    109, 117, 131, 195, 212, 256
Survival copula    104, 114, 132, 140, 166, 215
Swarm intelligence    22
Swiss Franc    208–210, 214, 276
Swiss National Bank    210

Tail dependence    168, 192, 212, 220, 233, 254
  Archimedean copula    170
  elliptical copula    172
  factor model    174
Tail risk    124, 131, 177, 216
Thai Baht    208, 210
Theorem
  central limit    VIII
  friendship    26
  Girsanov    144
  Gnedenko    45, 76
  GPBH    46, 115
  invariance    105
  Sklar    104, 107, 190
  Stone-Weierstrass    192
  Wilks    71, 93, 216, 224

US Dollar    208–210, 240, 276

Value-at-Risk    2, 5, 43, 46, 79, 100, 124, 128, 131, 168, 271, 276
Volatility    2, 231
  clustering    35, 38, 39, 43, 194, 217
Volume of transactions    41

Weibull distribution    50, 57, 67, 88
Wilks theorem    71, 93, 216, 224